Lecture Notes in Computer Science　　9700

Commenced Publication in 1973
Founding and Former Series Editors:
Gerhard Goos, Juris Hartmanis, and Jan van Leeuwen

More information about this series at http://www.springer.com/series/7408

Marco Bernardo · Rocco De Nicola
Jane Hillston (Eds.)

Formal Methods for the Quantitative Evaluation of Collective Adaptive Systems

16th International School on Formal Methods
for the Design of Computer, Communication,
and Software Systems, SFM 2016
Bertinoro, Italy, June 20–24, 2016
Advanced Lectures

 Springer

Editors
Marco Bernardo
Dipartimento di Scienze Pure e Applicate
Università di Urbino "Carlo Bo"
Urbino
Italy

Jane Hillston
School of Informatics
University of Edinburgh
Edinburgh
UK

Rocco De Nicola
IMT - School for Advanced Studies Lucca
Lucca
Italy

ISSN 0302-9743   ISSN 1611-3349   (electronic)
Lecture Notes in Computer Science
ISBN 978-3-319-34095-1   ISBN 978-3-319-34096-8   (eBook)
DOI 10.1007/978-3-319-34096-8

Library of Congress Control Number: 2016937952

LNCS Sublibrary: SL2 – Programming and Software Engineering

Printed on acid-free paper

This Springer imprint is published by Springer Nature
The registered company is Springer International Publishing AG Switzerland

Preface

This volume collects a set of papers accompanying the lectures of the 16th International School on Formal Methods for the Design of Computer, Communication and Software Systems (SFM). This series of schools addresses the use of formal methods in computer science as a prominent approach to the rigorous design of the above-mentioned systems. The main aim of the SFM series is to offer a good spectrum of current research in foundations as well as applications of formal methods, which can be of help to graduate students and young researchers who intend to approach the field. SFM 2016 was devoted to the quantitative evaluation of collective adaptive systems and covered topics such as self-organization in distributed systems, scalable quantitative analysis, spatio-temporal models, and aggregate programming. The eight papers of this volume represent the broad range of topics of the school.

The paper by Talcott, Nigam, Arbab, and Kappé proposes a framework called Soft Agents, formalized in the Maude rewriting logic system, to describe systems of cyber-physical agents that operate in unpredictable, possibly hostile, environments using locally obtainable information. Ghezzi's paper is a tutorial on how to design adaptable and evolvable systems that support safe continuous software deployment to guarantee correct operation in the presence of dynamic reconfigurations. Bortolussi and Gast study the limiting behavior of stochastic models of populations of interacting agents, as the number of agents goes to infinity, in the case that classical conditions ensuring the validity of mean-field results based on ordinary differential equations do not hold. The paper by Loreti and Hillston illustrates CARMA, a language recently defined to support specification and analysis of collective adaptive systems, and its tools developed for system design and analysis. Galpin's paper provides an overview of models of individuals and models of populations for collective adaptive systems, in which discrete or continuous space is explicitly represented. Ciancia, Latella, Loreti, and Massink also address spatial aspects of collective adaptive systems through a topology-inspired approach to formal verification of spatial properties, which is based on the logics SLCS and STLCS and their model-checking algorithms. The paper by Vandin and Tribastone shows how to efficiently analyze quantitative properties of large-scale collective adaptive systems by reviewing algorithms that reduce the dimensionality of models in a way that preserves modeler-defined state variables. Finally, Beal and Viroli present aggregate programming, a new paradigm for coping with an ever-increasing density of computing devices that raises the level of abstraction in order to allow programmers to reason in terms of collections of interacting devices.

We believe that this book offers a useful view of what has been done and what is going on worldwide in the field of formal methods for the quantitative evaluation of collective adaptive systems. This school was organized in collaboration with the EU FP7 project QUANTICOL, whose support we gratefully acknowledge. We wish to thank all the speakers and all the participants for a lively and fruitful school. We also wish to thank the entire staff of the University Residential Center of Bertinoro for the organizational and administrative support, as well as the Springer editorial office for the assistance with the editing of this book and the kind sponsorship.

June 2016

Marco Bernardo
Rocco De Nicola
Jane Hillston

Contents

Formal Specification and Analysis of Robust Adaptive Distributed Cyber-Physical Systems

Carolyn Talcott[1]([✉]), Vivek Nigam[2], Farhad Arbab[3,4], and Tobias Kappé[3,4]

[1] SRI International, Menlo Park, CA 94025, USA
carolyn.talcott@sri.com
[2] Federal University of Paraiba, João Pessoa, Brazil
vivek.nigam@gmail.com
[3] LIACS, Leiden University, Leiden, The Netherlands
tkappe@liacs.nl
[4] Centrum Wiskunde & Informatica, Amsterdam, The Netherlands
farhad@cwi.nl

Abstract. We are interested in systems of cyber-physical agents that operate in unpredictable, possibly hostile, environments using locally obtainable information. How can we specify robust agents that are able to operate alone and/or in cooperation with other agents? What properties are important? How can they be verified?

In this tutorial we describe a framework called Soft Agents, formalized in the Maude rewriting logic system. Features of the framework include: explicit representation of the physical state as well as the cyber perception of this state; robust communication via sharing of partially ordered knowledge, and robust behavior based on soft constraints. Using Maude functionality, the soft agent framework supports experimenting with, formally testing, and reasoning about specifications of agent systems.

The tutorial begins with a discussion of desiderata for soft agent models. Use of the soft agent framework for specification and formal analysis of agent systems illustrated in some detail by a case-study involving simple patrolling bots. A more complex case study involving surveillance drones is also discussed.

1 Introduction

Consider a future in which an explosion of small applications running on mobile devices combine and collaborate to provide powerful new functionality, not only in the realms such as smart vehicles, disaster response, home care, but also harnessing diverse communication mechanisms and robust people power for new kinds of cyber crowd sourcing tasks.

Complex cyber-physical agents are becoming increasingly ubiquitous, in part, due to increased computational performance of commodity hardware and the widespread adoption of standard mobile computing environments (e.g. Android).

The work was partially supported by ONR grant N00014–15–1–2202.

© Springer International Publishing Switzerland 2016
M. Bernardo et al. (Eds.): SFM 2016, LNCS 9700, pp. 1–35, 2016.
DOI: 10.1007/978-3-319-34096-8_1

For example, one can purchase (for a few hundred US dollars in the retail market) a four rotor "drone" capable of precise, controlled hovering that can be equipped with a portable Android phone that provides integrated communication (e.g. wifi ad hoc) and sensing (e.g. high resolution camera) capability as well as considerable processing power (e.g. multiple GPU cores) and memory.

Already there are impressive examples coming from both research and industrial settings. Here are a few.

Researchers in the Vijay Kumar Lab at the University of Pennsylvania [1–3] are experimenting with autonomous flying quadrotor robots that can be equipped with a variety of sensors such as IMUs (inertial measurement units), cameras, laser range scanners, altimeters and/or GPS sensors. Key capabilities being developed include navigation in 3-dimensional space, sensing other entities, and forming ad hoc teams. Potential applications include construction, search and rescue, first response, and precision farming.

The recent EU ASCENS project [4,5] focused on theoretical foundations and models for reliable and predictable system behavior while exploiting the possibilities of highly dynamic, autonomic components. The project included pragmatic case studies such as generation of a robot swarm with both autonomous and collective behavior.

A case study of an adaptive network consisting of smart phones, robots, and UAVs is reported in [6]. The temporal evolution of the macroscopic system state is controlled using a distributed logic [7,8], while the microscopic state is controlled by an algorithm based on 'artificial physics'. Communication is based on a partially-ordered knowledge sharing model for loosely coupled distributed computing. The ideas were tested via both simulation and in a parking lot using a cyber-physical testbed consisting of robots, quadcopters, and Android devices. This work was part of the SRI Networked Cyber-Physical Systems project [9].

Startups are developing applications for shipping, security, search and rescue, environmental monitoring, and agriculture to name a few. Of course, there are Google's driverless cars and Amazon's drone delivery system.

In October 2015 Stanford hosted *Drone swarms: the Buz of the future* [10]. The event included demonstrations (live and video) of a number of advanced autonomous robotic capabilities. The Knightscope [11] security robots roamed the busy plaza without bumping into people or other objects. Their normal job is surveillance of limited areas, looking for problems. Liquid Robotics [12] presented their *wave glider*, a surfboard size robot that is powered by the ocean, capable of multiple modes of communication and of carrying diverse sensors. Wave gliders are able to autonomously and safely navigate from the US to Australia, and able to call home when pirates try bot-napping gliders, in addition to collecting data.

According to a recent Fortune article [13] Burlington Northern Santa Fe (BNSF) Railway has gained FAA approval for a pilot(less) program to use drones to inspect its far-flung network of rails, over 32,500 miles. There are many technical obstacles to overcome, but use of drones has the potential for both greater efficiency and improved safety. Each rail section is inspected at least twice a week, and inspectors may have to deal with disagreeable insects, toxic vegetation, or poisonous snakes.

Formal executable models can provide valuable tools for exploring system designs, testing ideas, and verifying aspects of a systems expected behavior. Executable models are often cheaper and faster to build and experiment with than physical models, especially in early stages as ideas are developing. The value of using formal executable models in the process of designing and deploying a network defense system is illustrated in [14,32].

In the following we describe a *Soft Agent Framework* that provides infrastructure and guidance for building formal executable models of cyber-physical agent systems that we call *soft agent* systems. We present an executable specification in the Maude rewriting logic system [15] and illustrate the ideas with case studies.

Key features of soft agents include

- Simple goal specification (package A must get delivered from point X to point Y).
- Efficient (not necessarily optimal) agent trajectory planning.
- Distributed and local, there is no central planner with perfect global knowledge.
- Agents are resource constrained—limited communication range, energy (actions consume energy), lift power, etc.
- Robustness to unexpected/unplanned situations
  - agent actions (sensor readings, actuator actions, communication) can suffer delays/failure,
  - perturbations in environment. e.g., a gust of wind, a change of goal.

The soft agent framework combines ideas from several previous works: the use of soft constraints [16–18] and soft constraint automata [19] for specifying robust and adaptive behavior; partially ordered knowledge sharing for communication in disrupted environments [8,20–22], and the Real-time Maude approach to modeling timed systems [23]. A novel feature is the explicit representation of both cyber and physical aspects of a soft agent system. An environment component maintains the physical state of the agents and their surroundings. Each agent has a local knowledge base that stores the agents view of the world, which may or may not be the actual state of the world. To specify an agent system one must say what sensors an agent has and what actions it can perform. The 'physics' of these actions must also be specified: how an action affects the state of the agent and its surroundings (in particular how the sensor readings change). One must also specify how an agent decides what actions to carry out in different situations. Although the framework allows complete freedom in how this is done, we will focus on a soft constraint approach.

Soft constraints allow composition in multiple dimensions, including different concerns for a given agent, and composition of constraints for multiple agents. In [17] a new system for expressing soft constraints called Monoidal Soft Constraints is proposed. This generalizes the Semi-ring approach to support more complex preference relations. In [18] partially ordered valuation structures are explored to provide operators for combination of constraints for different features that respects the relative importance of the features. Steps towards a theoretical foundation for compositional specification of agent systems based on Soft Constraint Automata is presented in [24]. Although soft constraints provide an elegant and powerful foundation, as we will see there is much to do to develop principled ways to decompose problems and compose concerns with predictable results. Having compositional specification mechanisms is a step toward compositional reasoning which is import to manage the complexity of soft agent and similar systems. Systems operating in unpredictable environments, beyond our control are almost guaranteed to exhibit some kind of failure. Having a structured, compositional model will facilitate developing methods for determining the cause of failures and enabling (partial) recovery in many cases.

Plan. Section 2 discusses desiderata for a framework for Soft Agents. Section 3 presents the proposed framework and its formalization in Maude. Section 4 illustrates the application of soft-agents to a system of simple patrolling bots. Application to modeling more complex drone systems is also briefly summarized. Section 5 summarizes and discusses future directions.

2   Desiderata for Soft Agents

Cyber-physical agents must maintain an overall situation, location, and time awareness and make safe decisions that progress towards an objective in spite of uncertainty, partial knowledge and intermittent connectivity. The big question is: how do we design, build, and understand such systems? We want principles and tools for system design that achieve adaptive, robust functionality.

The primary desiderata for our Soft Agents are *localness*, *liveness* and *softness*. We explicitly exclude considering insincere or malicious agents in our current formulation.[1]

Localness. Cooperation and coordination should emerge from local behavior based on local knowledge. This is traditionally done by consensus formation algorithms. Consensus involves agreeing on what actions to take, which usually requires a shared view of the system state. In a distributed system spread over a large geographic area beyond the communication reach of individual agents, consensus can take considerable time and resources, but our agents must keep going. Thus consensus may emerge but can't be relied on, nor can it be be forced.

In less than ideal conditions what is needed is a notion of *sufficient consensus*: for any agent, consensus is sufficient when enough of a consensus is present so that agents can begin executing actions that are likely to be a part of a successful plan, given that there is some expectation for the environment to change.

Our partially ordered knowledge sharing (POKS) model for communication takes care of agreeing on state to the degree possible. In a quiescent connected

---

[1] This is a strong assumption, although not unusual. The soft agents framework supports modeling of an unpredictable or even "malicious" environment. We discuss the issue of trust or confidence in knowledge received as part of future work.

situation all agents will eventually have the same knowledge base. As communication improves, an soft agent system approaches a typical distributed system without complicating factors. This should increase the likelihood of reaching actual consensus, and achieving ideal behaviors.

A key question here is how a system determines the minimal required level of consensus? In particular what quality of connection/communication is required to support formation of this minimum level of consensus?

Safety and Liveness. Another formal property to consider concerns safety and liveness: often explained as something bad does not happen and something good will eventually happen. From a local perspective this could mean avoiding preventable disaster/failure as well as making progress and eventually sufficiently satisfying a given goal.

– An agent will never wait for an unbounded time to act.
– An agent can always act if local environment/knowledge/situation demands.
– An agent will react in a timely manner based on local information.
– An agent should keep itself "safe".

We note that the quality calculus [25, 26] provides language primitives to support programming to meet such liveness requirements and robustness analysis methods for verification. One of the motivations of the Quality Calculus was to deal with unreliable communication. It will be interesting to investigate how soft constraints and the quality calculus approach might be combined to provide higher level specification and programming paradigms.

Softness. We want to reason about systems at both the system and cyber-physical entity/agent level and systematically connect the two levels. Agent behavior must allow for uncertainty and partial information, as well as preferences when multiple actions are possible to accomplish a task, as often is the case.

Specification in terms of constraints is a natural way to allow for partiality. *Soft constraints* provide a mechanism to rank different solutions and compose constraints concerning different aspects. We refer to [17] for an excellent review of different soft constraint systems. Given a problem there will be system wide constraints characterizing acceptable solutions, and perhaps giving measures to rank possible behaviors/solutions. Depending on the system wide constraints, global satisfaction can be a safety requirement, a liveness/progress requirement, a cooperation requirement, to name a few. Rather than looking for a distributed solution of global constraints, each agent will be driven by a local constraint system. Multiple soft constraint systems maybe be involved (multiple agent classes) and multiple agents may be driven by the same soft constraint system. A key challenge is developing reasoning and analysis methods to determine conditions under which satisfaction of local constraints will lead to satisfaction of global constraints, monotonically.

3   The Soft Agent Formal Framework

The soft-agent framework provides infrastructure and context for developing *executable* specifications of cyber-physical agents that operate in unpredictable environments, and for experimenting with and analyzing the resulting system behaviors. The framework, specified in the rewriting logic language Maude [15], provides generic data structures for representing system state (cyber and physical), interface sorts and functions to be used to specify the environment, agent capabilities (physical) and behavior (cyber). The semantics, how a system evolves, is given by a small number of rewrite rules defined in terms of these sorts and functions. A soft agent model specializes the framework by refining the sort hierarchy, adding model specific data structures, and providing definitions of the interface functions.

What can we do with an executable specification? Once a model is defined, we can define specific agent system configurations and explore their behavior by rewriting and search. We can watch the system run by rewriting according to different builtin or user defined strategies to choose next steps. In this way, specific executions and their event traces can be examined. Search allows exploration of all possible executions of a given configuration (up to some depth), to look for desired or undesirable conditions. Metadata can be added to a configuration to collect quantitative information such as the lowest energy before charging, close encounters, time elapsed between events, and so on.

As an example we introduce our main case study, *Patrol bots*, informally. A patrol bot moves on a grid along a fixed track (fixed value of the y coordinate). One or more squares of the grid are charging stations. We want to specify patrol bot behavior such that in a system with one or more patrol bot agent: agents do not run out of energy, and they keep patrolling.

In a little more detail, a soft agent system configuration consists of an environment component and a collection of agents. Each agent has local knowledge which can be shared, selectively and opportunistically with other agents when they come within "hearing distance" (in contact). For example, patrol bots share location. In a more complex scenario agents could learn terrain features and share that information. Knowledge sharing replaces traditional message passing as a communication mechanism. The framework caches shared knowledge to support propagation through a network of agents as they move around and meet different agents. Knowledge comes with a partial order that is used by the framework to replace stale or superseded knowledge. Of course an agent can keep a history by aggregating superseded information locally, for example as a list of values.

A specific model will specify for each agent some ability to observe its physical state and local environment, and some ability to act locally to change its physical state and local environment. A patrol bot can move one square at a time in any direction (i.e., to an adjacent square), subject to staying on the grid, and not moving to a blocked or occupied square. Moving uses energy. If the bot is at a charging square it can charge for some time until its energy capacity is reached. A patrol bot has sensors to read its location on the grid, and its energy level.

An agent's state consists of several attributes including its local knowledge base, pending events (tasks for the agent, or actions to be carried out), and cached knowledge to share with other agents.[2] A patrol bot's local knowledge includes the grid dimensions and charging station locations. It may store its location and energy, or simply read the sensors when it needs that information. In a context with more than one patrolling bot, it may store information about the location of other bots.

The environment is modeled as a knowledge base that contains information about the physical state of each agent, along with non-agent specific information such as features of the terrain, or current weather.

How do agents work? Agents carry out tasks (the cyber/reasoning part) and schedule actions to be executed by the environment (the physical part). The typical process for an agent carrying out a task is the following:

- read and process shared knowledge it has received;
- read and process sensor data (local information about the agents state and local environment conditions);
- think and decide
  - new knowledge to share
  - actions to execute
  - next tasks
  - updates for the local knowledge base: sensor readings to remember, new information learned or computed; new goals, things to check, ...

In the Patrol Bot model there is just one task, `tick`. This triggers the processing described above. In selecting actions, if not at a charging station, a patrol bot must decide whether to continue on its track or to detour to the nearest charging station.

How does the system run? Agents schedule tasks with a delay, often just one time unit, Actions are scheduled with a delay (possibly zero) and a duration. If an agent has a ready task (zero delay) then it can be carried out. Once there are no more ready tasks, ready actions are executed concurrently, sharable knowledge is exchanged, and time passes until there are one or more ready tasks.

In a scenario with two patrol bots, each with a ready task, then both will execute their tasks (order doesn't matter) scheduling the next `tick` with one time unit delay. Suppose both also schedule moves to different locations with 0 delay. Then the moves will be executed concurrently, one unit of time will pass, and now the two bot will again process their `tick`s.

Making explicit models of the physical state and agents knowledge allow modeling a range of situations, including the case where an agent has an accurate view of its situation and a variety of cases with discrepancies by independently varying different parts of the model. For example as we will see, it is easy to add wind effects to the model and see how well the agents adapt.

---

[2] The shared cache is separated from the local knowledge base so an agent can choose what to share and what to keep to itself.

The knowledge sharing communication model supports opportunistic communication where a connected network is not possible. In an ideal situation, knowledge propagates quickly and agents can have good situation awareness. However, an agent should not wait to hear from others. If an agent senses that action is needed, it should figure out the an action that is safe and makes progress towards its goal if possible, based on local information.

3.1   Introduction to Rewriting Logic and Maude

Rewriting logic [27] is a logical formalism that is based on two simple ideas: states of a system are represented as elements of an algebraic data type, specified in an equational theory, and the behavior of a system is given by local transitions between states described by *rewrite rules*. An equational theory specifies data types by declaring constants and constructor operations that build complex structured data from simpler parts. Functions on the specified data types are defined by *equations* that allow one to compute the result of applying the function. A *term* is a variable, a constant, or application of a constructor or function symbol to a list of terms. A specific data element is represented by a term containing no variables. Assuming the equations fully define the function symbols, each data element has a canonical representation as a term containing only constants and constructors.

A rewrite rule has the form $t \Rightarrow t'$ *if* c where t and t' are terms possibly containing variables and c is a condition (a boolean term). Such a rule applies to a system in state s if t can be matched to a part of s by supplying the right values for the variables, and if the condition c holds when supplied with those values. In this case the rule can be applied by replacing the part of s matching t by t' using the matching values for the place holders in t'. The process of application of rewrite rules generates computations (also thought of as deductions).

We note that rewriting with rules is similar to rewriting with equations (traditional term rewriting), in that we match the lefthand side and replace the matched subterm by the instantiated righthand side. The difference is in the way the rewriting is used. Equations are used to define functions by providing a means of computation the value of a function application. This means that the equations of an equational theory should give the same result independent of the order in which they are applied. Furthermore, equational rewriting should terminate. In contrast, rules are used to describe change over time, rather than computing the value of a function. They often describe non-deterministic possibly infinite behavior.

Maude is a language and tool based on rewriting logic [15,28]. Maude provides a high performance rewriting engine featuring matching modulo associativity, commutativity, and identity axioms; and search and model-checking capabilities. Thus, given a specification S of a concurrent system, one can execute S to find one possible behavior; use search to see if a state meeting a given condition can be reached; or model-check S to see if a temporal property is satisfied, and if not to see a computation that is a counter example.

To introduce Maude notation and give some intuition about how concurrent systems are specified and analyzed in Maude we consider specification of a simple Vending Machine. A Maude specification consists of a collection of modules. A module has a name, a set of imports (possibly empty), declarations of sorts and operations (constants, constructors, functions), equations defining functions, and rewrite rules. The vending machine specification is given in a module named VENDING-MACHINE.

```
mod VENDING-MACHINE is
  sorts Coin Item  Marking .
  subsorts Coin Item < Marking .
  op null : -> Marking [ctor] .  *** empty marking
  op _ _ : Marking Marking -> Marking
           [ctor assoc comm id: null] .
  ops $ q : -> Coin [ctor] .  *** dollar, quarter
  ops a c : -> Item [ctor] .  *** apple, cake
  rl[buy-c]: $ => c .
  rl[buy-a]: $ => a q .
  rl[change]: q q q q => $ .
endm
```

First several sorts (think sets or data types) are declared (key word sort). The basic sorts are Coin and Item. They represent what you put in and get out of the machine. The sort Marking consists of multisets of items and coins. This is specified by the subsort (subset) declarations saying that coins and items are (singleton) markings; and the declaration of the union operator (_ _, key word op). The blanks indicate operator argument positions, thus union of two markings is represented by placing them side by side, just as one represents a string of characters. The operator attributes assoc, comm, and id:null declare union to be associative and commutative with identity null, the empty marking. (Text following *** is a comment.) After defining the data types to be used, some specific constants are declared: $ (dollar) and q (quarter) of sort Coin; and a (apple) and c (cake) of sort Item (the keyword ops is used when declaring multiple constants of the same sort). The key word ctor in the operator attributes indicates the a constructor. Constructors are used to *construct* data elements, while non-constructor operators are used to name constants and functions defined by equations. The idea is that, using the equations, each term can be rewritten to an equivalent term in canonical form built from constructor operations. Maude takes care of all this under the hood, allowing the user to think abstractly in terms of equivalence classes.

Finally there are three rewrite rules specifying the vending machine behavior. The rule labeled buy-c says that if you have a dollar you can buy a cake. More formally, any marking containing an occurrence of $ can be rewritten to one in which the $ is replaced by a c. Similarly the rule labeled buy-a says that with a dollar you can also get an apple and a quarter change. The rule labeled change says that when four quarters have accumulated they can be changed into a dollar. Note that if a dollar is present in a marking, there are two ways that the marking could be rewritten, each with a different outcome. If four quarters are also present, the change rule could be applied before or after one of the buy rules without affecting the eventual outcome.

To find one way to use three dollars, ask Maude to rewrite, and a quarter, an apple, and two cakes are the result.

```
Maude> rew $ $ $ .
result Marking: q a c c
```

Although there are several ways to rewrite three dollars, the Maude rewrite command uses a specific strategy for choosing rules to apply, and in this case chose to apply buy-c twice and buy-a once.

To discover more possibilities Maude can be asked to search for all ways of rewriting three dollars, such that the final state matches some pattern. For example, we can find all ways of getting at least two apples using the pattern

```
a a M:Marking
```

that is matched by any state that has at least two as.

```
Maude> search $ $ $ =>! a a M:Marking .
Solution 1 (state 8)
M:Marking --> q q c
Solution 2 (state 9)
M:Marking --> q q q a
```

There are two ways this can be done. In one solution the remainder of the state consists of a cake and two quarters, (indicated by M:Marking -> q q c in Solution 1). In the other solution, there is a third apple and three quarters.

We can ask Maude to show us a path (list of rules fired) corresponding to one of these solutions using the command

```
Maude> show path labels 8 .
buy-c
buy-a
buy-a
```

to find the path to state 8. One can also ask for the full path to state 8. In this case the sequence of states and rules applied will be printed.

3.2   Key Sorts and Functions for Agent State

Now we see how the key sorts and functions for soft agent state are specified in Maude. Agent state is represented using knowledge and events.

Knowledge. Identifiers are used to identify agents and to link knowledge and events to agents. The module ID-SET declares the sort Id of identifiers and defines IdSet to be multisets of identifiers, using the same mechanism that we explained in the Vending machine definition of Marking. The attributes assoc, comm, and id: none, say that the union operation (_ _) is associative and commutative with identity none. The subsort declaration Id < IdSet says that identifiers are also singleton identifier sets.

```
fmod ID-SET is
   sort Id .  sort IdSet .   subsort Id < IdSet .
   op none : -> IdSet [ctor] .
   op _ _ : IdSet IdSet -> IdSet [ctor assoc comm id: none] .

   var id : Id .  var ids : IdSet .
   op member : Id IdSet -> Bool .
   eq member(id, id ids) = true .
   eq member(id, ids) = false [owise] .
endfm
```

The boolean function, member, tests if an identifier is an element of a given set. Variables are declared with the key word var and give the variable name and its sort. The definition of member uses Maude's builtin matching modulo AC capability to check if the second argument can be written in the form id ids. If so, we have found id in the given set. If the match fails, then id is not a member of the given set. The equation with the owise attribute applies in this case.

The structure of knowledge is specified in the KNOWLEDGE module.

```
fmod KNOWLEDGE is
   inc ID-SET .  inc NAT-TIME-INF .

   sort KB .  subsort KItem < KB .
   op none : -> KB [ctor] .
   op __ : KB KB -> KB [ctor assoc comm id: none] .

   sorts PKItem TKItem .  *** Persistent, Timed
   subsort PKItem TKItem < KItem .

   sort Info .
   op _@_ : Info Time -> TKItem .

   *** knowledge partial order
   op _<<_ : KItem KItem -> Bool .
   eq ki << ki' = false [owise] .

      op addK : KB KB -> KB .
      ...
   endfm
```

The statements

```
inc ID-SET .   inc NAT-TIME-INF .
```

are import statements. They effectively include the declarations and operations of the modules ID-SET and NAT-TIME-INF into KNOWLEDGE. The module NAT-TIME-INF models discrete time as natural numbers and adds a notion of infinity which is a convenient for many purposes.

The main knowledge sorts are KItem (knowledge item) and KB (knowledge base, a multiset of knowledge items). PKItem (persistent knowledge items) and TKItem (time dependent knowledge items) are subsorts of KItem. The sort Info represents the content of a TKItem, and becomes a TKItem when timestamped (info @ t). Key to managing evolving knowledge is the partial order, <<, on knowledge items. ki << ki′ is intended to mean that ki′ supersedes ki. Often this is because ki′ has a later timestamp, but this is not necessarily so. The function addK(kb0,kb1) adds the knowledge base kb1 to kb0 using the partial order to drop any superseded knowledge.

The following extract from modules LOCATION-KNOWLEDGE and CLOCK-KNOWLEDGE specifies three forms of knowledge provided by the soft agent framework for use in specific agent models.

```
*** knowledge elements available to all models
  sort Class .
  op class : Id Class -> PKItem .
  op clock : Time -> KItem [ctor] .
  sort Loc .
  op noLoc : -> Loc [ctor] .
  op atloc : Id Loc -> Info [ctor] .

  vars t0 t1 : Time .
  var id : Id .    vars loc0 loc1 : Loc .
  eq clock(t0) << clock(t1) = t0 < t1 .
  ceq atloc(id,loc0) @ t0 << atloc(id,loc1) @ t1 =  t0 < t1 .
```

Soft agents are organized in classes, similar to object oriented models. The operator class defines a (partial) mapping from identifiers to classes. Soft agent classes are intended to capture fixed physical aspects that don't change over time. Thus the return sort for class is PKItem. For example, the intent is that an agent of class Bot (such as our patrol bot), that rolls on wheels, should not turn into an agent of class Drone, that flys. Information constructors could be added to capture information about behavior class that might change over time, for example to specify an agents *role* in a protocol or other coordinated activity.

The term atloc(id,loc0) @ t0 says that the agent with identifier id was at location loc2 at time t0. The operator clock is an exception to the convention that time sensitive items are obtained by time stamping information terms. A term clock(t0) in a local knowledge base says that the agents local time is t0.

The equations defining << for clocks and location formalize the decision that only the latest information should be kept. For example letting

```
lkb = clock(0) class(b(0),Bot) atloc(b(0),pt(0,0)) @ 0)
```

where b(0) is an identifier, pt(0,0) and pt(1,0) are grid locations, and Bot is a class, we have the following

```
addK(lkb, atloc(b(0),pt(1,0)) @ 1) =
   clock(0) class(b(0),Bot) (atloc(b(0),pt(1,0)) @ 1)
```

because $0 < 1$.

Events. Agents behavior is organized using events, formalized in the module EVENTS starting with the main sorts, Event and EventSet (multisets of events).

```
fmod EVENTS is inc KNOWLEDGE .

  sorts Event EventSet .    subsort Event < EventSet .
  op none : -> EventSet .
  op __ : EventSet EventSet -> EventSet [ctor assoc comm id: none] .

*** Tasks
  sort Task .
  op tick : -> Task [ctor] .  --- default tasks

*** Actions (must be annotated by ids)
  sorts Action ActSet .  subsort Action < ActSet .
  op none : -> ActSet [ctor] .
  op __ : ActSet ActSet -> ActSet [ctor assoc comm id: none] .
*** immediate events .
  sort IEvent .  subsort IEvent < Event .
  op rcv : KB -> IEvent [ctor] .
*** delayed events
  sort DEvent .  subsort DEvent < Event .
  op _@_ : Task Time -> DEvent .
  op _@_;_ : Action Time Time -> DEvent .
endfm
```

There are two kinds of event, immediate events (sort IEvent) and delayed events (sort DEvent). rcv(kb) is an immediate event, representing notification of newly arrived shared knowledge. Delayed events are built from tasks (sort Task) and actions (sort Action). Tasks are used to trigger agents to process information and schedule actions to be carried out. task @ delay is a delayed event specifying that task should be done after delay time units have passed. The generic task tick is provided by the framework. A specific agent model can introduce additional tasks as needed. act @ delay ; duration is a delayed event specifying that the action act is to be enacted starting after delay time units and lasting duration time units. For example in response to a tick, a patrol bot with identifier b(0) might schedule mv(b(0),E) @ 0 ; 1— move east, now, for one time unit. It might also schedule tick @ 2, to repeat the processing again in 2 time units. If the patrol bot is one square north of a charging station it could schedule

    (mv(b(0),S) @ 0 ; 1) (charge(b(0)) @ 1 ; 2) tick @ 3

to move to the charging station, charge for two time units and then repeat the task processing. We note that action terms must have an identifier, since ready actions from all agents are collected and executed concurrently. Thus it is necessary to know who is doing the action.

3.3   Agents and Configurations

Now we can formalize the structure of a soft agent (module AGENTS) and of agent systems (module AGENT-CONF).

```
fmod AGENTS is
    inc LOCATION-KNOWLEDGE .   inc CLOCK-KNOWLEDGE .
    inc EVENTS .

    sorts Attribute AttributeSet .
    subsort Attribute < AttributeSet .
    op none : -> AttributeSet  [ctor] .
    op _,_ : AttributeSet AttributeSet -> AttributeSet
             [ctor assoc comm id: none] .

    sort Agent  .
    op [_:_|_] : Id Class AttributeSet -> Agent [ctor] .
    op lkb`:_ : KB -> Attribute [ctor] .
    op ckb`:_ : KB -> Attribute [ctor] .
    op evs`:_ : EventSet -> Attribute [ctor] .
endfm
```

In the spirit of object oriented specification, an agent has an identifier (recall the module IDSET), a class, and a set of attributes. Three attributes are essential: lkb (the local knowledge base); ckb (the knowledge cache for sharing); and evs (pending events for this agent). A specific agent model might introduce additional attributes if needed. The initial state of a patrol bot might look like the following

```
    [ b(0) : Bot |
      lkb: class(b(0),Bot) (myDir(b(0),E) @ 0) (myY(b(0),0) @ 0)
           class(s(0),Station) (atloc(s(0),pt(2,1)) @ 0),
      ckb: none, evs: tick @ 1]
```

where Station is the class of charging stations, myY(b(0),0) says this bot should travel along $y = 0$ and myDir(b(0),E) says the bot is currently traveling east. In this case the local knowledge base does not store location and energy level since the corresponding sensors are read each time the bot processes a task.

A soft agent system configuration (sort `Conf`) is a multiset of configuration elements (sort `ConfElt`). The sort `ASystem` is used for top-level configurations.

```
fmod AGENT-CONF is
  inc AGENTS .
  sorts  ConfElt Conf .
  subsorts Agent  < ConfElt < Conf .
  op none : -> Conf .
  op __ : Conf Conf -> Conf [ctor assoc comm id: none] .

  sort Env . subsort Env < ConfElt .
  op [_|_] : Id KB -> Env [ctor] .

  op restrictKB : Id KB -> KB .
  ...

  sort ASystem .
  op '{_'} : Conf -> ASystem .
  op mte : Conf -> TimeInf .
endfm
```

An agent is a configuration element (subsort `Agent < ConfElt .`), and environment objects (sort `Env` are also configuration elements (subsort `Env < ConfElt .`). An environment object has an identifier and a knowledge base. A top-level system should have a unique environment object. The environment knowledge base includes the (physical) state of each agent as well as environment knowledge such as terrain, obstacles, or weather conditions. For example an environment knowledge base for a single patrol bot scenario could be the following

```
clock(3)
class(b(0),Bot)  (atloc(b(0),pt(2,0)) @ 3)  (energy(b(0),10)) @ 3)
class(s(0),Station)  (atloc(s(0),pt(2,1)) @ 0)
wind(N,pt(1,2)) @ 0)
```

where 3 time units have passed and the bot is now at grid location $(2,0)$ with energy 10. There is northerly wind at grid location $(1,2)$.

The function `restrictKB(id, kb)` selects the knowledge items in `kb` with identifier `id`. For example, restricting the above knowledge base to `b(0)` results in the following.

```
class(b(0),Bot)  (atloc(b(0),pt(2,0)) @ 3)  (energy(b(0),10)) @ 3)
```

The operator `{_}` is used to collect configuration elements to define a top-level system state of sort `ASystem`. The function `mte` computes how much time can elapse before some task is ready (0 delay).

Specific agent models can introduce new configuration elements to facilitate analysis. For example, simple flags can be defined that indicate a goal has been reached or an invariant has failed. Metadata elements can be defined that monitor execution state and collect information.

3.4 Rules

There are two rewrite rules: doTask that controls carrying out agent tasks, and timeStep, that carries out actions concurrently, propagates any sharable knowledge, and increments time.

The doTask rule uses the doTask function to determine how to update the agents state.

```
crl[doTask]:
 [id : cl | lkb : lkb,  evs : ((task @ 0) evs),  ckb : ckb, ats]
 [eid | ekb ]
=>
 [id : cl | lkb : lkb', evs : evs',  ckb : ckb', ats] [eid | ekb ]
if t := getTime(lkb)
/\ {ievs,devs} := splitEvents(evs,none)
/\ {lkb',evs',kb} kekset := doTask(cl,id,task,ievs,devs,ekb,lkb)
/\ ckb' := addK(ckb,kb)
[print "doTask:" id  "! time:" t "!!" evs' "\n" kekset].
```

The function getTime(lkb) gets the local time from the clock knowledge of a knowledge base, in this case the agents view of the current time. splitEvents splits the input event set into a pair consisting of the immediate events (ievs), and the delayed events (devs). Immediate events are to be processed by the agent. The only immediate events provided by the framework are rcv events, holding knowledge received from other agents by sharing. Delayed events are actions to be executed by the environment. The agent is free to add to or delete/cancel elements of its delayed events. The doTask function is declared as part of the framework,

```
op doTask : Class Id Task EventSet EventSet KB KB
          -> KBEventsKBSet.
```

but equations defining this function must be provided by each model. doTask is given the agents class (cl), its identifier (id), and the task to be addressed (task).[3] In addition it is given unprocessed knowledge acquired by knowledge sharing (ievs), the agents current scheduled tasks and actions (devs, delayed events), the agents local knowledge base (lkb), and the environment knowledge base from which it can extract the agents current sensor readings (ekb) and other local information. doTask returns a triple: the updated local knowledge base (lkb'), updated scheduled tasks and actions (evs'), and information items to be shared with other agents (kb), which is added to the agents cache to produce the updated cache knowledge base (ckb').

The timeStep rule concurrently performs actions using the function doEnvAct. It updates delays and durations in the event sets (counting down) using the function timeEffect. It propagates sharable knowledge using the

---

[3] In fact, we use a single task, tick, to schedule the next invocation of an agents cyberactivity. Information about what to do is kept in the local knowledge base.

function shareKnowledge. The function updateAConf is an auxiliary function used during model analysis to collect and update metadata (stored as configuration elements that are ignored by other framework functions that operate on configurations).

```
crl[timeStep]:
{ aconf }
=>
{ aconf2 }
if nzt := mte(aconf)
/\ t := getTime(envKB(aconf))
/\ ekb' := doEnvAct(t, nzt, envKB(aconf), effActs(aconf))
/\ aconf0 := updateEnv(ekb',timeEffect(aconf,nzt))
/\ aconf1 := shareKnowledge(aconf0)
/\ aconf2 := updateConf(aconf1)
 [print "eAct:" ekb' "\ntimeStep:" t "++" nzt] .
```

Note that both rules have a print attribute. The print attribute specifies a string to be printed (if the print option is turned on) each time the rule is applied. The keyword print is followed by one or more string literals and/or variables bound by rule matching. This is a handy way to produce an execution trace with just information you care about.

The function doEnvAct determines the effects of agents actions on the environment, including the agents physical state and interactions with neighbors. The function is given the current time (t), the maximum time that can elapse (nzt), the current situation (represented by the environment knowledge base, envKB(aconf)), the actions to be executed (effActs(aconf), actions, collected from all agents, that are no longer delayed), and the amount of time that has elapsed in the current execution round. doEnvAct returns the environment knowledge base resulting from (concurrent) execution of these actions along with the amount of time passed (t') which is as most nzt. This is used in case execution fails and less than the allotted time has passed.

doEnvAct proceeds by time increments (in the discrete case 1 time unit). The ready actions are concurrently executed for 1 time increment, each in the same initial environment using the function doUnitEnvAct. The combined environment resulting from the concurrent effects is used for the next time increment.[4]

Specifically, doEnvAct is defined by the conditional equations

```
ceq doEnvAct(t, nzt, ekb, evs, t')
  = doEnvAct(s(t), nzt monus 1, ekb', timeEffect(evs,1), s(t'))
  if ekb' := doUnitEnvAct(t, ekb, evs, none)
  /\ isOk(ekb) .
```

---

[4] It is possible that the concurrent actions are in conflict. In the current framework, we have two options. One is that time stops, and the user can investigate what went wrong. Or some actions are arbitrarily chosen to succeed and others fail. This is not unreasonable, since true simultaneity is rare.

```
eq doEnvAct(t, t'', ekb, evs, t') = {ekb,t'} [owise]
  --- applies if t'' is 0 or ekb' is not Ok

eq doUnitEnvAct(t, ekb, (act @ 0 ; nzt) evs, ekb')
  = doUnitEnvAct(t, ekb, evs,
     addK(ekb',doUnitEnvOneAct(act,ekb,t))) .
```

isOk recognizes impossible environments (like two agents in the same location). The function doUnitEnvAct simply iterates over the events, accumulating environment updates returned by doUnitEnvOneAct(act,ekb,t).

doUnitEnvOneAct is the heart of the matter. It models the physics of the action, act. Recall that actions are required to have an identifier component that determines the agent doing the action.

As an example, a move action mv(id,dir) will change the location of the agent identified by id, according to environmental conditions such as boundaries, terrain slope, or wind. The default is to move one unit in the specified direction dir.

4   Case Studies

In this section we present two soft agent case studies. The first is the simple Patrol Bot mentioned in Sect. 3 that we now formalize in some detail. The second is a more complex surveillance drone case study that we summarize briefly focusing on a set of formal analysis results. We begin by defining an extension of the soft agent framework to support use of soft constraints to specify agent behavior that is robust to modest disruptions.

4.1   Soft Constraints

Although the framework does not enforce a particular mechanism for defining the doTask function, we provide an example template for this function to facilitate use of soft constraint problem solving to determine what actions an agent might consider in a given situation.

The idea of soft constraints is to constrain possible values of a set of variables by mapping such assignments to a partially ordered domain and then selecting assignments with maximal value. Traditionally soft constraints use an algebraic structure such as c-Semirings as the valuation domain. Such structures have good properties with respect to different forms of composition [17,18,24]. In our examples, we do not need all the machinery of c-Semirings, so to simplify the formalization so we defined a theory VALUATION that captures what we do require of a valuation domain for our SCPs.

```
fth VALUATION is
  pr BOOL .  inc SOFT-AGENTS .
  sort Grade .
  op equivZero : Grade -> Bool .
```

```
op _<_ : Grade  Grade -> Bool .
op val : Id KB Action  -> Grade .
endfth
```

Specifically, there is a sort `Grade` (the values), a partial order `<` on `Grade`, a predicate, `equivZero`, that recognizes the minimal element(s) of `Grade`, and a valuation function `val` that evaluates an action from the point of view of the identified agent in the context of a knowledge base. Typically a specific instance of `val` will measure the effect of an action, carried out in a situation represented by the knowledge base–is it safe?, is it progressing towards some goal?, and so on. For example the evaluation of a patrol bots action with respect to energy (it should not run out of energy), might return 0 if the action leads to a state in which there is not sufficient energy to get to a charging station and 1 otherwise. This could be refined to return $1/2$ if the energy is sufficient to get to a charging station, but with less than $1/4$ the energy capacity to spare (formalizing a notion of caution).

The module `SOLVE-SCP` provides a simple mechanism for solving a soft constraint problem by finding the maximal solutions, i.e. the maximally graded actions. It is parameterized by a module `Z` that satisfies the theory `VALUATION`.

```
fmod SOLVE-SCP{Z :: VALUATION} is
   inc SOFT-AGENTS .

   sorts RankEle{Z} RankSet{Z} .
   op {_,_} : Z.Grade ActSet -> RankEle{Z} .

   subsort RankEle{Z} < RankSet{Z} .

   op updateRks : RankSet{Z} Action Z.Grade -> RankSet{Z} .
   op solveSCP : Id KB ActSet -> ActSet .
   eq solveSCP(id,kb,acts) = solveSCP!(id,kb,acts,none) .

   op solveSCP! : Id KB ActSet RankSet{Z} -> ActSet .
   eq solveSCP!(id,kb,none,rks) = getAct(rks) .
   ceq solveSCP!(id,kb,act actset,rks) =
         solveSCP!(id,kb,actset,rks1)
    if v0 := val(id,kb,act)
    /\ rks1 := updateRks(rks,act,v0) .
endfm
```

The function `solveSCP` (solve Soft Constraint Problem) takes an agent identifier, a knowledge base and an action set and returns the actions in the input action set with maximal grade according to the parameter theory. This function maintains a ranked action set where the ranks/grades are non zero and maximal among the actions considered so far. It uses the auxiliary function `updateRks` to update this set given the grade of an action. The function `equivZero` is used to ensure that the returned actions have non-zero grade. This method of solving soft constraint problems works well when there are only a few actions to consider. More efficient methods will be needed when situations get more complex.

To the degree possible we would like to derive valuation functions for agents from valuations with respect to different concerns. The choice of combination operation is not a simple task. Lexicographic combination of partial orders works in some cases [18], however it is not always possible to use a lexicographic ordering to obtain the desired combination. Initial steps towards a theory of composition of soft constraint problems/automata is presented in [24]. For the present the soft agent framework does not provide builtin compositions operations. Each agent model will need to develop its own valuation functions and combination methods. We will see examples in the case studies below.

4.2 doTask Template

We now introduce the specialization of the doTask function that uses solveSCP to determine the actions for an agent to consider. Recall that doTask returns a triple consisting of an update for the agents local knowledge base, an update for the agents event set, and knowledge to share. Then doTask is specified by the following conditional equation, that formalized the task process outlined in beginning of Sect. 3.

```
ceq doTask(cl,id,tick,ievs,devs,ekb,lkb) =
   if acts == none then
       {lkb2, devs (tick @ botDelay), none }
   else
       selector(doTask!(id,lkb2,devs (tick @ botDelay),acts))
    fi
 if lkb0 := handleS(cl,id,lkb,ievs)
 /\ lkb1 := getSensors(id,ekb)
 /\ lkb2 := proSensors(id,lkb0,lkb1)
 /\ acts0 := myActs(cl,id,lkb2)
 /\ acts := solveSCP(id,lkb2,acts0)  .
```

The functions handleS, getSensors, proSensors, and myActs are declared by the framework, but must be defined by each specific agent model. The function handleS processes the new shared knowledge (in ievs) producing an updated local knowledge base (lkb0). It could simply add the new knowledge, or could be more selective, do simple reasoning, or aggregate incoming data. The function getSensors reads relevant sensors (and local environment information) from the environment knowledge base (ekb), and the function proSensors processes the result, lkb1, updating lkb0 to produce the updated local knowledge lkb2. The function myActs returns a list of actions that are possible given the current situation, represented by lkb2. For example a charge action is only possible if the agent is at a charging station, and move actions may be constrained by boundaries or known obstacles.

The function doTask! constructs a result triple for each of the actions, act in acts, returned by solveSCP.

```
op doTask! : Id KB EventSet ActSet -> KBEventsKBSet .
ceq doTask!(id,lkb2,devs,act acts) =
    {lkb3,devs (act @ 0 ; 1),kbp} doTask!(id,lkb2,devs,acts)
 if kbp := tell(id,act,lkb2)
 /\ lkb3 := remember(id,act,lkb2) .
eq doTask!(id,lkb2,devs,none) = none .
```

doTask! uses remember(id,act,lkb2) to compute the local knowledge update, and tell(id,act,lkb2) to compute what new knowledge to share. By default, the updated events consist of devs, (tick @ botDelay) (to schedule the next execution of doTask), and act @ 0 ; 1 (scheduling the action to happen immediately for one time unit). This can be overridden as needed. The selector function, by default, returns all the result triples. Alternatively, one result could be picked arbitrarily (by Maude or by tossing a coin).

4.3   The Patrol Bot Case Study

The idea of patrol bots was introduced at the beginning of Sect. 3. Now we address the problem of specifying a patrol bot system, including knowledge, the possible actions of a patrol bot, the effects of these actions, and the soft constraint problem to be solved for deciding actions. Once individual patrol bots have been specified we define some scenarios to illustrate the use of the specifications to study possible system behaviors.

Recall that a patrol bot is moving on a grid, along some track (fixed y) continually going from one side to the other. Moving uses energy, so the patrol bot must recharge to avoid dying (so there must be a charging station somewhere on the grid). The patrol bot may drift off its path (faulty motor, or wind, . . .). It prefers to move along the assigned track so it should correct when it discovers it has drifted.

In addition to class, clock and location knowledge, patrol bot knowledge includes energy (a sensor reading), caution (how much reserve energy to keep), and its patrolling parameters myY,myDir.

```
**** Info constructors
  op energy : Id FiniteFloat -> Info [ctor] .
  op caution : Id FiniteFloat -> Info [ctor] .
  op myY : Id Nat -> Info [ctor] .    *** the bot's track
  op myDir : Id Dir -> Info [ctor] .  *** current direction
**** partial order--new information superceeds old
  eq energy(id,e0) @ t0 << energy(id,e1) @ t1 =  t0 < t1 .
  eq caution(id,e0) @ t0 << caution(id,e1) @ t1 =  t0 < t1 .
  eq myY(id,y0) @ t0 << myY(id,y1) @ t1 = t0 < t1 .
  eq myDir(id,dir0) @ t0 << myDir(id,dir1) @ t1 = t0 < t1 .
```

A patrol bot has two kinds of action: charge, to restore energy; and mv, to move one step in the given direction. The directions E,W,N,S stand for East, West North, and South (or left, right, up, and down).

```
*** actions
  sort Dir .    ops E W N S : -> Dir [ctor] .
  op mv : Id Dir -> Action [ctor] .
  op charge : Id -> Action [ctor] .
```

The effects of patrol bot actions depend on several global parameters, listed below.

```
*** global parameters
  ops gridX gridY : -> Nat [ctor] .          *** grid dimensions
  op chargeUnit : -> FiniteFloat [ctor] .    *** energy gained
  op maxCharge : -> FiniteFloat [ctor] .     *** energy capacity
  op botDelay : -> Nat [ctor] .              *** time between ticks
  op costMv : -> FiniteFloat [ctor] .        *** cost of moving
```

An agent's model of the effects of actions is given by the functions doMv and doAct.

```
  op doMv : Loc Dir -> Loc .
  op doAct : Id KB Action -> KB .
```

doMv simply returns the location after the move. In case the result would be off the grid, the initial location is returned. doAct updates the given knowledge base with the result of the action. Energy will be decreased in the case of a move action and increased by chargeUnit in the case of a charge action. The model assumes actions are carried out for one time unit, thus the new information is time stamped with a time that is one time unit in the future, the time the action completes.

The physical model of the effect of actions is given by doUnitEnvOneAct which is used in the timeStep rule. For actions that operate as expected, given the physical state (as reflected by the environment knowledge base), the doUnitEnvOneAct result is the same as doAct. This is the ideal case. In other cases doUnitEnvOneAct will differ. For example, charging stops when the maxCharge is reached. If there is an obstacle in the target of a move, then the agent doesn't move, but it does use energy trying. If there is wind, the final location may be different depending on the strength and direction of the wind.

doTask Auxiliary Functions. Recall that an agents behavior is specified by the function doTask that is defined in terms of auxiliary functions for handling shared knowledge and processing sensors to define the soft constraint problem to be solved. The function handleS processes newly received shared knowledge by simply adding it to the local knowledge base.

```
  op handleS : Class Id KB EventSet -> KB .
  eq handleS(cl,id,lkb,rcv(kb) ievs) = addK(lkb,kb) .
```

Sensors are read by restricting the environment knowledge base to the patrol bots identity.

```
op getSensors : Id KB -> KB .
eq getSensors(id,ekb) = restrictKB(id,ekb) .
```

Sensor processing adds sensed location and energy information to the local knowledge base. It also reverses the direction (reverseDir(dir)) if the current location is at one of the vertical edges (atVEdge(loc)).

```
op proSensors : Id KB KB -> KB .
ceq proSensors(id,lkb,ekb) = lkb1
  if ((myDir(id,dir) @ t0) clock(t) lkb0) := lkb
  /\ (atloc(id,loc) @ t1) (energy(id,e) @ t2) ekb0 := ekb
  /\ dir1 := (if atVEdge(loc) then reverseDir(dir) else dir fi)
  /\ lkb1 := addK(lkb0 (myDir(id,dir1) @ t) clock(t),
                  (atloc(id,loc) @ t1) (energy(id,e) @ t2)) .
```

By reversing the direction when the patrol bot reaches an edge it will always be able to move in the 'current' direction.

The possible actions in a given situation (myMvs(Bot,id,lkb)) include charging, if current location is that of a station and not fully charged, and moves in any direction that do not lead to an occupied location or a location that is off the grid (myMvs(Bot,id,lkb)).

```
op myActs : Class Id KB -> ActSet .
eq myActs(Bot, id,lkb) =
(myMvs(Bot,id,lkb)
(if canCharge(Bot,id,lkb) then charge(id) else none fi)) .
```

The functions remember and tell are used by the auxiliary doTask! to assemble triples from the set of actions returned by solveSCP.

```
op remember : Id Action KB -> KB .
eq remember(id,act,lkb) = lkb .
op tell : Id Action KB -> KB .
eq tell(id,act,(atloc(id,loc) @ t) lkb) = (atloc(id,loc) @ t) .
```

A patrol bot remembers all the information gathered to define the soft constraint problem. In fact, the location and energy could be forgotten as they are simply overridden during the next task processing. Only the location is shared. This is useful when there are several patrol bots, to avoid potential collisions or blocking.

Valuation Functions. What remains is to specify the valuation function used by a patrol bot to choose among available actions. The valuation function should ensure that the patrol bot does not run out of energy (assuming sufficient charge capacity and accessible charging stations). It should also ensure that the patrol bot continually patrols, i.e. it reaches one side, turns, reaches the other side, turns .... Given the two different *concerns* we define two valuation functions: val-energy for assuring the energy requirement and val-patrol for maximizing patrolling progress. val-energy is overloaded in that it can evaluate a knowledge base (val-energy(id,kb)) or an action in the context of a knowledge base (val-energy(id,kb,act)). In either case the valuation is relative to a specific patrol bot, identified by id. val-energy returns an element of TriVal which has three elements ordered by bot < mid < top.

```
op val-energy : Id KB Action -> TriVal .
eq val-energy(id,
          (energy(id,e) @ t0) (atloc(id,loc0) @ t)
          (atloc(st,loc1) @ t1) (class(st,Station)) kb,
          charge(id)) =
      (if (loc0 == loc1)
       then (if (e >= maxCharge) then bot else top fi)
       else bot fi) .
```

For a charge action, the value is top if the patrol bot is at a charging station and is not fully charged. Otherwise the value is bot. In the case of move actions, the resulting knowledge base is evaluated for energy safety.

```
ceq val-energy(id,kb,mv(id,dir)) =
          val-energy(id,doAct(id,kb,mv(id,dir)))
 if not (val-energy(id,doAct(id,kb,mv(id,dir))) == mid) .
```

If the result of knowledge base valuation is not mid then that result is returned.

```
ceq val-energy(id,kb,mv(id,dir)) =
        if towards(dir,loc,locb) then  mid else bot fi
  if val-energy(id,doAct(id,kb,mv(id,dir))) == mid
  /\ (atloc(id,loc) @ t) (atloc(st,locb) @ t1)
     (class(st,Station)) kb' := kb .
```

If the result of knowledge base valuation is mid then energy valuation prefers a move in the direction of a charging station.

```
op val-energy : Id KB  -> TriVal .
eq val-energy(id,(energy(id,e) @ t0) (atloc(id,loc) @ t)
     (atloc(st,locb) @ t1) (class(st,Station)) kb) =
  if e > cost2loc(loc,locb) + caution then top else
   (if e > cost2loc(loc,locb) then mid else bot fi) fi .
```

Energy valuation of a knowledge base uses the caution parameter that determines how much reserve energy it prefers. If the situation allows the agent to reach a charging station with out running out of energy, but the reserve energy is less than the caution parameter the value returned is mid. Otherwise the value is top if it is energy safe and bot if unsafe.

The valuation from a patrolling perspective, computed by val-patrol, has a range of 0.0 to 1.0. The value for charging is 1.0 assuming it will not be asked if charging is not feasible.

```
op val-patrol : Id KB Action -> Float .
eq val-patrol(id,
        (atloc(id,pt(x,y)) @ t) class(id,Bot)
        (myDir(id,dir) @ t0) (myY(id,y0) @ t1) kb,
         mv(id,dir1)) =
     (if (y0 < y)
      then (if   (dir1 == S) then 0.9 else 0.0 fi)
      else (if (y < y0)
            then (if   (dir1 == N) then 0.9 else 0.0 fi)
            else (if   (dir == dir1) then 0.9 else 0.0 fi) fi) fi).
eq val-patrol(id, kb, charge(id))   = 1.0 .
```

A move is given value 0.9 if the agent is off track and the move corrects, or if the move is in the current direction. From a patrolling perspective, charging is preferred if it is possible, otherwise moves that correct or move towards the current goal are preferred.

The combined valuation function val returns a pair: the first component is the energy valuation and the second component is the patrol valuation.

```
sort BUVal .
op {_,_} : TriVal Float -> BUVal .
op val : Id KB Action -> BUVal .
eq val(id,kb,act) =
    {val-energy(id,kb,act),val-patrol(id,kb,act)} .
op _<_ : BUVal BUVal -> Bool .
op equivZero : BUVal -> Bool .
eq {b1,u1} < {b2,u2} = (b1 < b2) or (b1 == b2 and u1 < u2) .
eq equivZero({b1,u1}) = (equivZero(b1)) .
```

The partial order on these pairs is the lexicographic order [18] with energy valuation (safety) given preference. A value pair is equivalent to zero just if the energy component is equivalent to zero. Thus energy consideration alone can veto an action. But an action with non-zero energy value is acceptable even if the patrol value is equivalent to zero (i.e., is 0.0).

Here are a few examples. We assume a knowledge base with a patrol bot moving east, along $y = 0$, in a 5×3 grid.

```
loc    energy caution act    val          comment
(2,0)  5.0    1.0     mv(E)  {top, 0.9}   min caution move E
                      mv(N)  {top, 0.0}
(2,0)  5.0    4.0     mv(E)  {bot, 0.9}    more caution, N wins
                      mv(N)  {mid, 0.0}
(2,1)  10.0   1.0     charge {top,1.0}
(2,1)  25.0   1.0     charge {bot,1.0} fully charged
```

With caution 1.0 and energy 5.0 moving east is preferred, but then the patrol bot will need to backtrack. With caution 4.0 and energy 5.0 moving north to the charging station is preferred over moving east in the patrolling direction since {bot, n} < {mid,n'} = true. Also, when at the charging station and fully charged, a charge action will not be considered since not(equivZero({mid,0}) = true.

4.4   Patrol Bot Scenarios

In this section we show how rewriting and search can be used to gain understanding of agent system behavior, the reasons for failures, and the effects of
changing parameters. To illustrate how environment effects can be introduced
we add potential for wind to blow a patrol bot off course. It is not intended to be
particularly realistic as a model of wind, but it does allow us to see how robust
the patrol bots can be by just varying the frequency of disruption and the level
of caution.

```
op wind : Dir Nat -> Info .
op windEffect : Loc KB -> Loc .
ceq windEffect(10, (clock(t)) (wind(dir,n) @ t0) ekb)
  = doMv(10,dir)
 if t rem n == 0 .
eq windEffect(10,ekb) = 10 [owise] .
```

wind(dir,n) @ t0 specifies wind effect in direction dir to be applied if
the current time is divisible by n. The point is that we don't want continual
wind. Periodic wind for different periods can test robustness without needing to
introduce machinery for probability distributions. The function windEffect is
applied to the result of a move in the function doUnitEnvOneAct that is the
core of the doEnvAct function used in the timeStep rule.

A Patrol bot system consists of one or more patrol bots each with their own
track and current direction, together with an environment. A system can be
instrumented for analysis in various ways. For example, to bound an execution
we define a new configuration element.

```
op bound : Nat -> ConfElt .
```

The auxiliary function updateConf that is applied during the timeStep rule
is used to decrement the bound each time step. When the bound reaches 0 the
configuration is replaced by a constant goalConf for which there are no rewrite
rules, so execution/search must stop. To analyze a system we define a notion of
critical configuration and carry out bounded search for such configurations.

```
ops criticalConf goalConf : -> ConfElt .
op critical : Conf -> Bool .

ceq critical([eId | (energy(id,ff) @ t) kb ] aconf) = true
 if equivZero(val(id,(energy(id,ff) @ t) kb)) .
eq critical(aconf) = false [owise] .

eq updateConf(bound(n) aconf) =
  if critical(aconf) then criticalConf aconf
   else (if (n == 0) then goalConf
        else bound(monus(n)) aconf fi) fi .
```

When a critical configuration is reached the constant criticalConf is added to the configuration by the function updateConf to simplify specifying the search command. For our examples, we define a critical configuration to be one in which the valuation for some agent is equivalent to zero using the boolean function critical.

We begin with a one patrol bot configuration, watch it run and search for critical configurations in a family of configurations parameterized by the level of caution and the frequency of wind. Then we will look at what happens when a second patrol bot is added. In all cases we will use the same global parameters:

```
eq gridZ = 5 .
eq gridY = 3 .
eq chargeUnit = 5.0 .
eq maxCharge = 20.0 .
```

To simplify notation we define an agent template

```
B(I,X,Y,Z,D,C)  =
  [b(i) : Bot |
    lkb : (clock(0) class(b(i), Bot)
           class(st(0), Station) (atloc(st(0), pt(2, 1)) @ 0)
           (atloc(b(I), pt(X,Y)) @ 0) (energy(b(I), 1.5e+1) @ 0)
           (myY(b(I),Z) @ 0) (myDir(b(I),D) @ 0)
  caution(b(I),C) @ 0),
      ckb : none,
      evs : (tick @ 1) ]
```

Thus B(I,X,Y,Z,D,C) is the state of a patrol bot at time 0 with identifier I, location pt(X,Y), track Z, direction D, and caution C. The family of agent systems with one patrol bot parameterized by level of caution, C, and Wind, Asys1(C,Wind), is given by

```
Asys1(C,Wind) =
{bound(200)
[eI | clock(0) class(b(0), Bot)
      (atloc(b(0), pt(0, 0)) @  0)   (energy(b(0), 15) @ 0)
      class(st(0), Station) (atloc(st(0), pt(2, 1)) @ 0)
      Wind  ]
 B(0,0,0,0,W,C)
}
```

The system Asys1(C,Wind) has a single patrol bot, with identifier b(0), location pt(0,0), moving west, with energy 15 and caution C. In the environment there is wind specification Wind. Note that the patrol bot will immediately turn and head east since it is at the western edge of the grid.

To watch the system run, we turn on printing of print attributes and rewrite for 20 steps. The following is a simplified version of what is printed. Following the eAct tag is the patrol bots location and energy at the end of the timeStep. Following the doTask tag is the patrol bot identifier, task, current time, and event set produced.

```
set print attribute on .
Maude> rew [20] updAkb(asys(200),b(0),caution(b(0),1.0) @ 0) .

eAct: clock(0) (atloc(b(0),pt(0,0)) @ 0) (energy(b(0),15) @ 0
timeStep: 0 ++ 1
doTask: b(0) ! tick time: 1 !! (tick @ 1) mv(b(0),E) @ 0 ; 1
eAct: clock(1) (atloc(b(0),pt(1,0)) @ 2) energy(b(0),14) @ 2
timeStep: 1 ++ 1
....
doTask: b(0) ! tick time: 4 !! (tick @ 1) mv(b(0),E) @ 0 ; 1
eAct: clock(4) (atloc(b(0),pt(4,0)) @ 5) (energy(b(0),11) @ 5)
timeStep: 4 ++ 1
**** the bot reversed direction
doTask: b(0) ! tick time: 5 !! (tick @ 1) mv(b(0),W) @ 0 ; 1
eAct: clock(5) (atloc(b(0),pt(3,0)) @ 6) (energy(b(0),10) @ 6)
timeStep: 5 ++ 1
. . . . .
```

Initially there is nothing to do since the patrol bot has a task with delay 1. Then doTask and timeStep alternate, with the patrol bot moving east until it reaches the edge. At time 5 it reverses direction and starts moving west. By adjusting what is printed one can observe just variables of interest and look for unexpected behavior.

Now we look for critical configurations starting with instances of the one patrol bot family using the search command

```
search [1] Asys1(C,Wind) =>+ {criticalConf aconf} .
```

for different values of Wind and C.

Table 1 summarizes the search results. We see that in ideal conditions, minimal caution seems good enough to ensure no critical configurations are reached. Minimal caution works for 'modest' wind conditions (N 17, S 13 or N 11, S 7).

Table 1. The columns Wind and C are values of the corresponding parameters. Found? indicates whether a critical configuration was found (in less that 200 time units). States is the number of states visited in the search, Rewrites is the number of rewrites, and Duratin is the search time in milliseconds. N n, S m stands for the wind items (wind(N,n) @ 0) (wind(S,m) @ 0).

Wind	C	Found?	States	Rewrites	Duration (ms)
none	1.0	no	406	59948	184
none	4.0	no	406	59948	184
N 17, S 13	1.0	no	1049	460717	474
N 11, S 7	1.0	no	1195	524411	564
N 7, S 5	1.0	yes (1)	2216	926342	935
N 7, S 5	4.0	no	6605	3022120	3207

A critical configuration is found at state 2215 with Wind N 7, S 5 and caution 1.0. The environment component of the found critical configuration is the following:

```
[eI | clock(191) class(b(0), Bot) class(st(0), Station)
        (atloc(b(0), pt(2,0)) @ 191) (energy(b(0), 1.0) @ 191)
        (atloc(st(0), pt(2, 1)) @ 0)
(wind(N,7) @ 0) wind(S, 5) @ 0]
```

Thus more caution (4.0) is needed when wind effects are more frequent. Recall that the effect specified by wind(dir,n) @ 0 is wind blowing in direction dir every n time units. Smaller values of n mean wind blowing more often and thus a greater chance to interfere with a patrol bots progress.

Caveat. The above searches were limited to time less than 200. This is already useful to find problems. In practice we may only need a given patrol instance to operate for a limited time, for example overnight or weekends. In this case bounded search is sufficient. If not, there are several ways to consider to extend the analysis. These will be discussed in Sect. 5.

The family of agent systems with two patrol bots parameterized by Wind and level of caution is given by

```
Asys2(C,Wind) =
{bound(200)
[eI | clock(0) class(b(0), Bot) class(b(1), Bot)
      (atloc(b(0), pt(0, 0)) @ 0)  (energy(b(0), 15) @ 0)
      (atloc(b(1), pt(4, 2)) @ 0)  (energy(b(1), 15) @ 0)
      class(st(0), Station) (atloc(st(0), pt(2, 1)) @ 0)
      Wind ]
 B(0,0,0,0,W,C)
 B(1,4,2,2,E,C)
}
```

The system Asys2(C,Wind) extends Asys1(C,Wind) with a second patrol bot, with identifier b(1), location pt(4,2), moving east, with energy 15 and caution C. The environment is also extended with class, location and energy knowledge about the second patrol bot.

The Table 2 summarizes results of searching for critical configurations in instances of the two patrol bot system using the search command

```
search [1] Asys2(C,Wind) =>+ {criticalConf aconf},
```

for different values of Wind and C.

The critical configuration (N 7, S 5, 4.0) is found at state 113. The environment component of the found critical configuration is the following:

```
[eI | clock(15) class(b(0), Bot) class(b(1), Bot)
      (atloc(b(0), pt(2, 0)) @ 15) (energy(b(0), 19) @ 15)
      (atloc(b(1), pt(2, 2)) @ 15) (energy(b(1), 1.0) @ 15)
      class(st(0), Station) (atloc(st(0), pt(2, 1)) @ 0)
```

We can get an idea of how the critical configuration arises in the two patrol bot scenario by using the command

```
show path 113 .
```

which shows the sequence of states and rules applied leading to the critical configuration (state 113). The following shows the clock and patrol bot location and energy information in environment components of the last three states.

```
state 89, ASystem: {bound(187)
[eI
| clock(13)
(atloc(b(0), pt(2, 1)) @ 11) (energy(b(0), 15) @ 13)
(atloc(b(1), pt(3, 1)) @ 13) (energy(b(1), 3.0) @ 13) ]
```

Patrol bot b(0) is at the station, charging, and b(1) is next to the station, presumably intending to enter.

```
state 101, ASystem: {bound(186)
[eI
| clock(14)
(atloc(b(0), pt(2, 1)) @ 11) (energy(b(0), 20) @ 14)
(atloc(b(1), pt(3, 1)) @ 14) (energy(b(1), 2.0) @ 14) ]
```

Now b(0) is fully charged, and b(1) is still waiting, but it has used one energy unit trying to enter the station.

```
state 113, ASystem: {criticalConf
[eI
| clock(15)
(atloc(b(0), pt(2, 0)) @ 15) (energy(b(0), 19) @ 15)
(atloc(b(1), pt(2, 2)) @ 15) (energy(b(1), 1.0) @ 15) ]
```

b(0) has left the station. b(1) could enter at the next time, but it has used up its energy. It seems that the patrol bot is not paying attention to the fact that the station is occupied. Perhaps it should just wait until the station is free before trying to move there.

Table 2. The table columns are the same as Table 1 for one patrol bot.

Wind	C	Found?	States	Rewrites	Duration (ms)
none	4.0	no	853	669244	684
N 11, S 7	4.0	no	64154	51976162	65039
N 7, S 5	4.0	yes (1)	114	101739	122
N 7, S 5	6.0	no	243096	209875128	504088

4.5   Surveillance Drone Case Study

Now we look at a more complex case study involving a surveillance problem: there are P points of interest and we need to continually have recent pictures of the area around each point. This case study is inspired by a project to develop drones with the ability to monitor health in agricultural fields. In this project formal models are being used to explore different strategies for robustly meeting the recency requirement we formalize an abstract version of the problem as follows. The points are distributed on a grid with dimensions $x_{max} \times y_{max}$. N drones are deployed to take pictures. As for patrol bots, the drones use energy to fly from one place to another, and to take pictures, and they have maximum energy of e_{max}. There is a charging station in the center of the grid. In addition to its charging service, the station serves as a knowledge exchange cache, so drones can share information with each other by sharing with the station. Drones use soft-constraints, which take into account the drone's position, energy, and picture status of the points, to rank their actions. They may perform any one of the best ranked actions. We use M to denote the maximal acceptable age of a picture. Thus a critical configuration is one in which a drone runs out of energy, or the latest picture at some point is older than M.

As for the patrol bot system, we searched for critical configurations. Here we used a time bound of $n = 4 \times M$. The search results are summarized in Table 3. We varied M and the maximum energy capacity of drones e_{max} (instead of varying caution).

Note that even when considering a large grid (20×20) and three drone, search finds critical configurations or covers the full bounded search space quite quickly (less than a minute). As expected the number of states and time to

Table 3. N is the number of drones, P the number of points of interest, $x_{max} \times y_{max}$ the size of the grid, M the time limit for photos, and e_{max} the maximum energy capacity of each drone. We measured st and t, which are, respectively, the number of states and time in seconds until finding a critical configuration if F (for fail), or until searching all traces with exactly $4 \times M$ time steps if S (for success, no critical configuration before the time bound is reached).

Exp 1: ($N = 1, P = 4, x_{max} = y_{max} = 10$)	
$M = 50, e_{max} = 40$	F, $st = 139, t = 0.3$
$M = 70, e_{max} = 40$	F, $st = 203, t = 0.4$
$M = 90, e_{max} = 40$	S, $st = 955, t = 2.3$

Exp 3: ($N = 2, P = 9, x_{max} = y_{max} = 20$)	
$M = 100, e_{max} = 500$	F, $st = 501, t = 6.2$
$M = 150, e_{max} = 500$	F, $st = 1785, t = 29.9$
$M = 180, e_{max} = 500$	S, $st = 2901, t = 49.9$
$M = 180, e_{max} = 150$	F, $st = 1633, t = 25.6$

Exp 2: ($N = 2, P = 4, x_{max} = y_{max} = 10$)	
$M = 30, e_{max} = 40$	F, $st = 757, t = 3.2$
$M = 40, e_{max} = 40$	F, $st = 389, t = 1.4$
$M = 50, e_{max} = 40$	S, $st = 821, t = 3.2$

Exp 4: ($N = 3, P = 9, x_{max} = y_{max} = 20$)	
$M = 100, e_{max} = 150$	F, $st = 3217, t = 71.3$
$M = 120, e_{max} = 150$	F, $st = 2193, t = 52.9$
$M = 180, e_{max} = 150$	S, $st = 2193, t = 53.0$
$M = 180, e_{max} = 100$	F, $st = 2181, t = 50.4$

search increases moderately with the increase of the number of drones and size of grid.

Although abstract, the surveillance drone model can help specifiers to decide how many drones to use and with which energy capacities. For example, in Exp 3, drones required a great deal of energy, namely 500 energy units. Adding an additional drone, Exp 4, reduced the energy needed to 150 energy units.

5 Conclusion and Future Perspectives

We have described a framework for modeling and reasoning about cyber-physical agent systems using executable models specified in the Maude rewriting logic language. Agents coordinate by sharing knowledge, and their behavior is specified by soft constraint problems (SCPs). Physical state and an agents perception of the state are modeled separately. These features are intended to help specify agents with some robustness, and to allow reasoning about agents that have only partial information about the system state, and about operation in an unpredictable environment. The soft agent framework does not provide methods for deciding what valuation domains and functions to use in defining SCPs. It does provide tools for formal exploration of parameter settings, weighting of valuation functions, degrees of caution, and so on.

The notion of soft-agent system is very similar to the notion of *ensemble* that emerged from the Interlink project [29] and that has been a central theme of the ASCENS (Autonomic Service-Component Ensembles) project [5]. In [30] a mathematical system model for ensembles is presented. Similar to soft agents, the mathematical model treats both cyber and physical aspects of a system. A notion of fitness is defined that supports reasoning about level of satisfaction. Adaptability is also treated. In contrast to the soft-agent framework which provides an executable model, the system model for ensembles is denotational. The two approaches are both compatible and complementary and there is intriguing potential for future extensions that could lead to a very expressive framework supporting both high-level specification and concrete design methodologies.

Soft Constraint Automata (SCA) is another approach to specifying soft agent systems [24]. It is an inherently compositional approach. Agents are composed from SCA for different aspects of their behavior, systems are composed from agent SCAs. The environment is treated as an agent, and it too is composed from smaller parts. Future plans include defining maps between SCAs and agent systems specified in the soft agent framework to be able take advantage of the benefits of each approach.

The soft agent framework presented here is just the beginning. In the following we discuss some of the remaining challenges and future directions.

We proposed agents that decide what to do by locally solving soft constraint problems (SCPs) and showed that this works in some simple cases. But, how much can be done usefully with local SCPs? When is higher level planning and coordination needed? Some specific questions to study include:

- Under what conditions are the local solutions good enough?
- Under what conditions would it not be possible?
- How much knowledge is needed for satisfactory solution/behavior? For example, for sufficient consensus.
- What frequency of decision making is needed so that local solutions are safe and effective?

Another challenge is compositional reasoning. Can we reason separately about different concerns by abstracting the rest of the system? Methods are needed to derive suitable decompositions and abstractions. We expect compositional approach based on Soft Constraint Automata [24] to lead to methods for compositional reasoning.

Since our objective is to reason about both cyber and physical aspects of a system, it is important to be able to have models that involve dense time. One of the reasons for the way time steps are formalized in the soft agent framework is to be able to specify actions using continuous functions (applied for some duration). We imagine that tasks will happen in discrete time, and that agents only observe continuously changing situations at discrete times. There are many details to worry about, including frequency of observation and frequency of adjusting controls, not to mention more complex interactions between elements of a model of the physical state. Importantly also, what are the right abstractions that manage complexity while remaining sufficiently faithful.

We showed how execution and bounded search provide simple tools for analyzing a soft agent system. But the guarantees provided are quite limited. One way to expand the analysis capability is to develop methods based on symbolic execution. Here large parts of a system are represented by variables, possibly subject to constraints. Search is carried out by unifying rules with symbolic states rather than matching rules to concrete states (this is called *narrowing*). If there are constraints, they accumulate and can be checked for satisfiability to prune impossible branches while avoiding the need to enumerate solutions. In this way whole families of systems can be analyzed at once. Backwards narrowing is another approach. Here one starts with a pattern representing a critical configuration and applies rules backwards to see if an initial state can be reached. If not, no instance of the critical configuration pattern is reachable. This approach has been used successfully in the MaudeNPA protocol analysis tool [31]. Another possibility is to identify systems where one can apply timeshift abstraction. Here one defines an equivalence relation on states at different times and attempts to show that modulo equivalence there are a finite number of states that repeat. This is a form of bounded induction. Another important direction is to develop efficient algorithms for model checking of soft constraint automata, taking advantage of compositionality.

Finally, an important issue that we have not mentioned is security. In the context of soft agents security has many aspects and subtilties. There are issues of trust or confidence of an agent in knowledge received, from another agent, or by reading its sensors. Security protocols may have space or time aspects. Like everything else, we probably want soft notions for security guarantees: trust for some purpose, secret for some small amount of time, and so on. It is clear that the soft agent framework needs to provide support for managing and reasoning about security. What are the right mechanisms? How much security should be built into knowledge sharing? How can we balance imposition of security precautions and need for agility, and open systems.

References

1. Robots that fly and cooperate. TED talk (2015). Accessed 07 March 2016
2. Das, J., Cross, G., Qu, C., Makineni, A., Tokekar, P., Mulgaonkar, Y., Kumar, V.: Devices, systems, and methods for automated monitoring enabling precision agriculture. In: IEEE International Conference on Automation Science and Engineering (2015)
3. Vijay Kumar lab. Accessed 11 March 2016
4. Wirsing, M., Hölzl, M., Koch, N., Mayer, P. (eds.): Software Engineering for Collective Autonomic Systems. The ASCENS Approach. LNCS, vol. 8998. Springer, Switzerland (2015)
5. Ascens: Autonomic service-component ensembles. Accessed 15 November 2014
6. Choi, J.-S., McCarthy, T., Kim, M., Stehr, M.-O.: Adaptive wireless networks as an example of declarative fractionated systems. In: Stojmenovic, I., Cheng, Z., Guo, S. (eds.) MOBIQUITOUS 2013. LNICST, vol. 131, pp. 549–563. Springer, Heidelberg (2014)
7. Kim, M., Stehr, M.O., Talcott, C.: A distributed logic for networked cyber-physical systems. In: Arbab, F., Sirjani, M. (eds.) FSEN 2011. LNCS, vol. 7141, pp. 190–205. Springer, Heidelberg (2012)
8. Stehr, M.-O., Talcott, C., Rushby, J., Lincoln, P., Kim, M., Cheung, S., Poggio, A.: Fractionated software for networked cyber-physical systems: research directions and long-term vision. In: Agha, G., Danvy, O., Meseguer, J. (eds.) Formal Modeling: Actors, Open Systems, Biological Systems. LNCS, vol. 7000, pp. 110–143. Springer, Heidelberg (2011)
9. Networked cyber physical systems. Accessed 11 March 2016
10. Drone swarms: The buzz of the future. Accessed 08 March 2016
11. Knightscope. Accessed 11 March 2016
12. Liquid robotics. Accessed 11 March 2016
13. Why BNSF railway is using drones to inspect thousands of miles of rail lines. Accessed 11 March 2016
14. Dantas, Y.G., Nigam, V., Fonseca, I.E.: A selective defense for application layer ddos attacks. In: SI-EISIC (2014)
15. Clavel, M., Durán, F., Eker, S., Lincoln, P., Martí-Oliet, N., Meseguer, J., Talcott, C.: All About Maude - A High-Performance Logical Framework. LNCS, vol. 4350. Springer, Heidelberg (2007)
16. Wirsing, M., Denker, G., Talcott, C., Poggio, A., Briesemeister, L.: A rewriting logic framework for soft constraints. In: Sixth International Workshop on Rewriting Logic and Its Applications (WRLA 2006). Electronic Notes in Theoretical Computer Science. Elsevier (2006)
17. Hölzl, M., Meier, M., Wirsing, M.: Which soft constraints do you prefer? In: Seventh International Workshop on Rewriting Logic and Its Applications (WRLA 2008). Electronic Notes in Theoretical Computer Science. Elsevier (2008)

18. Gadducci, F., Hölzl, M., Monreale, G.V., Wirsing, M.: Soft constraints for lexico-graphic orders. In: Castro, F., Gelbukh, A., González, M. (eds.) MICAI 2013, Part I. LNCS, vol. 8265, pp. 68–79. Springer, Heidelberg (2013)
19. Arbab, F., Santini, F.: Preference and similarity-based behavioral discovery of services. In: ter Beek, M.H., Lohmann, N. (eds.) WS-FM 2012. LNCS, vol. 7843, pp. 118–133. Springer, Heidelberg (2013)
20. Kim, M., Stehr, M.-O., Talcott, C.L.: A distributed logic for networked cyber-physical systems. Sci. Comput. Program. **78**(12), 2453–2467 (2013)
21. Choi, J.S., McCarthy, T., Yadav, M., Kim, M., Talcott, C., Gressier-Soudan, E.: Application patterns for cyber-physical systems. In: Cyber-Physical Systems Net-works and Applications (2013)
22. Stehr, M.-O., Kim, M., Talcott, C.: Partially ordered knowledge sharing and frac-tionated systems in the context of other models for distributed computing. In: Iida, S., Meseguer, J., Ogata, K. (eds.) Specification, Algebra, and Software. LNCS, vol. 8373, pp. 402–433. Springer, Heidelberg (2014)
23. Ölveczky, P.C., Meseguer, J.: Semantics and pragmatics of real-time maude. High. Order Symbolic Comput. **20**(1–2), 161–196 (2007)
24. Kappé, T., Arbab, F., Talcott, C.: A compositional framework for preference-aware agents (March 2016, submitted)
25. Nielson, H.R., Nielson, F., Vigo, R.: A calculus for quality. In: Păsăreanu, C.S., Salaün, G. (eds.) FACS 2012. LNCS, vol. 7684, pp. 188–204. Springer, Heidelberg (2013)
26. Nielson, H.R., Nielson, F.: Safety versus security in the quality calculus. In: Liu, Z., Woodcock, J., Zhu, H. (eds.) Theories of Programming and Formal Methods. LNCS, vol. 8051, pp. 285–303. Springer, Heidelberg (2013)
27. Meseguer, J.: Conditional rewriting logic as a unified model of concurrency. The-oret. Comput. Sci. **96**(1), 73–155 (1992)
28. The maude system. Accessed 15 November 2014
29. Hölzl, M., Rauschmayer, A., Wirsing, M.: Engineering of software-intensive sys-tems: state of the art and research challenges. In: Wirsing, M., Banâtre, J.-P., Hölzl, M., Rauschmayer, A. (eds.) Software-Intensive Systems. LNCS, vol. 5380, pp. 1–44. Springer, Heidelberg (2008)
30. Hölzl, M., Wirsing, M.: Towards a system model for ensembles. In: Agha, G., Danvy, O., Meseguer, J. (eds.) Formal Modeling: Actors, Open Systems, Biological Systems. LNCS, vol. 7000, pp. 241–261. Springer, Heidelberg (2011)
31. Escobar, S., Meadows, C., Meseguer, J.: Maude-NPA: cryptographic protocol analysis modulo equational properties. In: Aldini, A., Barthe, G., Gorrieri, R. (eds.) FOSAD 2009. LNCS, vol. 5705, pp. 1–50. Springer, Heidelberg (2009)
32. Dantas, Y.G., Lemos, M.O.O., Fonseca, I.E., Nigam, V.: Formal specification and verification of a selective defense for TDoS attacks. In: Lucanu, D. (ed.) Workshop on Rewriting Logic and Applications (2016)

Dependability of Adaptable and Evolvable Distributed Systems

Carlo Ghezzi[✉]

DEIB, DeepSE Group, Politecnico di Milano,
Piazza Leonardo da Vinci, 32, 20133 Milano, MI, Italy
carlo.ghezzi@polimi.it

Abstract. This article is a tutorial on how to achieve software evolution and adaptation in a dependable manner, by systematically applying formal modelling and verification. It shows how software can be designed upfront to tolerate different sources of uncertainty that cause continuous future changes. If possible changes can be predicted, and their occurrence can be detected, it is possible to design the software to be self-adaptable. Otherwise, continuous evolution has to be supported and continuous flow into operation has to be ensured. In cases where systems are designed to be continuously running, it is necessary to support safe continuous software deployment that guarantees correct operation in the presence of dynamic reconfigurations. The approaches we survey here have been mainly developed in the context of the SMScom project, funded by the European Commission –Programme IDEAS-ERC (http://erc-smscom.dei.polimi.it/.) – and lead by the author. It is argued that these approaches fit well the current agile methods for development and operations that are popularized as DevOps.

Keywords: Distributed, ubiquitous, pervasive systems · Cyber-physical systems · Environment uncertainty · Requirements · Software evolution · Dynamic reconfiguration

1 Introduction and Motivations

Modern software systems increasingly live in a dynamic and open world [3]. The goals to fulfil and the requirements to meet evolve over time. The environment in which the software is embedded often behaves in ways that cannot be predicted upfront during design. And if it can, it might later change after the software has been developed and became operational. This situation is often encountered in the design of *cyber-physical systems*, in which the physical and the cyber worlds are intertwined, through many kinds of devices behaving as sensors and actuators. Interaction with the physical world introduces a great variety of possible contingencies, like noise, vibrations, humidity, or temperature, which may unexpectedly affect the system's behavior. Sensors and actuators may also behave in a hard-to-predict manner, and this may change over time, e.g. their behavior may change because of the battery level. For these reasons, many different kinds

© Springer International Publishing Switzerland 2016
M. Bernardo et al. (Eds.): SFM 2016, LNCS 9700, pp. 36–60, 2016.
DOI: 10.1007/978-3-319-34096-8_2

of uncertainty may be present when the system is being designed and uncertainty may ultimately affect the system's ability to satisfy the requirements.

Design uncertainty is increasingly becoming the norm also for many other kinds of system. User-intensive, highly interactive systems depend on users' behaviors, which also may change over time. The widespread availability of virtual environments, providing infrastructure/software-as-a-service, which raise the level of abstraction for system designers, add their own sources of uncertainty that must be properly handled. Furthermore, modern systems are increasingly multi-owner. They depend upon parts (components, services) that are not under the developers' full control, but rather they are owned, managed, and operated by others. They may run on platforms that developers do not own and do not run; for example, they may run on a cloud. Yet, software designers are responsible for the service they provide to their clients, and the level of service they must guarantee has to satisfy the contractual agreements they subscribed with their customers.

Requirements volatility and environment uncertainty are two main causes that drive software evolution. Software evolution is not a new problem. It has been recognized as a key distinguishing factor of software with respect to other technologies since the early work pioneered by Belady and Lehman since the 1970's [4,17], although the phenomenon has reached today unprecedented levels of intensity. In the past software evolution was often viewed as a nuisance. The term *maintenance* was often used to capture the evolution of software needed to remedy inadequate requirements and wrong design choices. Evolution is instead intrinsic in software. Like evolution in nature, it has a positive connotation, which refers to the ability to adapt and improve in quality.

Today software is developed through evolutionary processes. Traditional predefined, monolithic, *waterfall* lifecycles are generally replaced by incremental, iterative, evolutionary, *agile* processes. Agility indicates a fast and flexible way to react to changes. At the same time, researchers developed approaches to embed in software capabilities to drive its own evolution, in an *autonomic* or *self-managed* manner [14].

Agile processes originated in the practitioners' world and have only been marginally investigated by researchers. As observed by [19], in their iconoclastic reaction to other approaches, the proponents of agile methods tend to dismiss some of the key principles of software engineering that lead to improved dependability. They dismiss requirements analysis —replaced by user stories— (formal) modelling —viewed as as a sterile exercise— and the value of formal verification —fully replaced by continuous testing. The importance of formal methods in the context of self-managed systems has also been largely underestimated by most initial research efforts.

The body of work we survey in this paper is fully reliant on formal methods to enable dependable software evolution. Within the vast area of software evolution, this article focuses mostly on two aspects:

1. *Non-functional requirements*: The system's evolution is dictated by the need to satisfy certain non-functional requirements in the presence of changes that

would otherwise lead to violations. Among non-functional requirements we focus in particular on those that can be modelled in a mathematically precise, quantitative way. This includes requirements on response time, reliability, power consumption. Often these can be expressed in a probabilistic manner.

2. *Self-adaptation*: We analyze when and how the system can be made capable of collecting and analyzing run-time data that hint at changes in the behavior of the environment that may lead to requirements violations and are amenable to reactions that may be decided autonomously.

3. *Dynamic software updates*: In the case of self-adaptation, the system must reconfigure itself dynamically, while it is operational. This requirement also holds for systems where updates are performed offline by software engineers and installed online while the system is offering service. This situation is becoming very common today because many systems are required to be continuously running and operation cannot be interrupted to accommodate new updates. Dynamic updates must be performed both safely and efficiently, to ensure timely reaction to changes.

The paper is structured as follows. Section 2 presents a general framework to understand and reason about evolution and adaptation. In particular, it allows us to articulate the complex interactions that may occur between the software and the environment in which it is embedded, and how dependency on the environment may affect dependability and drive adaptation. Section 3 introduces a case study. Section 4 introduces background material on modeling and verification. Section 5 discusses how models and verification may be brought to run time to support self-adaptation. Section 6 addresses the problem of safe dynamic software updates. Finally, Sect. 7 illustrates final considerations and points to future research.

2   Reference Framework

In this section we describe a framework to understand and reason about software and change, which was proposed by the foundational work on requirements engineering developed by Jackson and Zave [13,21]. Jackson and Zave observe that in requirements engineering one needs to carefully distinguish between two main concerns: the *world* and the *machine*. The machine is the system of interest that must be developed; the world (the environment) is the portion of the real-world affected by the machine. The ultimate purpose of the machine is always to be found in the world. The goals to be met and the *requirements* are ultimately dictated by the world and must be expressed in terms of the phenomena that occur in it. Some of these phenomena are shared with the machine: they are either controlled by the world and observed by the machine –through sensors– or controlled by the machine and observed by the world –through actuators. The machine is built exactly for the purpose of achieving satisfaction of the requirements in the real world. Its *specification* is a prescriptive statement of the relation on *shared phenomena* that must be enforced by the system to be developed. The machine that implements it must be correct with respect to the specification.

The task of software engineers is to develop first a specification and then an implementation for a machine that achieves requirements satisfaction. To this end, *domain* or *environment knowledge* plays an essential role. That is, the software engineer needs to understand the laws that govern the behavior of the environment and formulate the set of relevant assumptions that have to be made about the environment in which the machine is expected to work, which affect the achievement of the desired results. Quoting from [21],

> *"The primary role of domain knowledge is to bridge the gap between requirements and specifications."*

If R and S are the prescriptive statements that formalize the requirements and the specification, respectively, and D are the descriptive statements that formalize the domain knowledge, assuming that S and D are are both satisfied and consistent with each other, the designer's responsibility is ultimately to ensure that

$$S, D \models R$$

i.e., the machine's specification S we are going to devise must entail satisfaction of the requirements R in the context of the domain properties D. We call this the *dependability argument*.

Figure 1 provides a visual sketch of the Jackson/Zave approach. The domain knowledge D plays a fundamental role in establishing the requirements. We need to know upfront how the environment in which the software is embedded works, since the software to develop (the machine) can achieve the expected requirements only based on the assumptions on the behavior of the domain, described by D. Should the environment behave in a way that contradicts the statements in D, the current specification might lead to violation of R. The statements expressed by D may fail to capture the environment's behavior for two reasons: either because the domain analysis was initially flawed (i.e., the environment behaves according to different laws than the ones captured by D) or because changes occurred, which cause the assumptions made earlier to become invalid. An example of the latter case may be an exceptional and unexpected traffic of submitted user requests that may generate a denial of service.

It is possible to further breakdown D into two components: *domain laws* −*Dl*− and *domain assumptions* −*Da*. Laws indicate the physical or mathematical properties that have been *proved* for the domain, whose truth can only be invalidated by falsifying the theory. An example is the law of motion that says that the application of a force in a given direction to a body causes motion of the body in that direction. A designer relying on this property may specify that a command to a force actuator has to be issued by the software to satisfy the requirement that a body should be moved. This property holds and cannot be refuted. Assumptions instead are properties that are subject to some level of uncertainty and may be disproved. In some cases, they denote currently valid properties that may later change, as for example, traffic conditions changes. They represent the best of

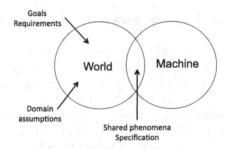

Fig. 1. The Jackson/Zave framework.

our knowledge at a given time. But because of design-time uncertainty and/or because variability in time, assumptions may become invalid.

Software evolution refers to changes that affect the machine, to enable it to respond to changes in the requirements and/or in the environment (we ignore in this paper the fact that implementation may be be incorrect, i.e., the running software violates its specification S). The term *adaptation* is used in this work to indicate the specific case of evolution dictated by changes in the environment, while self-adaptation indicates changes that can be handled autonomously by the machine.

The management of evolution in traditional software is performed off-line, during the *maintenance* phase. The traditional classification in perfective, adaptive, and corrective maintenance can also be explained by referring to the Jackson/Zave framework. Changes in the requirements, dictated by changes in the business goals of organizations or new demands by users, cause *perfective maintenance*. Environmental changes affecting domain assumptions, which may represent organizational assumptions or conditions on the physical context in which the software is embedded, cause *adaptive maintenance. Corrective maintenance* is instead caused by failure of the dependability argument and forces the specification to change.

According to the traditional paradigm, in order to undergo a maintenance intervention, software returns into its development stage, where changes are analyzed, prioritized, and scheduled. Changes are then handled by modifying the design and implementation of the application. The evolved system is then verified, typically via some kind of regression testing.

This paradigm does not meet the requirements of current application scenarios, which are subject to continuos changes in the requirements and in the environment, and which require rapid reaction to such changes. By following an *agile development* style, software development became incremental and iterative. By following the currently widely advocated *DevOps* culture, agility extends in a seamless manner to delivery and deployment, viewing development and operation as an integrated perpetual process.

Figure 2 illustrates our envisioned process that supports continuous development and operation, through two main, interacting loops: the development loop and the self-adaptation loop. The process incorporates the run-time feedback loop advocated by the autonomic computing proposal [14], which enables self-adaptation. Designers are in the loop and drive evolution. They get informed about the system's dynamic behavior by leveraging monitored data. They are required to initiate evolution whenever self-adaptation fails. Whenever they decide that components should be transferred to the running system to replace faulty functionalities, add functionalities, or enhance existing ones, they can instruct the operational environment to reconfigure itself dynamically in a completely safe, non-disruptive, and efficient way.

In this paper we embrace this holistic view and discuss the role that formal methods can play to support continuous evolution in a dependable manner, i.e., where the designer's focus is constantly driven by the need to formally guarantee satisfaction of the dependability argument.

The next section introduces a practical application domain and a case study that provide concrete motivations for this work. We subsequently show how the run-time adaptation loop can be structured and how safe dynamic reconfigurations can be supported. Finally we will conclude by discussing how we might progress to achieve the global picture of Fig. 2 and by outlining a research agenda.

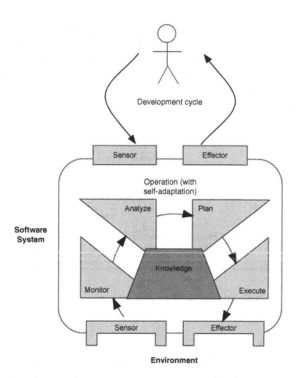

Fig. 2. The development and operation process.

3 A Case Study

Hereafter we illustrate a concrete example in which the approach described earlier is successfully applied. The example, which was originally introduced in [9], refers to a typical e-commerce application that sells merchandise on-line to end users by integrating several services offered by third-parties:

1. *Authentication Service.* This service manages the identity of users. It provides a *Login* and a *Logout* operation through which the system authenticates users.
2. *Payment Service.* This service provides a safe transactional payment service through which users can pay the selected merchandise via the *CheckOut* operation.
3. *Shipping Service.* This service is in charge of shipping goods to the customer's address. It provides two different operations: *NrmShipping* and *ExpShipping*. The former is a standard shipping functionality while the latter represents a faster and more expensive alternative. Finally, the system classifies the logged users as *NewCustomer* (NC) or *ReturningCustomer* (RC), based on their usage profile.

The case study illustrates a situation that has become quite common, which is abstracted by Fig. 3. It is a *user-intensive* application, where end-users interact with the application in a hard-to-predict and time variable manner. For example, usage patterns may vary during the different periods of the year, and may have seasonal peaks (for example, around Christmas holidays). Moreover, the behavior of integrated services may be subject to variability, and even deviations from the expected quality of service. Figure 4 provides a high-level view of the flow of interaction between users and the e-commerce application, expressed as an activity diagram.

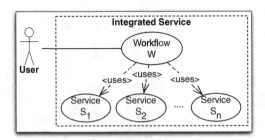

Fig. 3. A class of applications

This application has to guarantee a certain quality of service to customers. In particular, here we focus on *reliability*. Services may in fact fail to provide an answer by timing out incoming requests in situations where the load exceeds their capacity.

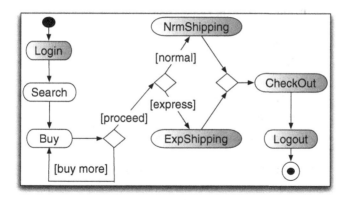

Fig. 4. Operational description of the specification via an activity diagram

Reliability requirements can be typically expressed in probabilistic terms: for example, the probability that a user-triggered transaction completes successfully must be higher than a given value. In the case of the e-commerce application, fulfilment of a reliability requirement clearly depends on certain assumptions about the environment, such as the reliability of the third-party services that are integrated into the application and usage profiles (e.g., the ratio between new and returning customers), which may affect the satisfaction of specific requirements that may refer to the different categories.

As mentioned, environment phenomena of these kinds are quite hard to predict when the system is initially designed. Even in cases where the expected failure rate of services may be stated in the contract with the service provider, values are subject to uncertainty and may very likely change over time (for example, due to a new release of the service). Likewise, usage profiles are hard to predict upfront and are very unstable.

Let us assume that the e-commerce application must satisfy the following reliability requirements:

- R1: *"Probability of success is greater then 0.8"*
- R2: *"Probability of a ExpShipping failure for a user recognized as ReturningCustomer is less then 0.035"*
- R3: *"Probability of an authentication failure is less then 0.06"*

Let us further assume that development time domain analysis tells us that expected usage profile can be reasonably described as in Table 1. The notation $P(x)$ denotes the probability of "x". Table 2 instead summarizes the results of domain analysis concerning the external services integrated in the e-commerce application. $P(Op)$ here denotes the probability of failure of service operation Op. The environment assumptions expressed in Tables 1 and 2 may derive from different sources. For example, reliability properties of third-party services may be published as part of the service-level agreement with service providers. Usage profiles may instead be derived from previous experience of the designers or knowledge extracted from previous similar systems.

Table 1. Domain assumptions on usage profiles

Description	Value
P(User is a RC)	0.35
P(RC chooses express shipping)	0.5
P(NC chooses express shipping)	0.25
P(RC searches again after a buy operation)	0.2
P(NC searches again after a buy operation)	0.15

Table 2. Domain assumptions on external services

Description	Value
P(Login)	0.03
P(Logout)	0.03
P(NrmShipping)	0.05
P(ExpShipping)	0.05
P(CheckOut)	0.1

4   Modeling and Verification Preliminaries

As we discussed earlier, the software engineer's goal is to derive a specification S which leads to satisfaction of requirements R, assuming that the environment behaves as described by D. From the activity diagram in Fig. 4 and the information contained in the tables regarding environment assumptions, we can derive an enriched state-machine model that summarizes a formal description both of the application and of the environment. The state machine transitions describe the possible sequences of interactive operations, according to the protocol specified by the activity diagram of Fig. 4. Domain assumptions are modeled as probabilities that label the transitions. The model also represents failure and success states for the external services.

Formally, the model in Fig. 5 is a Discrete Time Markov Chain (DTMC). It contains one state for every operation performed by the system plus a set of auxiliary states representing potential failures associated with auxiliary operations (e.g., state 5) or specific internal logical states (e.g., state 2).

Once a formal model is provided, like the DTMC in Fig. 5, it is possible to formally verify whether requirements are satisfied, provided they are expressed in suitable language for which a verification procedure exists. In the case of DTMCs, requirements may be formalized using the probabilistic temporal logic language PCTL, and then checked against the model using a probabilistic model checker, like PRISM [12,16]. By doing so on our example, we obtain the following results:

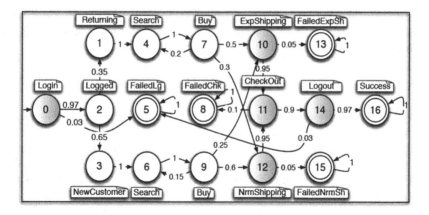

Fig. 5. DTMC model for the case study

- *Probability of success = 0.804*
- *Probability of a ExpShipping failure for a user recognized as ReturningCustomer =o 0.031*
- *Probability of an authentication failure (i.e., Login or Logout failures) = 0.056*

which ensure satisfaction of the requirements.

Hereafter we explain how this can be done, by first briefly reviewing DTMCs and then introducing PCTL.

4.1 Discrete Time Markov Chains

DTMCs are defined as state-transition systems augmented with probabilities. *States* represent possible configurations of the system. *Transitions* among states occur at discrete time and have an associated probability. DTMCs are discrete stochastic processes with the Markov property, according to which the probability distribution of future states depend only upon the current state.

Formally, a (labeled) DTMC is tuple (S, S_0, P, L) where

- S is a finite set of states
- $S_0 \subseteq S$ is a set of initial states
- $P : S \times S \to [0, 1]$ is a stochastic matrix, where $\sum_{s' \in S} P(s, s') = 1 \ \forall s \in S$. An element $P(s_i, s_j)$ represents the probability that the next state of the process will be s_j given that the current state is s_i.
- $L : S \to 2^{AP}$ is a labeling function which assigns to each state the set of *Atomic Propositions* which are true in that state.

For reasons that will become clear later, we implicitly extend this definition by also allowing transitions to be labeled with variables (with values in the range 0..1) instead of constants. A state $s \in S$ is said to be an *absorbing state* if $P(s, s) = 1$. If a DTMC contains at least one absorbing state, the DTMC itself is said to be an *absorbing DTMC*.

In an absorbing DTMC with r absorbing states and t transient states, rows and columns of the transition matrix P can be reordered such that P is in the following *canonical form*:

$$\mathbf{P} = \begin{pmatrix} Q & R \\ 0 & I \end{pmatrix}$$

where I is an r by r identity matrix, 0 is an r by t zero matrix, R is a nonzero t by r matrix and Q is a t by t matrix.

Consider now two distinct transient states s_i and s_j. The probability of moving from s_i to s_j in exactly 2 steps is $\sum_{s_x \in S} P(s_i, s_x) \cdot P(s_x, s_j)$. Generalizing, for a k-steps path and recalling the definition of matrix product, it follows that the probability of moving from any transient state s_i to any other transient state s_j in exactly k steps corresponds to the entry (s_i, s_j) of the matrix Q^k. As a natural generalization, we can define Q^0 (representing the probability of moving from each state s_i to s_j in 0 steps) as the identity t by t matrix, whose elements are 1 iff $s_i = s_j$ [10].

Due to the fact that R must be a nonzero matrix, and P is a stochastic matrix, Q has uniform-norm strictly less than 1, thus $Q^n \to 0$ as $n \to \infty$, which implies that eventually the process will be absorbed with probability 1.

In the simplest model for reliability analysis, the DTMC will have two absorbing states, representing the correct accomplishment of the task and the task's failure, respectively. The use of absorbing states is commonly extended to modeling different failure conditions. For example, different failure states may be associated with the invocation of different external services. Once the model is in place, we may be interested in estimating the probability of reaching an absorbing state or in stating the property that the probability of reaching an absorbing failure state should be less than a certain threshold. In the next section we discuss how these and other interesting properties of systems modeled by a DTMC can be expresses and how they can be evaluated.

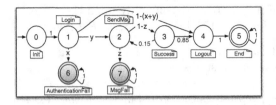

Fig. 6. DTMC example.

Let us consider the simple example of DTMC in Fig. 6, which represents a system sending authenticated messages over the network. States 5, 6, and 7 are absorbing states; states 6 and 7 represent failures associated respectively to the authentication and to message sending. We use variables as transition labels to indicate that the value of the corresponding probability is unknown, and may change over time.

In matrix form, the same DTMC would be characterized by the following matrices Q and R:

$$Q = \begin{pmatrix} 0 & 1 & 0 & 0 & 0 \\ 0 & 0 & y & 0 & 1-x-y \\ 0 & 0 & 0 & 1-z & 0 \\ 0 & 0 & 0.15 & 0 & 0.85 \\ 0 & 0 & 0 & 0 & 0 \end{pmatrix}$$

$$R = \begin{pmatrix} 0 & 0 & 0 \\ 0 & x & 0 \\ 0 & 0 & z \\ 0 & 0 & 0 \\ 1 & 0 & 0 \end{pmatrix}$$

This is a toy example that we use hereafter instead of the more complex original case study to exemplify the approach within a constrained space.

4.2 Formally Specifying Requirements

Formal languages to express properties of systems modeled through DTMCs have been studied in the past and several *model checkers* have been designed and implemented to support property analysis. Through model checking one can verify that a given model (representing domain assumptions and the specification) satisfies the requirements, provided they are formalized in a language, such as PCTL, for which a verification procedure exists. In particular, PCTL [2] –which is briefly introduced hereafter– proved to be useful to express a number of interesting reliability properties.

PCTL extends the branching-time temporal logic language CTL [2] to deal with probabilities. Instead of the existential and universal quantification of CTL, PCTL provides the probabilistic operator $\mathcal{P}_{\bowtie p}(\cdot)$, where $p \in [0,1]$ is a probability bound and $\bowtie \in \{\leq, <, \geq, >\}$.

PCTL is defined by the following syntax:

$$\Phi ::= true \mid a \mid \Phi \wedge \Phi \mid \neg \Phi \mid \mathcal{P}_{\bowtie p}(\varphi)$$
$$\varphi ::= X \, \Phi \mid \Phi \, U \, \Phi \mid \Phi \, U^{\leq t} \, \Phi$$

Formulae Φ are named *state formulae* and can be evaluated over a boolean domain (true, false) in each state. Formulae ψ are named *path formulae* and describe a pattern over the set of all possible paths originating in the state where they are evaluated.

The satisfaction relation for PCTL is defined for a state s as:

$$s \models true$$
$$s \models a \qquad \text{iff} \quad a \in L(s)$$
$$s \models \neg\Phi \qquad \text{iff} \quad s \not\models \Phi$$
$$s \models \Phi_1 \wedge \Phi_2 \quad \text{iff} \quad s \models \Phi_1 \text{ and } s \models \Phi_2$$
$$s \models \mathcal{P}_{\bowtie p}(\psi) \quad \text{iff} \quad Pr(s \models \psi) \bowtie p$$

A formal definition of how to compute $Pr(s \models \psi)$ is presented in [2]. The intuition is that its value corresponds to the fraction of paths originating in s and satisfying ψ over the entire set of paths originating in s. The satisfaction relation for a path formula with respect to a path π originating in s ($\pi[0] = s$) is defined as:

$$\pi \models X\Phi \qquad \text{iff} \quad \pi[1] \models \Phi$$
$$\pi \models \Phi U\Psi \qquad \text{iff} \quad \exists j \geq 0.(\pi[j] \models \Psi \wedge (\forall 0 \leq k < j.\pi[k] \models \Phi))$$
$$\pi \models \Phi U^{\leq t}\Psi \quad \text{iff} \quad \exists 0 \leq j \leq t.(\pi[j] \models \Psi \wedge (\forall 0 \leq k < j.\pi[k] \models \Phi))$$

PCTL is an expressive language that allows reliability-related properties to be specified. A taxonomy of all possible reliability properties is out of the scope of this paper. The most important case is a *reachability property*. A reachability property states that a state where a certain characteristic property holds is eventually reached from a given initial state. In most cases, the state to be reached is an absorbing state. Such state may represent a *failure state*, in which a transaction executed by the system modeled by the DTMC eventually (regrettably) terminates, or a *success state*. Reachability properties are expressed as $\mathcal{P}_{\bowtie p}(true \; U \; \Phi)$[1], which expresses the fact that the probability of reaching any state satisfying Φ has to be in the interval defined by constraint $\bowtie p$. Φ is assumed to be a simple state formula that does not include any nested path formula. In most cases, it just corresponds to the atomic proposition that is true in an absorbing state of the DTMC. In the case of a failure state, the probability bound is expressed as $\leq x$, where x represents the upper bound for the failure probability; for a success state it would be instead expressed as $\geq x$, where x is the lower bound for success.

PCTL allows more complex properties than plain reachability to be expressed. Such properties would be typically domain-dependent, and their definition is delegated to system designers. For example, referring to the example in Fig. 6, we express the following reliability requirements:

- **R1**: *"The probability that a MsgFail failure happens is lower than 0.001"*
- **R2**: *"The probability of successfully sending at least one message for a logged in user before logging out is greater than 0.001"*
- **R3**: *"The probability of successfully logging in and immediately logging out is greater than 0.001"*
- **R4**: *"The probability of sending at least 2 messages before logging out is greater than or equal to 0.001"*

---

[1] Note that this is often expressed as $\mathcal{P}_{\bowtie p}F\Phi$, using the *finally* operator.

Notice that **R1** is an example of reachability property. Also notice that these requirements have different sets of initial states: **R1**, **R3**, and **R4** must be evaluated starting from state 0 (i.e., $S_0 = \{0\}$) while **R2** must be evaluated starting from state 1. Formalization of requirements R1-R3 using PCTL is left as an exercise.

5 Supporting Self-adaptation via Run-Time Verification

Let us refer to the process model represented in Fig. 2, which shows the interplay between the run-time adaptation and the off-line evolution feedback loops. To support dependable self-adaptation, we root the analysis phase taking place during operation (see Fig. 2) into model checking. The model that represents both the software system and the environment –such as the one shown in Fig. 6– is kept alive at run time and is updated according to the data gathered by monitoring, which can be used to infer possible environment changes through a machine learning component. In our case study, new values for the probabilities of certain transitions representing service failures may be inferred by monitoring the failure rate of service invocations. Likewise, user profiles may be inferred by monitoring log-in customers' data. Inference can be based on standard statistical approaches, like the Bayesian learning method we used in [6]. Once the model is updated, the properties of interest can be checked. Violation of a given property is a trigger for self-adaptation, which is successful if changes of the implementation can be found that can eliminate the problem through a dynamic reconfiguration.

The key concepts upon which this approach is based are that (1) the models of interest are kept at run time and continuously updated, and (2) model checking provides continuous verification support to detect the need for adaptive reactions. Reactions are often subject to hard real-time constraints: they must lead to a valid software reconfiguration before the violation of requirements leads to unacceptable mishaps. The conventional model checking techniques are not really suitable for use at run time. They require the model checker to be run from scratch after any model change. It is thus necessary to re-think model checking algorithms to make them suitable for run-time use.

Our work has focused on making DTMC model checking for PCTL *incremental*. An incremental approach avoids re-analysis of the entire model by pre-computing the effects of changes. To achieve this goal, we make the assumption that changes are local and not disruptive. This is a reasonable assumption in most practical cases, assuming that the source model for the update is a reasonable approximation of the target. For DTMC models this assumption boils down to the hypothesis that the structure of the model does not change: only transition parameters may change. Furthermore, although in principle all such parameters may change, the solution we found works very efficiently if the number of transition parameters that may change is a small fraction of all transitions.

In the next section we present an incremental approach to probabilistic model checking that is based on parameterization. Changeable transition probabilities are treated as variables and a mathematical procedure computes a symbolic analytic expression for the properties we want to verify at run time. The underlying idea is that computation of the analytic expression, which takes place at design time, can be computationally expensive, but then evaluation of the pre-computed analytic expression, which occurs at run time, can be very efficient.

5.1 Run-Time Efficient Parametric Model Checking

The most commonly studied property for reliability analysis concerns the probability of reaching a certain state, which typically represents the success of the system or some failure condition. Both success and failure are modeled by absorbing states. The reachability formula in this case has the following form: $\mathcal{P}_{\bowtie p}Fl$, where l is the label of the target absorbing state. Hereafter we focus our discussion on how to pre-compute at design time a reachability formula for an absorbing state of a DTMC. All the details and the extension of the approach to cover all PCTL can be found in [7,8].

We assume that a DTMC can contain both numerically and symbolically labeled transitions. Since the sum of probabilities of all transitions exiting any given state must be 1, in the case where one transition is a variable, we require that all transitions exiting the state be also variable. We refer to such state as *variable state*.

For an absorbing DTMC, the matrix $I - Q$ has an inverse N and $N = I + Q + Q^2 + \cdots = \sum_{i=0}^{\infty} Q^i$ [10]. The entry n_{ij} of N represents the expected number of times the Markov chain reaches state s_j, given that it started from state s_i, before getting absorbed. Instead, q_{ij} represents the probability of moving from the transient state s_i to the transient state s_j in exactly one step.

Given that $Q^n \to 0$ when $n \to \infty$ (as discussed in Sect. 4.1), the process will always be absorbed with probability 1 after a large enough number of steps, no matter from which state it started off. Hence, our interest is to compute the probability distribution over the set of absorbing states. This distribution can be computed in matrix form as:

$$B = N \times R$$

where r_{ik} is the probability of being absorbed in state s_k given that the process started in state s_i.

B is a $t \times r$ matrix and it can be used to evaluate the probability of each termination condition starting from any DTMC state as an initial state. In particular the element b_{ij} of the matrix B represents the probability of being absorbed into state s_j given that the execution started in state s_i.

The design-time computation of an entry b_{ij} requires mixed symbolic and numeric computation, since variable states may be traversed to reach state s_j. Let us evaluate the complexity of such computation. Inverting matrix $I - Q$ by means of the Gauss-Jordan elimination algorithm [1] requires t^3 operations. The computation of the entry b_{ij} once N has been computed requires t more products, thus the total complexity is $t^3 + t$ arithmetic operations on polynomials. The computation could be further optimized by exploiting the sparsity of $I - Q$. Notice that the symbolic nature of the computation makes the design-time phase quite costly [11].

The complexity can be significantly reduced if the number of variable components c is small and the matrix describing the DTMC is sparse, as very frequently happens in practice. Let $W = I - Q$. The elements of its inverse N are defined as follows:

$$n_{ij} = \frac{1}{det(W)} \cdot \alpha_{ji}(W)$$

where $\alpha_{ji}(W)$ is the cofactor of the element w_{ji}. Thus:

$$b_{ik} = \sum_{x \in 0..t-1} n_{ix} \cdot r_{xj} = \frac{1}{det(W)} \sum_{x \in 0..t-1} \alpha_{xi}(W) \cdot r_{xj}$$

Computing b_{ik} requires the computation of t determinants of square matrices with size $t - 1$. Let τ be the average number of outgoing transitions from each state ($\tau << n$ by assumption). Each of the determinants can be computed by means of Laplace expansion. Precisely, by expanding first the c rows representing the variable states (each has τ symbolic terms), we need to compute at most τ^c determinants and then linearly combine them. Each submatrix of size $t - c$ does not contain any variable symbol, by construction, thus its determinant can be computed with $(t-c)^3$ operations among constant numbers (LU-decomposition), thus much faster than the corresponding ones among polynomials. The final complexity is thus:

$$\tau^c \cdot (t - c)^3 \sim \tau^c \cdot t^3$$

which significantly reduces the original complexity and makes the design-time pre-computation of reachability properties feasible in a reasonable time, even for large values of t.

As a term of comparison, the computation of reachability properties performed by probabilistic model-checkers is based on the solution of a system of n equations in n variables [2], which has, in a sequential computational model, a complexity equal to n^3 [5].

Summing up, we discussed the computation of properties in the form $\mathcal{P}_{\bowtie p}(F s_k)$, where s_k is an absorbing state, starting in any initial transient state of the system[2]. With this procedure, it is possible to obtain closed formulae for a number of interesting reliability properties.

For example, evaluating **R1** on our toy system, that is the probability of reaching the state *MsgFail* failure in any number of execution steps corresponds to evaluating b_{07} as:

$$\textbf{R1:} \quad \frac{(yz)}{(0.85 + 0.15z)} \leq 0.001$$

The approach can be extended to computing the probability of successfully reaching a non-absorbing state. This extension supports verification of properties like *"the probability of reaching state s_j without reaching any failure"* or *"the*

---

[2] Actually we discussed the computation of the probability associated with the property, to which the constraint $\bowtie p$ has to be applied.

probability of a successfully performing a certain operation or service". In our example, the probability of reaching the *Logout* state 7 after any number of steps is expressed by the following formula: $f_{04} = \frac{0.85 - 0.85x + 0.15z - 0.15xz - yz}{0.85 + 0.15z}$. This extension, as well as the ones needed to cover the entire PCTL are presented in [8].

6   Achieving Safe Dynamic Software Update

Once the need for a change in running software is identified, an alternative solution has to be found and then instantiated. A number of different approaches have been proposed to address the problem, focusing on changes at different levels of granularity. In this section we assume that the implementation has a distributed component-based architecture, where components interact via remote invocations. We do not address here the issue of how the alternative solution may be identified, but instead focus on how the architectural update may be instantiated at run time in a safe way, while the system is running.

Traditional approaches to software update are static. They require (1) to shut down the currently running version, (2) deploy the new version, and (3) restart the system. This allows safe replacement if off-line verification has proved that the new version satisfies the new requirements, but cannot be applied in the increasingly common cases where the system cannot be shut down and the update must be performed while the system is running.

Dynamic software update must satisfy two main requirements. It has to have *low disruption*, i.e. it must have low overhead and minimize the the delay with which the system is updated. It also has to be *safe*, i.e. it must not lead the system into an unexpected erroneous state.

The rest of this section summarizes the work presented in [18], where different criteria for dynamic update are assessed and a new criterion, called *version consistency* is proposed. This criterion leads to a safe and efficient dynamic update approach for distributed component-based architectures.

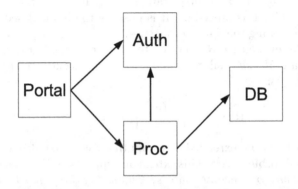

Fig. 7. Our example system.

Let us consider, as an example, the architecture shown in Fig. 7. A portal component (Portal) interacts with an authentication component (Auth) and a business processing component (Proc), while Proc interacts with both Auth and a database component (DB). This means that Portal statically depends on (i.e., can invoke) Proc and Auth, and Proc depends on Auth and DB.

A component can host (execute) transactions. A transaction is a sequence of actions that completes in bounded time. Actions include local computations and message exchanges. A transaction T can be initiated by an outside client or by another transaction T'. T is called a *root transaction* in the former case and a *sub-transaction* (of T') in the latter case. The term $sub(T', T)$ denotes that T is a direct sub-transaction of T'. The set $ext(T) = \{x | x = T \vee sub^+(T, x)\}$ is the *extended transaction set* of T, which contains T and all its direct and indirect sub-transactions. The extended transaction set of a root transaction models the concept of *distributed transaction* that can span over multiple components. The host component of transaction T is denoted as h_T. Transactions are also always notified of the completion of their sub-transactions. This implies that a transaction T cannot end before its sub-transactions T_i. All other exchanged messages between h_T and h_{T_i} —because of T_i— are temporally scoped between the two corresponding messages that initiate the sub-transaction and notify its completion.

Figure 8 shows a usage scenario for the example system. The Portal first gets an authentication token from Auth and then uses it to require services

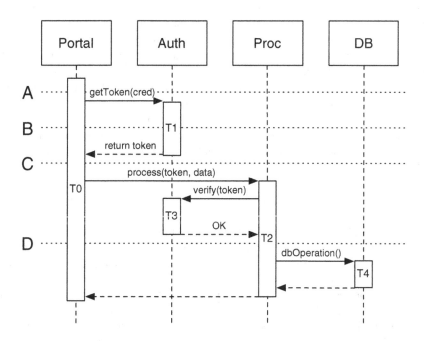

Fig. 8. Detailed scenario.

from Proc. Proc verifies the token through Auth and then starts computing, and interacting with DB. If we consider the root transaction T_0 at Portal, its extended transaction set is $ext(T_0) = \{T_0, T_1, T_2, T_3, T_4\}$, where T_1 at Auth is in response to the getTocken request, T_2 at Proc in response to process, T_3 at Auth in response to verify, and T_4 at DB for T_2's request of database operations.

A dynamic update can be specified as an operation that substitutes one or more components of the original configuration with new versions. We assume components to be stateless; i.e., there is no need to transfer the state from one component to its replacement during the update. We also assume the update to be correct, i.e., the update satisfies the requirements in the current environment conditions. The update leads to a dynamic reconfiguration, where new bindings are established between the existing components and a newly installed component. We assume that re-binding is performed as an atomic operation.

Let $\mathbb{S}$ be the current specification of the requirements to be satisfied by the system, and let $\mathbb{S}'$ be the updated specification that must be satisfied after the update. The dynamic reconfiguration is defined to be correct if:

- The transactions that end before the update satisfy $\mathbb{S}$;
- The transactions that begin after the update satisfy $\mathbb{S}'$;
- The transactions that begin before the update, and end after it, satisfy either $\mathbb{S}$ or $\mathbb{S}'$.

In our example, suppose that Auth has to be updated to exploit a stronger encryption algorithm and prevent weaknesses in system security. Although the new algorithm is incompatible with the old one, the other components need not to be updated because all encryption/decryption operations are done within Auth. If the update is allowed to happen any time, however, it may be impossible to ensure correctness. An obvious restriction on *when* the update can happen is that components targeted for update must be *idle*, that is, they are not hosting transactions. This constraint is a necessary but not sufficient condition for safe dynamic update. In fact, if we consider the scenario of Fig. 8, and substitute Auth when idle, but after serving getTocken, the resulting system would behave incorrectly since the security token would be created with an algorithm and validated by another.

It can be proved that correctness of arbitrary runtime updates is undecidable, even if the corresponding off-line update is correct and the on-line update only happens when components are idle. However, it is possible to derive automatically checkable sufficient correctness conditions.

In a seminal paper, Kramer and Magee [15] proposed a criterion called *quiescence* as a sufficient condition for a component to be safely replaced in dynamic reconfigurations. Their approach models a distributed system as a directed graph, whose nodes represent components and edges represent static dependencies. A node can initiate transactions on itself, or initiate two-party transactions on another node if there is an edge between the two nodes. A node's state can only be affected by transactions. Every two-party transaction is a sequence of message exchanges between the two nodes. A (dependent) transaction T can "contain" other (consequent) transactions T_i: the completion of T depends on the completion of all the T_i. Transactions always complete in bounded time and the initiator is always notified about their termination.

Definition 1 (Quiescence). *A node is quiescent if:*

1. *It is not currently engaged in a transaction that it initiated;*
2. *It will not initiate new transactions;*
3. *It is not currently engaged in servicing a transaction;*
4. *No transactions have been or will be initiated by other nodes which require service from this node.*

A component node satisfying the first two conditions is said to be *passive*. A node is required to respond to a passivate command from the configuration manager by driving itself into a passive state in bounded time. The last two conditions further make the node independent of all existing or future transactions, and thus it can be manipulated safely. To drive a node into a quiescent status, in addition to passivating it, all the nodes that statically depend on it must also be passivated to ensure the last two conditions.

According to this approach, a node cannot be quiescent before completion of all the transactions initiated by statically dependent nodes. This means that the actual update could be deferred significantly. In our example, Auth cannot be quiescent before the end of the transactions initiated by Portal and Proc. Moreover, all the other nodes that could potentially initiate transactions, which require service from Auth, directly or indirectly, are passivated, and their progress blocked till the end of the update. Again, in our example Portal and Proc are to be passivated. This means that the this approach can introduce significant disruption in the service provided by the system.

To reduce disruption, Vandewoude et al. [20] proposed an alternative criterion, called *tranquillity*. The idea is that there is no need for waiting a transaction to complete if it will not further request the service provided by the node targeted for update, even if the node has been involved in the transaction. Symmetrically, it is also permitted to update a node even if some on-going transactions will require the service provided by the node in the future, but they have not interacted with it yet.

Definition 2 (Tranquillity). *A node is tranquil if:*

1. *It is not currently engaged in a transaction that it initiated;*
2. *It will not initiate new transactions;*
3. *It is not actively processing a request;*
4. *None of its adjacent nodes are engaged in a transaction in which it has both already participated and might still participate in the future.*

If applied to our example, however, tranquillity would lead to unsafe updates. In fact, after Auth returns the token to Portal, it will not participate in the session initiated by Portal anymore. Before the request for verification is sent, Auth has not participated in the session initiated by Proc. So Auth is tranquil at time Ⓐ.

However, if Auth is updated at this time a failure may occur since the token was issued by the old version of Auth with an incompatible encryption algorithm. This failure would not happen if the system was either entirely in the old or in the new configuration.

To conclude, we can say that the quiescence is a general and safe criterion, but it can be disruptive. Tranquillity is less disruptive, but it can be applied safely in a restricted set of cases assumption, otherwise it can be unsafe.

Version consistency is a new criterion introduced in [18], which tries to get the best of the previous two proposals and achieves safety while reducing disruption. The criterion can be stated as follows:

Definition 3 (Version Consistency). *Transaction T is version consistent iff $\nexists T_1, T_2 \in ext(T) \mid h_{T_1} \in \omega \wedge h_{T_2} \in \omega'$. A dynamic reconfiguration of a system is version consistent if all its transactions are kept version consistent.*

This means that a dynamic reconfiguration of a system is correct if it happens at a time instant where all its transactions, including those started before and ended after the update, are kept version consistent. This is because of the correctness of the old and new configurations and the fact that any version-consistent transaction is served—along with all its sub-transactions—as if it entirely completed within the old or the new configuration, no matter when the update actually happens. Also note that a transaction that ends before (starts after) the update cannot have a direct or indirect sub-transaction hosted by the new (old) version of a component being updated.

For our example, if the update of Auth happens after transaction T_0 begins but before it sends a getToken request to Auth, all transactions in $ext(T_0)$ (i.e., all transactions in Fig. 8) are served in the same way as if the update happened before they all began. If it happens at any time after Auth replies to the verify request issued by Proc (time Ⓑ), all transactions in $ext(T_0)$ are served the same way as if the update happened after they all ended. However, if it happens at time Ⓐ, then $h_{T_1} =$ Auth, but $h_{T_3} =$ Auth$'$. As both T_1 and $T_3 \in ext(T_0)$, T_0 would not be version-consistent.

Since version consistency is not directly checkable, we need to identify a condition that is checkable on a component (or a set of components) and that ensures that its (their) runtime update does not break version consistency.

Dynamic dependences are the means to define such a condition, and they can easily be added to the diagram of Fig. 7 through properly-labelled edges besides those that represent the static dependencies. A static-labelled edge represents both a static dependence and the communication channel between the two components; future and past edges represent dynamic dependences. Future and past edges are also labelled with the identifier of a root transaction. We use $C \xrightarrow[T]{future(past)} C'$ to denote a future(past) edge labelled with the identifier of root transaction T, from component C to component C'. This means that because of T, some transactions in $ext(T)$ hosted by C will use (has used) the service provided by C' by initiating sub-transactions on it.

Definition 4 (Valid Configuration). *A valid configuration with dynamic dependences, hereafter configuration, must satisfy the following constraints:*

1. (LOCALITY) *For each* **future** *or* **past** *edge* $C \xrightarrow[T]{future(past)} C'$ *there is a* **static** *edge between* C *and* C';

2. (FUTURE-VALIDITY) *A* **future** *edge* $C \xrightarrow[T]{future} C'$ *must be in place before the first sub-transaction* $T' \in ext(T)$, *where* $T' \neq T$, *is initiated, and continues to exist at least until no transactions hosted by* C *will initiate further* $T'' \in ext(T)$ *on* C';

3. (PAST-VALIDITY) *A* **past** *edge* $C \xrightarrow[T]{past} C'$ *must be in place at the end of any transaction* $T' \in ext(T)$ *initiated by a transaction hosted by* C *on* C' *and continues to exist at least until the end of* T.

Figure 9 shows some configurations of the example system. Active components that are executing a transaction are marked with a $*$, and numbers correspond to the order with which edges are added.

The configuration of Fig. 9 (A) corresponds to time point Ⓐ in Fig. 8: transaction T_0 is executing on Portal, which is $*$-annotated. The dynamic edges indicate that to serve transactions in $ext(T_0)$, Portal might use Auth and Proc in the future, and also Proc might use Auth and DB. Figure 9 (B) corresponds to time Ⓑ and says that a transaction in $ext(T_0)$ (T_1) is currently running on

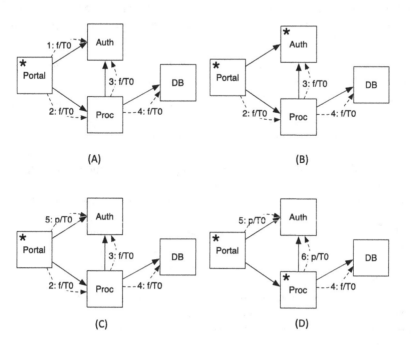

Fig. 9. Some configurations of the example system with explicit dynamic dependencies.

Auth, but no further transaction in $ext(T_0)$ hosted on Portal will initiate any sub-transaction on Auth anymore because there is no T_0-labelled future edge between the two nodes. Figure 9 (C), which corresponds to time Ⓒ in Fig. 8, indicates that Auth might have hosted transactions in $ext(T_0)$ initiated by Portal in the past, and might host further transactions in $ext(T_0)$ initiated by Proc in the future. Figure 9 (D) corresponds to time Ⓓ in Fig. 8 and shows that Auth, although it might have hosted transactions in $ext(T_0)$, is not hosting and will not host these transactions anymore.

Given a valid configuration, we can identify a locally checkable condition that is sufficient for the version consistency of dynamic reconfigurations.

Definition 5 (Freeness). *Given a configuration Σ, a component c is said to be free of dependencies with respect to a root transaction T iff c is not hosting any transaction in $ext(T)$ and there does not exist a pair of T-labelled future/past edges entering c. c is said to be free in Σ iff it is free with respect to all the root transactions in the configuration.*

In our example, Auth is free with respect to T_0 in the configurations of Fig. 9 (A) and (D), but not in the one of Fig. 9 (C) since there exist two f/T_0 and p/T_0 edges that enter Auth. Moreover, since Auth is active, it as also not free in Fig. 9. Intuitively, for a valid configuration Σ, the freeness condition for a component c —with respect to a root transaction T— means that the distributed transaction modeled by $ext(T)$ either has not used c yet (otherwise there should be a past edge), or it will not use c anymore (otherwise there should be a future edge). This leads to the following proposition, which is not proved here[3].

Proposition 1. *Given a valid configuration Σ of a system, a dynamic update of a component c is version consistent if it happens when c is free in Σ.*

Without entering into details, for which we refer to [18], our solution proposes a distributed algorithm for efficiently managing dynamic dependencies that: (1) keeps the configuration valid and (2) ensures version consistency with limited disruption. Dynamic dependencies are maintained in a distributed way. Each component only has a local view of the configuration that includes itself and its direct neighbors. A component is responsible for the creation and removal of the outgoing dynamic edges, but it is also always notified of the creation and removal of the incoming ones. This is achieved by exchanging management messages that keep the consistency among the views of neighbor components.

The management of dynamic dependencies may slightly delay the execution of the actual transactions, but it guarantees that no transaction will be blocked forever. The underlying message delivery is assumed to be reliable, and the messages between two components are kept in order. Dynamic edges are labelled with the identifiers of the corresponding root transactions to allow for the management of the dynamic edges of a root transaction independently of those of other transactions.

---

[3] A proof can be found in [18].

To assess version consistency we used simulation to evaluate its disruption for a wide set of randomly generated component-based distributed systems that varied in the number of components, service time, and network latency. The results showed that dynamic updates based on version consistency are on average more than 50 % less disruptive than those based on quiescence.

7 Conclusions

The objective of this paper was to give a high-level view of the problems involved in supporting software evolution without compromising its correctess, i.e., continuous requirements satisfaction. After setting the problem of software evolution in the context of Jackson and Zave's framework [13], we digged into the problem of achieving self-adaptation via models and verification at run time. Focusing on requirements that ask for probabilistic models and properties, we have shown how probabilistic model checking can be brought to run time to drive self-adaptation. We have then focused on another important problem that must be solved to support both self-adaptation and also, more generally, any kind of dynamic reconfiguration that is a consequence of evolution.

The approaches presented in this paper are a first step in the direction of integrating development and operation (DevOps), conceived as two interacting feedback loops that are funded on mathematically precise models and continuous formal verification. Models and verification are necessary in both loops, and they must be handled in an iterative and incremental manner. Agile development is often hostile to modeling and verification, sometimes they are even viewed as *deprecated upfront activities* [19]. Requirements are replaced by user stories. Although they realize that continuous verification is necessary, verification is simply equated to testing. Likewise, modeling and verification are often conceived as heavy-weight monolithic processes. For example, verification of partial and incomplete models is seldom supported, while incremental development intrinsically goes through incomplete descriptions. Verification is seldom incremental, support to understanding the effect of changes and reasoning on them is rarely provided. The two worlds, however, should get together, and this urgently calls for a sustained research agenda that goes widely beyond the initial steps presented in the paper.

References

1. Althoen, S.C., McLaughlin, R.: Gauss-Jordan reduction: a brief history. Am. Math. Monthly **94**(2), 130–142 (1987)
2. Baier, C., Katoen, J.-P.: Principles of Model Checking. The MIT Press, Cambridge (2008)
3. Baresi, L., Di Nitto, E., Ghezzi, C.: Toward open-world software: issue and challenges. Computer **39**(10), 36–43 (2006)
4. Belady, L.A., Lehman, M.M.: A model of large program development. IBM Syst. J. **15**(3), 225–252 (1976)

5. Bojanczyk, A.: Complexity of solving linear systems in different models of computation. SIAM J. Numer. Anal. **21**(3), 591–603 (1984)
6. Epifani, I., Ghezzi, C., Mirandola, R., Tamburrelli, G.: Model evolution by runtime adaptation. In: Proceedings of the 31st International Conference on Software Engineering, pp. 111–121. IEEE Computer Society (2009)
7. Filieri, A., Ghezzi, C., Tamburrelli, G.: Run-time efficient probabilistic model checking. In: Proceedings of the 33rd International Conference on Software Engineering (2011)
8. Filieri, A., Tamburrelli, G., Ghezzi, C.: Supporting self-adaptation via quantitative verification and sensitivity analysis at run time. IEEE Trans. Softw. Eng. **42**(1), 75–99 (2016)
9. Ghezzi, C., Tamburrelli, G.: Reasoning on non-functional requirements for integrated services. In: Proceedings of the 17th International Requirements Engineering Conference, pp. 69–78. IEEE Computer Society (2009)
10. Grinstead, C., Snell, J.: Introduction to probability. Amer Mathematical Society, Providence (1997)
11. Hahn, E.M., Hermanns, H., Zhang, L.: Probabilistic reachability for parametric markov models. In: Păsăreanu, C.S. (ed.) Model Checking Software. LNCS, vol. 5578, pp. 88–106. Springer, Heidelberg (2009)
12. Hinton, A., Kwiatkowska, M., Norman, G., Parker, D.: PRISM: a tool for automatic verification of probabilistic systems. In: Hermanns, H., Palsberg, J. (eds.) TACAS 2006. LNCS, vol. 3920, pp. 441–444. Springer, Heidelberg (2006)
13. Jackson, M., Zave, P.: Deriving specifications from requirements: an example. In: ICSE 1995: Proceedings of the 17th international conference on Software engineering, pp. 15–24, New York, NY, USA. ACM (1995)
14. Kephart, J.O., Chess, D.M.: The vision of autonomic computing. IEEE Comput. **36**(1), 41–50 (2003)
15. Kramer, J., Magee, J.: The evolving philosophers problem: dynamic change management. IEEE Trans. Softw. Eng. **16**(11), 1293–1306 (1990)
16. Kwiatkowska, M., Norman, G., Parker, D.: Prism 2.0: a tool for probabilistic model checking. In: Proceedings of First International Conference on the, Quantitative Evaluation of Systems, QEST 2004, pp. 322–323 (2004)
17. Lehman, M.M., Belady, L.A. (eds.): Program Evolution: Processes of Software Change. Academic Press Professional Inc., Cambridge (1985)
18. Ma, X., Baresi, L., Ghezzi, C., Manna, V.P.L., Lu, J.: Version-consistent dynamic reconfiguration of component-based distributed systems. In: ESEC/FSE 2011: The 19th ACM SIGSOFT Symposium on the Foundations of Software Engineering and the 13rd European Software Engineering Conference, pp. 245–255. ACM (2011)
19. Meyer, B.: Agile!: The Good, the Hype and the Ugly. Springer Science and Business Media, Berlin (2014)
20. Vandewoude, Y., Ebraert, P., Berbers, Y., D'Hondt, T.: Tranquility: a low disruptive alternative to quiescence for ensuring safe dynamic updates. IEEE Trans. Softw. Eng. **33**(12), 856–868 (2007)
21. Zave, P., Jackson, M.: Four dark corners of requirements engineering. ACM Trans. Softw. Eng. Methodol. **6**(1), 1–30 (1997)

Mean-Field Limits Beyond Ordinary Differential Equations

Luca Bortolussi[1,2,3](✉) and Nicolas Gast[4]

[1] DMG, University of Trieste, Trieste, Italy
lbortolussi@units.it
[2] MOSI, Saarland University, Saarbrücken, Germany
[3] CNR-ISTI, Pisa, Italy
[4] Inria, University of Grenoble Alpes, CNRS, LIG, 38000 Grenoble, France

Abstract. We study the limiting behaviour of stochastic models of populations of interacting agents, as the number of agents goes to infinity. Classical mean-field results have established that this limiting behaviour is described by an ordinary differential equation (ODE) under two conditions: (1) that the dynamics is smooth; and (2) that the population is composed of a finite number of homogeneous sub-populations, each containing a large number of agents. This paper reviews recent work showing what happens if these conditions do not hold. In these cases, it is still possible to exhibit a limiting regime at the price of replacing the ODE by a more complex dynamical system. In the case of non-smooth or uncertain dynamics, the limiting regime is given by a differential inclusion. In the case of multiple population scales, the ODE is replaced by a stochastic hybrid automaton.

Keywords: Population models · Markov chain · Mean-field limits · Differential inclusions · Hybrid systems

1 Introduction

Many systems can be effectively described by stochastic population models, for instance biological systems [51], epidemic spreading [1], queuing networks [41]. These systems are composed of a set of objects, agents, or entities interacting together. Each individual agent is typically described in a simple way, as finite state machines with few states. An agent changes state spontaneously or by interacting with other agents in the system. All transitions happen probabilistically and take a random time to be completed. By choosing exponentially distributed times, the resulting stochastic process is a continuous-time Markov chain with a finite state space. Many numerical techniques exist to compute probabilities of such chains [3], and they are part of state-of-the art stochastic model checking tools like PRISM [38] or MRMC [36].

These techniques, however, are limited in their applicability, as they suffer from the state-space explosion: the state-space grows exponentially with the

M. Bernardo et al. (Eds.): SFM 2016, LNCS 9700, pp. 61–82, 2016.
DOI: 10.1007/978-3-319-34096-8_3

number of agents and even simple agents, when present in large quantities, can generate a huge state space which is far beyond the capabilities of current tools.

This results in the need for approximation techniques to estimate the probabilities and the behaviours of the system. A classic way is to resort to stochastic simulation, which scales better but is still a computationally intensive process for large populations. Precisely in this regime of large populations, mean field analysis offers a viable, often accurate, and much more efficient alternative. The basic idea of mean field is that, when counting the number of agents that are in a given state, the fluctuations due to stochasticity become negligible as the number of agents N grows. For large N, the system becomes essentially deterministic.

A series of results, *e.g.*, [4,7,37], have established that when the state space of each agent is finite and the dynamics is sufficiently smooth, the system's behaviour converges as N goes to infinity to a limiting behaviour that is described by system of ordinary differential equations (ODE). The dimension of this system of ODE is equal to the number of states of the individual agents, but independent of the population size N. The dimension of the differential equation is typically small, hence the numerical integration of these equations is extremely fast. These results show that the intensity of the fluctuations goes to zero as $1/\sqrt{N}$. This approach is used in many domains, including computer-based systems [29,31, 43,50], epidemic or rumour propagation [17,34] or bike-sharing systems [24]. Is it also used to construct approximate solutions of stochastic model checking problems [10–13].

However, these limiting results have two main shortcomings. First, these models cannot deal with discontinuities on the rates of interaction between agents, or uncertainty in model parameters in an obvious way. Second, being able to approximate the number of agents by a continuous variable requires all populations to be large. These limitations are essentially due to restricting the attention to a limiting regime that can be expressed in terms of smooth ODEs.

In this document, we show that by enlarging the set of possible limiting regimes, it is possible to extend the classical framework in multiple directions. We first begin in Sect. 2 by a concise introduction to classical mean field models and their ODE limits. This section requires basic knowledge of CTMCs and ODEs. We then show in Sect. 3 how discontinuities and uncertainties can be treated uniformly and consistently considering mean field limits in terms of differential inclusions. We then tackle the presence of multiple population scales in Sect. 4. We show that when the number of agents in some populations go to infinity while others remain finite, the mean field limit is naturally expressed as a stochastic hybrid automaton, where continuous-deterministic and discrete-stochastic dynamics coexist and modulate each other. Last, we mention other related work and extensions of this framework, for instance to cooperative games, in Sect. 5.

2   The Classical Mean Field Framework

In this section, we will introduce the fundamental mean field approximation. We assume the reader familiar with basic concepts of Markov Chains in

Continuous Time, see e.g. [23,44] for an introduction, and with ordinary differential equations.

We start in Sect. 2.1 by introducing a framework to describe the class of systems amenable of mean field analysis, namely Markov population processes (see also the chapter on spatial representations [26]). We illustrate these concepts in Sect. 2.2 by means of a classic epidemic spreading model. In Sect. 2.3, we describe the basic mean-field theorems.

2.1 Population Continuous-Time Markov Chains

Population continuous-time Markov chains (PCTMCs) describe a set of interacting agents, which can have different internal states. Interactions involve a small number of agents, and can happen randomly in time, according to an exponential distribution with system-dependent rate. We describe these systems in terms of counting variables and transition classes, following the conventions of [7,33].

More specifically, a PCTMC $\mathcal{M}$ model is a tuple $(\mathbf{X}, \mathcal{T}, \mathbf{x}_0, N)$, where

- $\mathbf{X} = (X_s)_{s \in \mathcal{S}} \in \mathbb{R}^{|\mathcal{S}|}$ is the population vector. The state space of an agent is $\mathcal{S}$ and $X_s \in \mathbb{N}$ counts the number of agents in state $s \in \mathcal{S}$. The state space of the model is a finite or countable subset of $\mathbb{R}^{|\mathcal{S}|}$.
- $\mathcal{T}$ is the set of transition classes, each of the form $\eta = (\phi_\eta(\mathbf{X}), \mathbf{v}_\eta, f_\eta(\mathbf{X}))$, where
  - $\phi_\eta(\mathbf{X}) \in \{0, 1\}$ is a guard predicate, representing a subset of $\mathcal{S}$ in which the transition is active;
  - $\mathbf{v} \in \mathbb{R}^n$ is an update vector, encoding the relative change of $\mathbf{X}$ induced by the firing of the transition η: the new state will be $\mathbf{X} + \mathbf{v}_\eta$;
  - $f_\eta(\mathbf{X})$ is the rate function, giving the rate at which an η transition is fired as a function of the state space of the system. Typically, $f_\eta(\mathbf{X})$ is a (locally) Lipschitz continuous function of the population variables.
- $\mathbf{x}_0 \in \mathbb{R}^{|\mathcal{S}|}$ is the initial state of the system.
- N is the population size.

Each PCTMC model $\mathcal{M}$ defines a CTMC $\mathbf{X}(t)$ on the state space $\mathcal{S}$. This chain is characterised by the infinitesimal generator matrix Q [23], whose off-diagonal entries are given by

$$Q_{\mathbf{x}_1, \mathbf{x}_2} = \sum_{\eta \in \mathcal{T} \text{ s.t. } \mathbf{x}_2 = \mathbf{x}_1 + \mathbf{v}_\eta} \phi_\eta(\mathbf{x}_1) f_\eta(\mathbf{x}_1).$$

An important concept related to population models is the *system size*, N. Typically, system size is the total (initial) population. In some domains, though, like biochemical networks or ecological models, N may represent another measure of size, like the volume or the area in which the dynamics described by a PCTMC happens. We refer to [7] for a deeper discussion of this.

2.2   Example: SIR Epidemic Spreading

As a simple and illustrative example, we consider the spreading of a disease in a community of N agents (which can be humans, animals, computers). The state space of an agent is $\mathcal{S} = \{S, I, R\}$. This model is one of the classical examples of a Markov population process and is referred to as the SIR model.

The contagion happens when a susceptible agent (X_S) enters in contact with another agent who turns out to be infected (at rate $k_{si}X_S X_I/N$) or enters in contact with an external source of the disease (at rate $k_i S$). Infected individuals spontaneously recover at rate k_r, and become Recovered (X_R) and immune from the disease. This immunity, however, can be lost with rate k_s.

Formally, the model can be described as a PCTMC with three variables (X_S, X_I, X_R), each taking values in the integers $\{0, \ldots, N\}$, as no birth or death events are considered. The model has four transition classes, all having a guard predicate evaluating to true $(=1)$ in all states:

- Internal infection: $(\mathbf{true}, \mathbf{e}_I - \mathbf{e}_S, k_{si}X_S X_I/N)$;
- External infection: $(\mathbf{true}, \mathbf{e}_I - \mathbf{e}_S, k_i X_S)$;
- Recovery: $(\mathbf{true}, \mathbf{e}_R - \mathbf{e}_I, k_r X_I)$;
- Immunity loss: $(\mathbf{true}, \mathbf{e}_S - \mathbf{e}_R, k_s X_R)$;

2.3   Classic Mean Field Equations

Mean field theory answers the following question about population models: what happens when the population is very large? More specifically, it can be shown that, for a large class of models, the dynamics of the system greatly simplifies as the system size goes to infinity. The classic theorem, dating back to the 1970s [37], shows that trajectories of suitably rescaled processes for large populations look deterministic, and in fact converge to the solution of an ordinary differential equation (ODE).

An important operation in the path to mean field is to normalise population processes, dividing variables by the system size, and updating accordingly the transitions. This allows one to compare different models, as they will now have the same scale, intuitively these are population densities. Roughly, the deterministic behaviour appears because fluctuations around the mean of a population process grow as $\sqrt{N}$, hence while normalising, i.e. dividing by N variables, fluctuations will be of magnitude $1/\sqrt{N}$, and will thus go to zero.

More formally, consider a population model $\mathcal{M}^N$, where we make explicit the dependence on the system size, and define its normalised version $\hat{\mathcal{M}}^N$ as follows:

- Population variables (and initial conditions) are rescaled by N: $\hat{\mathbf{X}}^N = \mathbf{X}/N$;
- Transition rates and guard predicates are expressed in the normalised variables, by substituting $N\hat{\mathbf{X}}^N$ for $\mathbf{X}$ in the functions: $\hat{f}^N(\hat{\mathbf{X}}^N) = f(N\hat{\mathbf{X}})$ and $\hat{\phi}^N(\hat{\mathbf{X}}^N) = \phi(N\hat{\mathbf{X}})$;
- Update vectors are rescaled by N, too: $\hat{\mathbf{v}}^N = \mathbf{v}/N$;

The CTMC associated with the normalised model will be denoted by $\hat{\mathbf{X}}^N(t)$.

Example. Consider the SIR model of Sect. 2.2. Its normalised version, for a population of N agents, has the following four transition classes:

- Internal infection: $\hat{\mathbf{v}}_{si}{}^N = (\mathbf{e}_I - \mathbf{e}_S)/N$, $\hat{f}^N_{si}(\hat{\mathbf{X}}) = Nk_{si}\hat{X}_S\hat{X}_I$;
- External infection: $\hat{\mathbf{v}}_i{}^N = (\mathbf{e}_I - \mathbf{e}_S)/N$, $\hat{f}^N_i(\hat{\mathbf{X}}) = Nk_i\hat{X}_S$;
- Recovery: $\hat{\mathbf{v}}_r{}^N = (\mathbf{e}_R - \mathbf{e}_I)/N$, $\hat{f}^N_r(\hat{\mathbf{X}}) = Nk_r\hat{X}_I$;
- Immunity loss: $\hat{\mathbf{v}}_s{}^N = (\mathbf{e}_S - \mathbf{e}_R)/N$, $\hat{f}^N_s(\hat{\mathbf{X}}) = Nk_s\hat{X}_R$.

As we can see, all transition rates depend linearly on system size. When this happens, rates are called density dependent [7], a condition that usually guarantees the applicability of the mean field results.

Drift. The main quantity required to define mean field equations is the drift. The drift is the average direction of change of the population model, conditional on being in a certain state at some time t. The drift of the normalised model is

$$F(\mathbf{x}) = \sum_{\eta \in \mathcal{T}} \hat{\phi}_\eta(\mathbf{x})\hat{f}_\eta(\mathbf{x})\hat{\mathbf{v}}_\eta, \tag{1}$$

Usually, mean field is defined under some additional restrictions on the population model:

(C1) Guards are true for any $\mathbf{x} \in S$, hence the indicator function can be safely removed from the drift: $F^N(\mathbf{x}) = \sum_{\eta \in \mathcal{T}} \hat{f}^N_\eta(\mathbf{x})\hat{\mathbf{v}}^N_\eta$.
(C2) F is a Lipschitz continuous function.

Note that by definition of the rescaled model, the drift $F(\mathbf{x})$ does not depend on N, because the update vectors are rescaled by $1/N$ while the transition rates are rescaled by N. When the drift $F^N(\mathbf{x})$ does depend on N, condition (C2) can be replaced in all theorems by a condition (C2'): $F^N(\mathbf{x})$ converges uniformly as $N \to \infty$ to a Lipschitz continuous function $F(\mathbf{x})$.

Note that in the SIR model, Conditions (C1) and (C2) are satisfied. For the second one, in particular, we can see that by multiplying a normalised update vector, e.g. $\hat{\mathbf{v}}_i{}^N$, by the corresponding rate, e.g. $\hat{f}^N_i(\hat{\mathbf{X}}) = Nk_i\hat{X}_S\hat{X}_I$, the dependency on system size cancels out, so that the drift $F^N(\mathbf{x})$ is independent of N. Lipschitz continuity[1] is easily proved. See [7] for a deeper discussion of these conditions. In the following, we will discuss how to weaken these assumptions.

The following theorem can then be proved (see [4,7,20]):

Theorem 1. *Assume the above conditions C1 and C2 hold and that $\hat{\mathbf{X}}_0$ converges to $\mathbf{x}_0$ almost surely (resp. in probability) as N goes to infinity. Let $\mathbf{x}$ be the solution of the ODE:*

$$\frac{d}{dt}\mathbf{x}(t) = F(\mathbf{x}(t)) \qquad \mathbf{x}(0) = \mathbf{x}_0. \tag{2}$$

---

[1] In fact, Lipschitz continuity is satisfied only locally, but this enough for mean field convergence to work.

Then, for any $T > 0$,

$$\lim_{N\to\infty}\sup_{t\leq T}\|\hat{\mathbf{X}}^N(t) - \mathbf{x}(t)\| = 0 \qquad \textit{almost surely (resp. in probability).}$$

The theorem essentially states that trajectories of the PCTMC, for large N, will be indistinguishable from the solution of the mean field ODE restricting to any finite time horizon $T > 0$. This can be seen as a functional version of the law of large numbers. An example of the theorem at work, for the SIR model, can be seen in Fig. 1. For the SIR model, the mean field approximation is given by the following system of ODEs:

$$\begin{aligned}
\dot{x}_S &= -k_i x_S - k_{si} x_S x_I + k_s x_R \\
\dot{x}_I &= k_i x_S + k_{si} x_S x_I - k_r x_I \\
\dot{x}_R &= k_r x_I - k_s x_R
\end{aligned} \tag{3}$$

We simulate the model for population size of $N = 10$, $N = 100$ and $N = 1000$ agents and report the evolution of the numbers of susceptible or infected agents as a function of time. The parameters are $k_i = k_{si} = k_s = k_r = 1$ and the initial conditions are $X_I(0) = X_S(0) = 2X_R(0) = 2N/5$. Each plot contains three curves: a sample path of one simulation, the mean field (ODE) approximation and an average over 10^4 simulations.

This figure illustrates how large the population size N has to be for the approximation to be accurate. We observe that the mean field approximation describes correctly the overall dynamics of the PCTMC for $N = 100$ and $N = 1000$. In fact, it can be shown that the rate of convergence in Theorem 1 is of the order of $1/\sqrt{N}$ [4,20] but with bounds that often an underestimation of the real convergence speed. In practice, we often observe that the convergence is quicker than this bound. This is particularly true when one considers the average stochastic value: $\mathbb{E}\,[\mathbf{X}(t)]$. To illustrate this fact, we simulated the SIR model 10^4 times to compute the values $\mathbb{E}\,[\mathbf{X}(t)]$. We report the evolution of $\mathbb{E}\,[X_S(t)]$ and $\mathbb{E}\,[X_I(t)]$ with time in Fig. 1. We observe than, already for $N = 10$, the ODE $\mathbf{x}(t)$ is extremely close to the value of $\mathbb{E}\,[\mathbf{x}(t)]$ computed by simulation. For $N = 100$ and $N = 1000$, the curves are indistinguishable.

Notice that in Theorem 1, the restriction to finite time horizons is fundamental, as convergence at steady state does not necessarily hold. An example is given by the SIR model when $k_i = 0$, $k_{si} = 3$, $k_s = k_r = 1$ and $x_R(0) = x_S(0) = x_I(0) = 1/3$. This initial condition is a fixed point of the mean field ODE which therefore predicts an endemic equilibrium. In the PCTMC model, however, for any N, the epidemic always extinguishes (i.e. eventually $X_I(t) = 0$) because there is no external infection in this situation. For a deeper discussion of this issue, see the next section as well as [7].

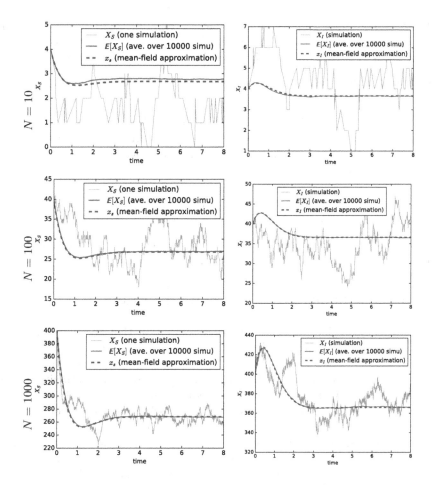

Fig. 1. Simulation of the SIR model: comparison between ODE and simulation for various values of N. We observe that the simulation converges to the ODE as N goes to infinity. Moreover, even for $N = 10$, the average simulation is very close to the ODE.

In fact, steady state results can be obtained at the price of adding two additional conditions:

(C3) For any N, the PCTMC has a steady-state distribution π^N. The sequence of distributions π^N is tight.[2]

(C4) The ODE (2) has a unique fixed point x^* to which all trajectories converge.

The condition C3 is a natural condition that is in general true for PCTMC models. For example, if the population CTMC of size N has a finite number of states, it has a steady-state distribution. Moreover, the tightness of the measure is true if $\mathbf{X}$ is almost surely bounded.

---

[2] A sequence of distributions is π^N tight if their probability does not escape to infinity, i.e. for each ϵ there is a compact set K such that $\pi^N(K) \geq 1 - \epsilon$ for each N.

However, condition C4 is a condition that is often difficult to check for a given set of ODEs. Proving it requires to exhibit a Lyapunov function witnessing global attractiveness of the unique equilibrium point.

Theorem 2. *Assume that the above conditions C1, C2, C3 and C4 hold. Then, π^N converges weakly to the Dirac measure x^*.*

Similarly to Theorem 1, it can be shown that under mild additional conditions, the speed of convergence of the steady-state distribution π^N to x^* is also $1/\sqrt{N}$. For example, it is shown in [53] that this holds when x^* is exponentially stable, in which case we have

$$\sqrt{\mathbb{E}\left[\|\mathbf{X}^N(\infty) - x^*\|^2\right]} = O(1/\sqrt{N}),$$

where $\mathbf{X}^N(\infty)$ denotes a random point distributed according to the stationary measure π^N.

3 Non-continuous Dynamics and Uncertainties

The classical mean field models make the assumption that the drift F, given by (1), is Lipschitz-continuous. Yet, this is not the case in many practical problems. For example, this occurs when a transition η has a guard predicate ϕ_η that is true only in a sub-part of the domain. This causes a discontinuity in the drift between the two regions where the predicate is true or false. In this case, the ODE $\dot{x} = F(x)$ is often not well-defined. A classical way of overcoming this difficulty is to replace the ODE by a differential inclusion $\dot{x} \in F(x)$.

In this section, we give some general discussion of differential inclusions. We then show how the classical mean field results can be generalised to differential inclusion dynamics in the case of discontinuous dynamics. Last, we show in Sect. 3.3 how this framework can be used to deal with uncertainties in the parameters. This section collects results from [6,28].

3.1 The Differential Inclusion Limit

Let $\mathcal{M} = (\mathbf{X}, \mathcal{T}, \mathbf{x}_0, N)$ be a population model as defined in Sect. 2.1. Each transition class has the form $\eta = (\phi(\mathbf{X}), \mathbf{v}_\eta, f_\eta(\mathbf{X}))$. Recall that the drift of the stochastic system, defined in Eq. (1), is

$$F(\mathbf{x}) = \sum_{\eta \in \mathcal{T}} \hat{\phi}_\eta(\mathbf{x}) \hat{f}_\eta(\mathbf{x}) \hat{\mathbf{v}}_\eta, \tag{4}$$

The classical mean field results presented in the previous section (Theorems 1 and 2) apply when the guard predicates ϕ_η always evaluate to true. This condition guarantees that, if the functions f_η are Lipschitz continuous, the function F is also Lipschitz continuous. This ensures that the ODE $\dot{x} = F(x)$ is well-defined: from any initial condition, it has a unique solution. However, this no longer holds when guard predicates can take the two values true and false. In this case, the drift F is not continuous and Theorems 1 and 2 no longer apply.

One of the reasons for the inapplicability of those theorems is that when F is not a Lipschitz-continuous function, the ODE $\dot{x} = F(x)$ does not necessarily have a solution. For example, let $F : \mathbb{R} \to \mathbb{R}$ be defined by $F(x) = -1$ if $x \geq 0$ and $F(x) = 1$ if $x < 0$. The ODE $\dot{x} = F(x)$ starting in 0 has no solution. A natural way to overcome this limitation is to use differential inclusions.

Let G be a multivalued map, that associates to each $x \in \mathbb{R}^S$ a set $G(x) \subset \mathbb{R}^S$. A trajectory $\mathbf{x}$ is said to be a solution of the differential inclusion $\dot{x} \in G(x)$ starting in x_0 if:

$$x(t) = x_0 + \int_0^t g(s)ds, \quad \text{where for all } s : g(s) \in G(x(s)).$$

The sufficient condition for the existence of at least one solution $\mathbf{x} : [0, \infty) \to \mathbb{R}^d$ of a differential inclusion from any initial condition x_0 is the following (see [2]):

(C5) For all $x \in \mathbb{R}^d$, $G(x)$ is closed, convex and non-empty; $\sup_{x \in \mathbb{R}^d} |G(x)| < \infty$ and G is upper-hemicontinuous.[3]

Theorem 3 [28]. *Let $\mathcal{M} = (\mathbf{X}^N, \mathcal{T}, \mathbf{x}_0, N)$ be a population model with drift F such that there exists a function G satisfying (C5) such that for all $x \in \mathbb{R}^d$: $F(x) \in G(x)$. Let $\mathcal{S}$ be the set of solutions of the differential inclusion $\dot{x} \in G(x)$ starting in x_0. Then, for all T, almost surely*

$$\lim_{N \to \infty} \inf_{\mathbf{x} \in \mathcal{S}_{x_0}} \sup_{t \in [0,T]} \left\| \mathbf{X}^N(t) - x(t) \right\| = 0.$$

In other words, as N grows, the distance between the stochastic process $\mathbf{X}$ and the set of solutions of the differential inclusion goes to 0. If this set has a unique solution, then $\mathbf{X}^N$ converges to this solution.

Theorem 3 is a generalisation of Theorem 1 that relaxes the condition (C1) by using a larger drift function G. We will see in the next section a natural way to choose G when the drift is not continuous. The price to be paid by this generalisation is composed of two drawbacks. First, differential inclusions can have multiple solutions. Theorem 3 implies that $\mathbf{X}^N$ gets closer to the solutions of the differential inclusion but does not indicate towards which solutions the process will converge. Second, the speed of convergence of $\mathbf{X}^N$ to $\mathcal{S}$ is unknown, apart in the special case of one-sided Lipschitz drift, for which the distance between $\mathbf{X}^N$ and $\mathcal{S}$ decays in $1/\sqrt{N}$ (see [28]). In general, the convergence appears to be slower in the case of non-continuous dynamics (see Fig. 2).

The steady-state of a non-continuous PCTMC can also be approximated by using the same approximation. In fact, the results of Theorem 2 can also be directly generalised to the case of non-continuous dynamics: if the differential inclusion has a unique point x^* to which all trajectories converge, then the steady-state distribution of $\mathbf{X}^N$ concentrates on x^* as N goes to infinity (see [6, 28]).

---

[3] G is upper-hemicontinuous if for all $x, y \in \mathbb{R}^d$, $x_n \in \mathbb{R}^d$, $y_n \in F(x_n)$, $\lim_{n \to \infty} x_n = x$ and $\lim_{n \to \infty} y_n = y$, then $y \in F(x)$.

3.2 Application to Discontinuous Dynamics

A natural way to define a multivalued map that satisfies (C5) is to consider the multivalued map $\bar{F}$, defined by

$$\bar{F}(x) = \bigcap_{\varepsilon>0} \text{convex\_hull} \left(\bigcup_{x':\|x-x'\|\le\varepsilon} F(x') \right). \tag{5}$$

It is shown in [28] that if F is bounded, then $\bar{F}$ satisfies (C5). By Theorem 3, this implies that, regardless of the properties of the original drift F, the trajectories of the stochastic system $\mathbf{X}^N$ converge to the solution of the differential inclusion $\dot{x} \in \bar{F}(x)$.

When F is continuous at a point $x \in \mathbb{R}^d$, the $\bar{F}(x) = F(x)$. When F is not continuous in $x \in \mathbb{R}^d$, $\bar{F}(x)$ is multivalued. To give a concrete example, let us consider the SIR model of Sect. 2.2 in which we add an additional transition corresponding to the treatment of some infected people. This treatment is applied when the proportion of infected people is greater than 0.3 and changes an infected person into a susceptible individual at rate k_t. This corresponds to a transition class:

– Treatment: $\eta = (\mathbf{1}_{X_I \ge 0.3}, \mathbf{e}_S - \mathbf{e}_I, k_t X_I)$

Adding this transition to the original ODE (3), the drift is given by

$$F(x) = \begin{pmatrix} -k_i x_S - k_{si} x_S x_I + k_s x_R + k_t \mathbf{1}_{x_I \ge 0.3} x_I \\ k_i x_S + k_{si} x_S x_I - k_r x_I - k_t \mathbf{1}_{x_I \ge 0.3} x_I \\ k_r x_I - k_s x_R \end{pmatrix}, \tag{6}$$

where the guard predicate leads to the term $\mathbf{1}_{x_I \ge 0.3}$.

This drift is not continuous in x. In fact, it can be shown that because of this discontinuity, the corresponding ODE has no solution on $[0, \infty)$ starting from $x_0 = (.4, .4, .2)$. The corresponding $\bar{F}$ defined by Eq. (5) is then given by

$$\bar{F}(x) = \begin{pmatrix} -k_i x_S - k_{si} x_S x_I + k_s x_R + k_t \mathbf{1}_{x_I > 0.3} x_I + k_t[0, x_I] \mathbf{1}_{x_I = 0.3} \\ k_i x_S + k_{si} x_S x_I - k_r x_I - k_t \mathbf{1}_{x_I > 0.3} x_I - k_t \mathbf{1}_{x_I = 0.3}[0, x_I] \\ k_r x_I - k_s x_R \end{pmatrix},$$

where the notation $a + [b, c]$ denotes the set $[a + b, a + c]$.

It can be shown that the differential inclusion $\dot{x} \in \bar{F}(x)$ has a unique solution, $\mathbf{x}$. Hence, Theorem 3 applies to show that $\mathbf{X}^N$ converges to $\mathbf{x}$ as N goes to infinity. To illustrate this fact, we simulated the modified SIR model with the treatment policy and report the results in Fig. 2. We observe that, as stated by Theorem 3, $\mathbf{X}^N$ converges to $\mathbf{x}$ as N goes to infinity. In this case, the convergence appears to be slower than for the Lipschitz-continuous case. This is especially visible when looking at the average values $\mathbb{E}[X_S]$ and $\mathbb{E}[X_I]$: in Fig. 1, we observe that for the Lipschitz-continuous case, $\mathbb{E}[X_S]$ is almost equal to x_s already for $N = 10$. In the non-continuous case, Fig. 2 indicates that $\mathbb{E}[X_I]$ does converge to x_I but at a much slower rate.

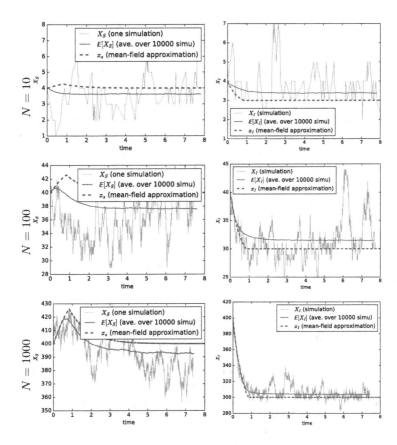

Fig. 2. Simulation of the SIR model: comparison between differential inclusions and simulation for various values of N. We observe that the simulation converges to the solution of the differential inclusion as N goes to infinity. Moreover, even for $N = 10$, the average simulation is very close to the solution of the differential inclusion.

3.3   Imprecise and Uncertain Models

Stochastic models are one way of representing uncertainties in a system but they depend on parameters whose precise values are not always known. The differential inclusion framework is also well adapted to study models with imprecise or unknown parameters. Following [6], we distinguish two ways to model uncertainties in models of complex systems:

– **Imprecise scenario:** Some parameters ϑ can depend on features of the environment external to the model. We fix a set Θ of possible values for ϑ and assume that ϑ depends on time t and can take any value of Θ at any time instant, i.e. that $\vartheta_t \in \Theta$.
– **Uncertain scenario:** In a simpler scenario, a parameter ϑ is assumed fixed, but its precise value not known precisely. In this case, we just assume that $\vartheta \in \Theta$, where Θ is the possible set of values of ϑ, as above.

An imprecise or uncertain PCTMC model is a tuple $(\mathbf{X}, \mathcal{T}, x_0, \Theta, N)$, where Θ is a set of parameters. The difference with classical PCTMC is that the rate function $f_\eta(\mathbf{X}, \theta)$ of each transition class $\eta \in \mathcal{T}$ depends on a parameter $\theta \in \Theta$.

The differential inclusion framework can be used to study the limits of imprecise and uncertain PCTMC. For the uncertain scenario, there is a differential inclusion $\dot{x} \in F(x, \vartheta)$ associated with each parameter ϑ. Denoting by S_ϑ the set of solutions of this differential inclusion, Theorem 3 shows that, as N goes to infinity, any sequence of uncertain trajectories $\mathbf{X}^N$ converges to the set $S_{\text{uncertain}} = \bigcup_{\vartheta \in \Theta} S_\vartheta$. The differential inclusion corresponding to the imprecise scenario is $\dot{x} \in \bigcup_{\vartheta \in \Theta} F(x, \vartheta)$. Denoting by $S_{\text{imprecise}}$ the set of solutions of this differential inclusion, Theorem 3 shows that, as N goes to infinity, any imprecise trajectory $\mathbf{X}^N$ converges to $S_{\text{imprecise}}$.

Some numerical methods are developed in [6] to compute or approximate the set of solutions of differential inclusions corresponding to the imprecise and the uncertain model. In particular, we obtain the most precise results by describing the set of reachable values of x at time t as a maximisation problem. Then Pontryagin's maximum principle [46] can be used to numerically compute the solution.

4 Hybrid Mean Field Limits

In the previous sections, we considered scenarios where all populations of the model are large and grow with the system size. This allows one to prove that their density has vanishing fluctuations around the mean, given by the solution of the mean field equation. However, in many practical cases, there may be multiple population scales in a model, typically in the form of some entities being present in small numbers, independent of the total population size. Examples can be found in genetic regulatory networks, where genes are present in a fixed quantity, typically one or few copies, or more generally in the presence of a centralised form of control [8]. This suggest that in these scenarios we need to consider mean field models in which the continuous and deterministic limit dynamics of parts of the system coexists with the discrete and stochastic dynamics of other parts. Mathematically, this behaviour is captured by stochastic hybrid systems (SHS) [8,21], which will be introduced in the next subsection.

4.1 Stochastic Hybrid Systems

We introduce a model of SHS, essentially borrowing from the treatment of [8, 15] of a class of stochastic hybrid processes known as Piecewise-Deterministic Markov Processes (PDMP) [21].

We consider two sets of variables, the *discrete variables* $\mathbf{Z} = Z_1, \ldots, Z_k$ and the *continuous variables* $\mathbf{Y} = Y_1, \ldots, Y_m$. The former describes populations that remain discrete also in the mean-field limit, while the second describes populations that will be approximated as continuous. We call $E = E_d \times E_c$ the hybrid state space of the SHS, with $E_d \subset \mathbb{N}^k$ a countable set of possible values for $\mathbf{Z}$, and with $E_c \subset \mathbb{R}^m$ the continuous state space in which variables $\mathbf{Y}$ can take values. Each possible value that the vector $\mathbf{Z}$ can take is called a discrete mode, and can be identified with a node in a graph describing the transitions of the discrete states. This graph-based point of view is taken in the definition of stochastic hybrid automata, see e.g. [15].

The evolution of the continuous state is governed by an m-dimensional *vector field* $F(\mathbf{Z}, \mathbf{Y})$, depending on the continuous and the discrete variables. Hence, the continuous variables will evolve following the solution of the differential equation defined by F, which can be different in each discrete mode $\mathbf{z}$. Such a mode-specific continuous dynamics is one of the characteristic features of SHS.

The dynamics of the discrete state is governed by a stochastic Markovian dynamics, specified by two quantities: a *rate function* $\lambda(\mathbf{Z}, \mathbf{Y})$, depending both on discrete and continuous variables, and a *jump or reset kernel* $R(\mathbf{Z}, \mathbf{Y}, \cdot)$, specifying for each $\mathbf{Z}, \mathbf{Y}$ a distribution on E, giving the state in which the system will find itself after a jump of the discrete transition. For the purpose of this chapter, we can restrict ourselves to finitely supported reset kernels, defined by a finite set of pairs of update vectors $\{(\mathbf{v}_j^d, \mathbf{v}_j^c) \mid j = 1, \ldots, h\}$ and associated probability functions $p_j(\mathbf{Z}, \mathbf{Y})$, giving the likelihood of jumping from state $\mathbf{Z}, \mathbf{Y}$ to state $\mathbf{Z} + \mathbf{v}_j^d, \mathbf{Y} + \mathbf{v}_j^c$, if a stochastic event fires when the system is in state $\mathbf{Z}, \mathbf{Y}$. This, in turn, happens after an exponentially distributed delay with rate $\lambda(\mathbf{Z}, \mathbf{Y})$.

Discrete and continuous dynamics in a SHS are intertwined. The system starts in a given state $\mathbf{z}_0, \mathbf{y}_0$, and its continuous state evolves following the solution of the initial value problem $\frac{d}{dt}\mathbf{Y}(t) = F(\mathbf{z}_0, \mathbf{Y}(t))$, $\mathbf{Y}(0) = \mathbf{y}_0$. This continuous flow will go on until a discrete event will happen, at a random time governed by an exponential distribution with rate $\lambda(\mathbf{Z}, \mathbf{Y}(t))$. Note that, as $\mathbf{Y}$ will change value in time following the flow of the vector field, the rate of a discrete jump is also time-dependent. When a discrete transition happens, say in state $\mathbf{z}, \mathbf{y}$, then the system will jump to the state $\mathbf{z} + \mathbf{v}_j^d, \mathbf{y} + \mathbf{v}_j^c$ with probability $p_j(\mathbf{Z}, \mathbf{Y})$. Note that both the discrete mode and the value of continuous variables can change. From this new state, the system continues to evolve following the dynamics given by the vector field in the new discrete mode. The overall dynamics is given by an alternation of periods of continuous evolution interleaved by discrete jumps. For a proper mathematical formalisation of this process, we refer the interested reader to [8, 21].

4.2   From PCTMC to SHS

In this section we will show how to construct a SHS from a PCTMC $(\mathbf{X}, \mathcal{T}, \mathbf{x}_0, N)$, and how to guarantee the asymptotic correctness of the method. The starting point is a partition of the variables $\mathbf{X}$ of the PCTMC into two distinct classes: discrete and continuous. We will denote discrete variables with $\mathbf{Z}$ and continuous ones with $\mathbf{Y}$, so that $\mathbf{X} = \mathbf{Z}, \mathbf{Y}$. Transitions $\mathcal{T}$ also have to be separated in two classes: discrete $\mathcal{T}_d$ and continuous $\mathcal{T}_c$. Intuitively, continuous transitions and variables will define the continuous dynamics, and discrete transitions and variables the discrete one. The only request is that continuous transitions do not affect discrete variables, i.e. for each $\eta \in \mathcal{T}_c$, $\mathbf{v}_\eta[\mathbf{Z}] = \mathbf{0}$, where $\mathbf{v}_\eta[\mathbf{Z}]$ denotes the vector $\mathbf{v}_\eta$ restricted to the components of $\mathbf{Z}$.

Remark 1. The choice of how to partition variables and transitions into discrete and continuous is not obvious, and depends on the model under consideration. Often, this is easily deduced from model structure, e.g. due to the presence of conservation laws with a small number of conserved agents. A further help comes from the request to make explicit in the rates and updates the dependency on system size N. An alternative is to define rules to automatically switch between a discrete or a continuous representation of variables and transitions, depending on the current state of the model. We refer the interested reader to the discussion in [8] for further details.

Normalisation of Continuous Variables. To properly formalise hybrid mean-field limits, we need to perform a normalisation operation on the continuous variables, taking system size into account (hence we will use the superscript N from now on). This can be obtained as in Sect. 2.3, by introducing the normalised variables $\hat{\mathbf{Y}}^N = \mathbf{Y}/N$, and expressing rates, guards, and update vectors with respect to these normalised variables. Note that normalised update vectors $\hat{\mathbf{v}}_\eta^N$ are divided by N only in the continuous components, as the discrete variables are not rescaled. Transitions, after normalisation, have to satisfy some scaling conditions:

(Tc) Continuous transitions $\eta \in \mathcal{T}_c$ are such that $\hat{f}_\eta^N(\mathbf{Z}, \hat{\mathbf{Y}})/N \rightarrow f_\eta(\mathbf{Z}, \hat{\mathbf{Y}})$, as $N \rightarrow \infty$, uniformly on $\hat{\mathbf{Y}}$ for each $\mathbf{Z}$. The limit function is required to be (locally) Lipschitz continuous. Furthermore, their non-normalised update is independent of N. Guards can depend only on discrete variables: $\phi_\eta = \phi_\eta(\mathbf{Z})$.

(Td) Discrete transitions $\eta \in \mathcal{T}_d$ are such that their rate function $\hat{f}_\eta^N(\mathbf{Z}, \hat{\mathbf{Y}})$ converges (uniformly in $\hat{\mathbf{Y}}$ for each $\mathbf{Z}$) to a continuous function $f_\eta(\mathbf{Z}, \hat{\mathbf{Y}})$. Their normalised jump vector $\hat{\mathbf{v}}_\eta^N$ has also to converge to a vector $\hat{\mathbf{v}}_\eta$ as N diverges. Guards can depend only on discrete variables: $\phi_\eta = \phi_\eta(\mathbf{Z})$.

Note that, for discrete transitions, we consider the change in the normalised continuous variables, and we admit that the update vectors can depend on N. In particular, the update for continuous variables can be linear in N, thus resulting in a non-vanishing jump in the density, in the large N limit. This means that limit discrete transitions may also induce jumps on continuous variables.

Construction of the Limit SHS. Given a family of PCTMC models $(\mathbf{X}, \mathcal{T}, \mathbf{x}_0, N)$, indexed by N, with a partition of variables into $\mathbf{Z}, \mathbf{Y}$ and transitions into $\mathcal{T}_c, \mathcal{T}_d$ we can normalise continuous variables and formally define the SHS associated with it:

– The vector field defining the continuous dynamics of the SHS is given by the following drift:
$$F(\mathbf{Z}, \hat{\mathbf{Y}}) = \sum_{\eta \in \mathcal{T}_c} \mathbf{v}_\eta I\{\phi_\eta(\mathbf{Z})\} f_\eta(\mathbf{Z}, \hat{\mathbf{Y}})$$

- The jump rate of the SHS is given by

$$\lambda(\mathbf{Z}, \hat{\mathbf{Y}}) = \sum_{\eta \in \mathcal{T}_d} I\{\phi_\eta(\mathbf{Z})\} f_\eta(\mathbf{Z}, \hat{\mathbf{Y}})$$

- The reset kernel is specified by the pair of update vectors $(\hat{\mathbf{v}}_\eta[\mathbf{Z}], \hat{\mathbf{v}}_\eta[\hat{\mathbf{Y}}])$ and by the probability

$$p_\eta(\mathbf{Z}, \hat{\mathbf{Y}}) = \frac{I\{\phi_\eta(\mathbf{Z})\} f_\eta(\mathbf{Z}, \hat{\mathbf{Y}})}{\lambda(\mathbf{Z}, \hat{\mathbf{Y}})},$$

  for each transition $\eta \in \mathcal{T}_d$.
- The initial state is $\mathbf{z}_0, \hat{\mathbf{y}}_0$.

SIR Model with Vaccination. We consider now an extension of the SIR model of Sect. 2.2, with the possibility of starting a vaccination campaign of susceptible individuals. The model has an additional variable, $X_V \in \{0, 1\}$, which is going to be the only discrete variable of the system and encodes if the vaccination is in force or not. We further have two additional transitions:

- Vaccination of susceptible: $(\texttt{true}, \mathbf{e}_R - \mathbf{e}_S, k_v X_S X_V)$;
- Activation of the vaccination policy: $(X_V = 0, \mathbf{e}_V, k_a X_I/N)$;

The first transition, which will be a continuous transition, models the effect of vaccination, moving agents from susceptible to recovered state. Note that the rate depends on X_V, hence the transition is in force only if $X_V = 1$. We could alternatively specify the same behaviour by introducing a guard in the transition, depending only on discrete variables, which would result in a rate active only in a subset of discrete modes. The second transition will be kept discrete and model the activation of the vaccination policy. Its rate depends on the density of infected individuals (the higher the infected, the higher the activation rate). The guard on X_V allows the activation transition to be in force only when the vaccination is inactive.

In Fig. 3 (left), we show a trajectory of the system for $N = 100$, and compare it with a trajectory of the limit SHS. Parameters are $k_{si} = 0$, $k_i = 1$, $k_r = 0.1$, $k_s = 0.01$, $k_v = 2$, $k_a = 2$, $X_S(0) = 0.95N$, $X_I(0) = 0.05N$, $X_R(0) = 0$. As we can see, around time $t = 2.5$ there is a sudden drop on the number of susceptible individuals, caused by the beginning of vaccination. The similarity between the PCTMC and the SHS trajectories is a clear hint on the existence of an underlying convergence result.

Mean-Field Convergence Results. Consider a family of PCTMC model $(\mathbf{X}, \mathcal{T}, \mathbf{x}_0, N)$, and denote by $(\mathbf{Z}^N(t), \hat{\mathbf{Y}}^N(t))$ the normalised CTMC associated with it, for size N, where we made explicit the partition of variables into discrete and continuous. Denote by $(\mathbf{z}(t), \mathbf{y}(t))$ the limit SHS, constructed according to the recipe of this section. We can then prove [8] the following theorem:

Theorem 4. *Assume $(\mathbf{z}_0^N, \hat{\mathbf{y}}_0^N) \rightarrow (\mathbf{z}_0, \mathbf{y}_0)$ and that the transitions of the PCTMC model satisfy conditions (Tc) and (Td). Then*

$$(\mathbf{Z}^N(t), \hat{\mathbf{Y}}^N(t)) \Rightarrow (\mathbf{z}(t), \mathbf{y}(t)),$$

for all times $t \geq 0$, where $\Rightarrow$ denotes weak convergence.[4]

This theorem states that the distribution of the PCTMC will look like the distribution of the SHS for large N. In particular, if E_d is finite (i.e. there is a finite number of discrete modes) and E_c is compact, then all conditional and unconditional moments of the distribution converge.

In Fig. 3 (right), we see the theorem at work in the epidemic with vaccination example. The figure compares the empirical cumulative distribution of the density of infected individuals at time $t = 10$. The curves look quite similar already for $N = 100$, and are almost identical for $N = 1000$.

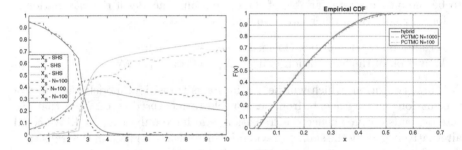

Fig. 3. Left: comparison of a simulated trajectory of the PCTMC model, for $N = 100$, with a simulated trajectory of the limit SHS. Right: comparison of the empirical cumulative distribution of $\hat{X}_I$ at time $t = 10$ of the limit SHS and the PCTMC models for $N = 100$ and $N = 1000$ (Color figure online).

4.3   Extensions of the Hybrid Limit Framework

The hybrid mean field limit presented in the previous section can be extended in many ways, as discussed in [8]. Here we will sketch them briefly, referring the interested reader to [8] for further details.

One possible direction to enrich the framework is to consider forced transitions. In the context of SHS, these are discrete jumps happening as soon as a condition on the system variables becomes true. Typically, they are introduced by constraining the continuous state space E_c (in a mode dependent way), and forcing a jump to happen as soon as the trajectory of the continuous variables hits the boundary ∂E_c [21]. Then a jump is done according to the reset kernel R,

---

[4] In fact, weak convergence holds for $(\mathbf{z}, \mathbf{y})$ as processes in the Skorokhod space of cadlag functions, see [8]. For a definition of weak convergence, see [5].

whose definition has to be extended on the boundary ∂E_c. Hence, discrete jumps may happen at stochastic times, or when the condition for a forced jump is met.

In the PCTMC setting, introducing forced transitions requires us to allow transitions with an infinite rate and with a non-trivial guard, firing as soon as their guard becomes true. Their guards, then, can be used to constrain the continuous state space. Hence, E_c will be defined in each mode as the interior of the complement of the region obtained by taking the union of the satisfaction sets of all the guards of forced transitions. The reset kernel in a point of the boundary ∂E_c will then be defined by the active immediate transitions at that point, choosing uniformly among the active transitions.[5]

As an example, consider again the SIR model with vaccination, but assume its activation and deactivation is threshold-based: when the density of infected becomes greater than a threshold I_{high}, the vaccination is started, while if it falls below I_{low}, the vaccination is stopped. In the PCTMC model, we would have the following two transitions in place of the stochastic one discussed previously:

- Activation of the vaccination policy: $(X_I \geq NI_{high}, \mathbf{e}_V, \infty)$;
- Deactivation of the vaccination policy: $(X_I \leq NI_{low}, -\mathbf{e}_V, \infty)$.

Theorem 4 is readily extended to the presence of instantaneous transitions, under some additional technical conditions on the vector fields (called transversal crossing), see [8] for details. The validity of the result for the SIR model with vaccination is illustrated in Fig. 4, where we compare simulations of the PCTMC model and the limit SHS, for thresholds $I_{high} = 0.3$ and $I_{low} = 0.2$.

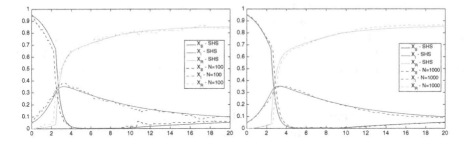

Fig. 4. Left: comparison of a simulated trajectory of the PCTMC model with instantaneous transitions (normalised variables), for $N = 100$ (left) and $N = 1000$ (right), with a simulated trajectory of the limit SHS (Color figure online).

Other extensions of the hybrid mean field include dealing with guards in deterministic and stochastic transitions. These introduce discontinuities in the

---

[5] In [8], weights are introduced to solve non-determinism between instantaneous transitions. Furthermore, the possibility of seeing a chain of instantaneous events firing is taken into account. Termination of this chain is discussed in [15] (where it is proved undecidable for countable state spaces), and in [27], where sufficient and testable conditions for termination are given.

model, and require further technical conditions for the limit theorems to hold. As for guards in the continuous transitions, this in fact requires one to introduce in the hybrid context the mean field techniques based on differential inclusions of Sect. 3.

5   Related Work and Examples

Load Balancing and Discontinuous Dynamics. The use of mean field approximation is popular for studying load-balancing policies in server farms. In such a system, an object represents a server and its state is typically the number of jobs that are waiting in its queue. A popular randomised load balancing policy is the *two-choice* policy: for each incoming packet two servers are picked at random and the job is allocated to the least-loaded of the two. This policy has been successfully analysed by classical mean field techniques in [43] where it is shown that it leads to an important gain of performance compared to a purely random allocation. The classical approach then fails when one considers a centralised load balancing policy such as *join the shortest queue* because these policies lead to discontinuous dynamics. As demonstrated in [24,28,49], these problems can be modelled and resolved by using differential inclusions.

Heterogeneous Systems and Uncertainties. Another problem where differential inclusions can help is the case of heterogeneous systems. In such cases, there is a large number of objects each having distinct parameters. One example is the caching model of [31] in which an object i has a popularity p_i. One possibility to solve the problem is to consider a system of $N \times S$ ODEs, where N is the number of objects and S the dimension of the state space. This method scales linearly in the number of objects but might still be problematic for large populations. An alternative is to consider upper and lower bounds on the dynamics [48] or to study a PDE approximation when the number of objects is large, see for example [25].

Hybrid Mean Field Limits. The use of hybrid approximation of population models is quite common in the area of systems and synthetic biology, where genes and often mRNA molecules are present in such low numbers (genes usually in one copy) that classic mean field assumptions are not correct and can lead to models failing to capture important features of the system like bursting protein expression [39]. In addition to the investigation of hybrid limits, carried out independently for general population models [8,9] and more specifically for gene networks [19], considerable work has been done in hybrid simulation [45] and in developing moment closure techniques for hybrid limits [32].

Mean Field Games. Game theory studies the decisions taken by competing rational agents. Recently, the notion of mean field games has been introduced in [35,40] to model decisions in systems composed of a large number of agents. In a mean field game, each agent tries to minimise an objective function that depends on the average behaviour of the population of agents but not on the action of a precise agent. This simplifies the analysis of Nash equilibria that are replaced by mean field equilibria. This theory is used for modelling purposes but also to solve optimisation problems in a decentralised way [30,52]. It can be shown that in certain cases, mean field equilibria are the limit of a sub-class of Nash equilibria [22].

Modelling Languages Supporting a Mean Field Semantics. In the last ten years, there has been a considerable interest in extending stochastic modelling languages, in particular stochastic process algebras, in order to derive automatically mean field equations. Examples are e.g. [14,18,47]. This work has also been extended to generate hybrid semantics in [8,15,16]. See also the chapter of this book on CARMA [42].

6 Conclusion

In this document, we reviewed the notion of mean field limits of a stochastic population process and presented two extensions. The classical mean field results show that, under some conditions, a stochastic population process converges to a deterministic system of ODEs as the number of objects of the population grows. We have shown that, by replacing the system of ODEs by either a differential inclusion or a hybrid system, it is possible to enlarge the set models for which a mean field limit exists. We illustrated these notions by using a classical SIR example. The last section gives a few pointers to papers in which these frameworks can be applied or generalised.

Acknowledgements. L.B. and N.G. acknowledge partial support from the EU-FET project QUANTICOL (nr. 600708).

References

1. Andersson, H., Britton, T.: Stochastic Epidemic Models and Their Statistical Analysis. Springer, Heidelberg (2000)
2. Aubin, J., Cellina, A.: Differential Inclusions. Springer, Heidelberg (1984)
3. Baier, C., et al.: Model-checking algorithms for continuous-time Markov chains. IEEE Trans. Softw. Eng. **29**(6), 524–541 (2003). http://ieeexplore.ieee.org/xpls/abs_all.jsp?arnumber=1205180
4. Benaim, M., Le Boudec, J.-Y.: A class of mean field interaction models for computer and communication systems. Perform. Eval. **65**(11), 823–838 (2008)
5. Billingsley, P.: Probability and Measure. English. Wiley, Hoboken (2012). ISBN: 9781118122372 1118122372
6. Bortolussi, L., Gast, N.: Mean field approximation of imprecise population processes. QUANTICOL Technical report TR-QC-07-2015 (2015)
7. Bortolussi, L., et al.: Continuous approximation of collective systems behaviour: a tutorial. Perform. Eval. **70**(5), 317–349 (2013). ISSN: 0166-5316, doi:10. 1016/j.peva.2013.01.001, http://www.sciencedirect.com/science/article/pii/ S0166531613000023

8. Bortolussi, L.: Hybrid behaviour of Markov population models. In: Information and Computation (2015)
9. Bortolussi, L.: Limit behavior of the hybrid approximation of stochastic process algebras. In: Al-Begain, K., Fiems, D., Knottenbelt, W.J. (eds.) ASMTA 2010. LNCS, vol. 6148, pp. 367–381. Springer, Heidelberg (2010). http://link.springer.com/chapter/10.1007/978-3-642-13568-2_26. Accessed 11 June 2015
10. Bortolussi, L., Hillston, J.: Model checking single agent behaviours by uid approximation. Inf. Comput. **242**, 183–226 (2015). ISSN: 0890-5401, doi:10.1016/j.ic.2015.03.002
11. Bortolussi, L., Lanciani, R.: Fluid model checking of timed properties. In: Sankaranarayanan, S., Vicario, E. (eds.) FORMATS 2015. LNCS, vol. 9268, pp. 172–188. Springer, Heidelberg (2015)
12. Bortolussi, L., Lanciani, R.: Model checking Markov population models by central limit approximation. In: Joshi, K., Siegle, M., Stoelinga, M., DArgenio, P.R. (eds.) QEST 2013. LNCS, vol. 8054, pp. 123–138. Springer, Heidelberg (2013)
13. Bortolussi, L., Lanciani, R.: Stochastic approximation of global reachability probabilities of Markov population models. In: Horvath, A., Wolter, K. (eds.) EPEW 2014. LNCS, vol. 8721, pp. 224–239. Springer, Heidelberg (2014)
14. Bortolussi, L., Policriti, A.: Dynamical systems and stochastic programming: to ordinary differential equations and back. In: Priami, C., Back, R.-J., Petre, I. (eds.) Transactions on Computational Systems Biology XI. LNCS, vol. 5750, pp. 216–267. Springer, Heidelberg (2009)
15. Bortolussi, L., Policriti, A.: (Hybrid) automata, (stochastic) programs: the hybrid automata lattice of a stochastic program. J. Logic Comput. **23**, 761–798 (2013). http://dx.doi.org/10.1093/logcom/exr045
16. Bortolussi, L., Policriti, A.: Hybrid dynamics of stochastic programs. Theor. Comput. Sci. **411**(20), 2052–2077 (2010). ISSN: 0304-3975
17. Chaintreau, A., Le Boudec, J.-Y., Ristanovic, N.: The age of gossip: spatial mean field regime. In: Proceedings of the ACM SIGMETRICS, vol. 37, issue 1, pp. 109–120. ACM (2009)
18. Ciocchetta, F., Hillston, J.: Bio-PEPA: a framework for the modelling and analysis of biological systems. Theor. Comput. Sci. **410**(33), 00185, 3065–3084 (2009). http://www.sciencedirect.com/science/article/pii/S0304397509001662. Accessed 25 Nov 2013
19. Crudu, A., et al.: Convergence of stochastic gene networks to hybrid piecewise deterministic processes. Ann. Appl. Probab. **22**(5), 00015, 1822–1859 (2012). http://projecteuclid.org/euclid.aoap/1350067987. Accessed 05 Nov 2013
20. Darling, R., Norris, J.R., et al.: Differential equation approximations for Markov chains. Probab. Surv. **5**, 37–79 (2008)
21. Davis, M.H.A.: Markov Models and Optimization. Chapman & Hall, London (1993)
22. Doncel, J., Gast, N., Gaujal, B.: Mean-Field Games with Explicit Interactions. Working paper or preprint, February 2016. https://hal.inria.fr/hal-01277098
23. Durrett, R.: Essentials of Stochastic Processes. Springer, Heidelberg (2012). ISBN: 9781461436157
24. Fricker, C., Gast, N.: Incentives and redistribution in homogeneous bike-sharing systems with stations of finite capacity. EURO J. Trans. Logistics, 1–31 (2014)
25. Fricker, C., Gast, N., Mohamed, H.: Mean field analysis for inhomogeneous bike sharing systems. DMTCS Proc. **01**, 365–376 (2012)
26. Galpin, V.: Spatial representations, analysis techniques. In: SFM (2016)

27. Galpin, V., Bortolussi, L., Hillston, J.: HYPE: hybrid modelling by composition of flows. Formal Aspects Comput. **25**(4), 503–541 (2013)
28. Gast, N., Gaujal, B.: Markov chains with discontinuous drifts have differential inclusion limits. Perform. Eval. **69**(12), 623–642 (2012)
29. Gast, N., Gaujal, B.: Mean field limit of non-smooth systems and differential inclusions. ACM SIGMETRICS Perform. Eval. Rev. **38**(2), 30–32 (2010)
30. Gast, N., Le Boudec, J.-Y., Tomozei, D.-C.: Impact of demand-response on the efficiency, prices in real-time electricity markets. In: Proceedings of the 5th International Conference on Future Energy Systems, pp. 171–182. ACM (2014)
31. Gast, N., Van Houdt, B.: Transient and steady-state regime of a family of list-based cache replacement algorithms. In: ACM SIGMETRICS 2015 (2015)
32. Hasenauer, J., et al.: Method of conditional moments (MCM) for the chemical master equation: a unified framework for the method of moments and hybrid stochastic-deterministic models. J. Math. Biol. **69**, 687–735 (2013). ISSN: 0303-6812, 1432–1416, doi:10.1007/s00285-013-0711-5, http://link.springer.com/10.1007/s00285-013-0711-5. Accessed 31 July 2014
33. Henzinger, T., Jobstmann, B., Wolf, V.: Formalisms for specifying Markovian population models. Int. J. Found. Comput. Sci. **22**(04), 823–841 (2011). http://www.worldscience.com/doi/abs/10.1142/S0129054111008441
34. Hu, L., Le Boudec, J.-Y., Vojnoviae, M.: Optimal channel choice for collaborative ad-hoc dissemination. In: 2010 Proceedings of the IEEE INFOCOM, pp. 1–9. IEEE (2010)
35. Huang, M., Malhame, R.P., Caines, P.E., et al.: Large population stochastic dynamic games: closed-loop McKean-Vlasov systems and the Nash certainty equivalence principle. Commun. Inf. Syst. **6**(3), 221–252 (2006)
36. Katoen, J.-P., Khattri, M., Zapreevt, I.S.: A Markov reward model checker. In: Second International Conference on the Quantitative Evaluation of Systems, pp. 243–244 (2005). Accessed 18 Jan 2014
37. Kurtz, T.: Solutions of ordinary differential equations as limits of pure jump Markov processes. J. Appl. Probab. **7**, 49–58 (1970)
38. Kwiatkowska, M., Norman, G., Parker, D.: PRISM 4.0: verification of probabilistic real-time systems. In: Gopalakrishnan, G., Qadeer, S. (eds.) CAV 2011. LNCS, vol. 6806, pp. 585–591. Springer, Heidelberg (2011). http://link.springer.com/chapter/10.1007/978- 3-642-22110-1_47. Accessed 18 Jan 2014
39. Krn, M., et al.: Stochasticity in gene expression: from theories to phenotypes. Nat. Rev. Genet. **6**(6), 451–464 (2005). ISSN: 1471-0056, 1471–0064, doi:10.1038/nrg1615, http://www.nature.com/doifinder/10.1038/nrg1615. Accessed 09 Feb 2016
40. Lasry, J.-M., Lions, P.-L.: Mean field games. Jpn. J. Math. **2**(1), 229–260 (2007)
41. Le Boudec, J.-Y.: Performance Evaluation of Computer and Communication Systems. EPFL Press, Lausanne (2010)
42. Loreti, M.: Modeling and analysis of collective adaptive systems with CARMA and its tools. In: SFM (2016)
43. Mitzenmacher, M.: The power of two choices in randomized load balancing. IEEE Trans. Parallel Distrib. Syst. **12**(10), 1094–1104 (2001)
44. Norris, J.R.: Markov Chains. English. Cambridge University Press, Cambridge (1998). ISBN: 978-0-511-81063-3 0-511-81063-6

45. Pahle, J.: Biochemical simulations: stochastic, approximate stochastic and hybrid approaches. Briefings Bioinform. **10**(1), 53–64 (2008). ISSN: 1467-5463, 1477-4054, doi:10.1093/bib/bbn050, http://bib.oxfordjournals.org/cgi/doi/10./bib/bbn050. Accessed 14 July 2014

46. Todorov, E.: Optimal control theory. In: Bayesian Brain: Probabilistic Approaches to Neural Coding, pp. 269–298 (2006)

47. Tribastone, M., Gilmore, S., Hillston, J.: Scalable differential analysis of process algebra models. IEEE Trans. Softw. Eng. **38**(1), 205–219 (2012). http://ieeexplore.ieee.org/xpls/abs_all.jsp?arnumber=5567115. Accessed 24 Nov 2013

48. Tschaikowski, M., Tribastone, M.: Approximate reduction of heterogenous nonlinear models with differential hulls. IEEE Trans. Autom. Control **61**(4), 1099–1104 (2016). doi:10.1109/TAC.2015.2457172

49. Tsitsiklis, J.N., Xu, K., et al.: On the power of (even a little) resource pooling. Stochast. Syst. **2**(1), 1–66 (2012)

50. Van Houdt, B.: A mean field model for a class of garbage collection algorithms in flash-based solid state drives. In: Proceedings of the ACM SIGMETRICS, SIGMETRICS 2013, Pittsburgh, PA, USA, pp. 191–202. ACM (2013). ISBN: 978-1-4503-1900-3, doi:10.1145/2465529.2465543, http://doi.acm.org/10.1145/2465529.2465543

51. Wilkinson, D.: Stochastic Modelling for Systems Biology. Chapman & Hall, Florida (2006)

52. Yang, T., Mehta, P.G., Meyn, S.P.: A mean-field control-oriented approach to particle filtering. In: American Control Conference (ACC), pp. 2037–2043. IEEE (2011)

53. Ying, L.: On the rate of convergence of mean-field models: Stein's method meets the perturbation theory. arXiv preprint arXiv:1510.00761 (2015)

Modelling and Analysis of Collective Adaptive Systems with CARMA and its Tools

Michele Loreti[1(✉)] and Jane Hillston[2]

[1] Dipartimento di Statistica, Informatica,
Applicazioni "G. Parenti", Università di Firenze, Florence, Italy
michele.loreti@unifi.it
[2] Laboratory for Foundations of Computer Science,
University of Edinburgh, Edinburgh, UK

Abstract. Collective Adaptive Systems (CAS) are heterogeneous collections of autonomous task-oriented systems that cooperate on common goals forming a collective system. This class of systems is typically composed of a huge number of interacting agents that dynamically adjust and combine their behaviour to achieve specific goals.

This chapter presents CARMA, a language recently defined to support specification and analysis of collective adaptive systems, and its tools developed for supporting system design and analysis. CARMA is equipped with linguistic constructs specifically developed for modelling and programming systems that can operate in open-ended and unpredictable environments. The chapter also presents the CARMA Eclipse plug-in that allows CARMA models to be specified by means of an appropriate high-level language. Finally, we show how CARMA and its tools can be used to support specification with a simple but illustrative example of a socio-technical collective adaptive system.

1 Introduction

In the last forty years *Process Algebras* (see [3] and the references therein), or *Process Description Languages* (PDL), have been successfully used to model and analyse the behaviour of concurrent and distributed systems. A Process Algebra is a formal language, equipped with a rigorous semantics, that provides models in terms of processes. These are agents that perform actions and communicate (interact) with similar agents and with their environment.

At the beginning, Process Algebras were only focussed on *qualitative aspects* of computations. However, when complex and large-scale systems are considered, it may not be sufficient to check if a property *is satisfied or not*. This is because random phenomena are a crucial part of *distributed systems* and one is also interested in verifying *quantitative aspects* of computations.

This motivated the definition of a new class of PDL where *time* and *probabilities* are explicitly considered. This new family of formalisms have proven to be particularly suitable for capturing important properties related to performance and quality of service, and even for the modelling of biological systems.

© Springer International Publishing Switzerland 2016
M. Bernardo et al. (Eds.): SFM 2016, LNCS 9700, pp. 83–119, 2016.
DOI: 10.1007/978-3-319-34096-8_4

Among others we can refer here to PEPA [19], MTIPP [18], EMPA [4], Stochastic π-Calculus [23], Bio-PEPA [9], MODEST [5] and others [8,17].

The ever increasing complexity of systems has further changed the perspective of the system designer that now has to consider a new class of systems, named *Collective adaptive systems* (CAS), that consist of massive numbers of components, featuring complex interactions among components and with humans and other systems. Each component in the system may exhibit autonomic behaviour depending on its properties, objectives and actions. Decision-making in such systems is complicated and interaction between their components may introduce new and sometimes unexpected behaviours.

CAS operate in open and non-deterministic environments. Components may enter or leave the collective at any time. Components can be highly heterogeneous (machines, humans, networks, etc.) each operating at different temporal and spatial scales, and having different (potentially conflicting) objectives.

CAS thus provide a significant research challenge in terms of both representation and reasoning about their behaviour. The pervasive yet transparent nature of the applications developed in this paradigm makes it of paramount importance that their behaviour can be thoroughly assessed during their design, prior to deployment, and throughout their lifetime. Indeed their adaptive nature makes modelling essential and models play a central role in driving their adaptation. Moreover, the analysis should encompass both functional and non-functional aspects of behaviour. Thus it is vital that we have available robust modelling techniques which are able to describe such systems and to reason about their behaviour in both qualitative and quantitative terms. To move towards this goal, it is important to develop a theoretical foundation for CAS that will help in understanding their distinctive features. From the point of view of the language designers, the challenge is to devise appropriate abstractions and linguistic primitives to deal with the large dimension of systems, to guarantee adaptation to (possibly unpredicted) changes of the working environment, to take into account evolving requirements, and to control the emergent behaviours resulting from complex interactions.

To design this new language for CAS we first have identified the *design principles* together with the *primitives* and *interaction patterns* that are needed in CAS design. Emphasis has been given placed on identifying the appropriate abstractions and linguistic primitives for modelling and programming collective adaptation, locality representation, knowledge handling, and system interaction and aggregation.

To be effective, any language for CAS should provide:

- Separation of knowledge and behaviour;
- Control over abstraction levels;
- Bottom-up design;
- Mechanisms to take into account the environment;
- Support for both global and local views; and
- Automatic derivation of the underlying mathematical model.

These design principles have been the starting point for the design of a language, developed specifically to support the specification and analysis of CAS, with the particular objective of supporting quantitative evaluation and verification. We named this language CARMA, Collective Adaptive Resource-sharing Markovian Agents [7,20].

CARMA combines the lessons which have been learned from the long tradition of stochastic process algebras, with those more recently acquired from developing languages to model CAS, such as SCEL [12] and PALOMA [13], which feature attribute-based communication and explicit representation of locations.

SCEL [12] (Software Component Ensemble Language), is a kernel language that has been designed to support the programming of autonomic computing systems. This language relies on the notions of *autonomic components* representing the collective members, and *autonomic-component ensembles* representing collectives. Each component is equipped with an interface, consisting of a collection of attributes, describing different features of components. Attributes are used by components to dynamically organise themselves into ensembles and as a means to select partners for interaction. The stochastic variant of SCEL, called StocS [22], was a first step towards the investigation of the impact of different stochastic semantics for autonomic processes, that relies on stochastic output semantics, probabilistic input semantics and on a probabilistic notion of knowledge. Moreover, SCEL has inspired the development of the core calculus AbC [1,2] that focuses on a minimal set of primitives that defines attribute-based communication, and investigates their impact. Communication among components takes place in a broadcast fashion, with the characteristic that only components satisfying predicates over specific attributes receive the sent messages, provided that they are willing to do so.

PALOMA [13] is a process algebra that takes as its starting point a model based on located Markovian agents each of which is parameterised by a location, which can be regarded as an attribute of the agent. The ability of agents to communicate depends on their location, through a perception function. This can be regarded as an example of a more general class of attribute-based communication mechanisms. The communication is based on a multicast, as only agents who enable the appropriate reception action have the ability to receive the message. The scope of communication is thus adjusted according to the perception function.

A distinctive contribution of the language CARMA is the rich set of communication primitives that are offered. This new language supports both unicast and broadcast communication, and locally synchronous, but globally asynchronous communication. This richness is important to enable the spatially distributed nature of CAS, where agents may have only local awareness of the system, yet the design objectives and adaptation goals are often expressed in terms of global behaviour. Representing these rich patterns of communication in classical process algebras or traditional stochastic process algebras would be difficult, and would require the introduction of additional model components to represent buffers, queues, and other communication structures. Another feature of CARMA is the explicit representation of the environment in which processes interact, allowing rapid testing of a system under different open world scenarios. The environment in CARMA models can evolve at runtime, due to the feedback from the system, and it further modulates the interaction between components, by shaping rates and interaction probabilities.

The focus of this tutorial is the presentation of the language and its discrete semantics, which are presented in the FuTS style [11]. The structure of the chapter is as follows. Section 2 presents the syntax of the language and explains the organisation of a model in terms of a collective of agents that are considered in the context of an environment. In Sect. 3 we give a detailed account of the semantics, particularly explaining the role of the environment. The use of CARMA is illustrated in Sect. 4 where we describe a model of a simple bike sharing system, and explain the support given to the CARMA modeller in the current implementation. Section 5 considers the bike sharing system in different scenarios, demonstrating the analytic power of the CARMA tools. Some conclusions are drawn in Sect. 6.

2   CARMA: Collective Adaptive Resource-Sharing Markovian Agents

CARMA is a new stochastic process algebra for the representation of systems developed according to the CAS paradigm [7, 20]. The language offers a rich set of communication primitives, and the exploitation of attributes, captured in a store associated with each component, to enable attribute-based communication. For most CAS systems we anticipate that one of the attributes could be the location of the agent [15]. Thus it is straightforward to model those systems in which, for example, there is limited scope of communication or, restriction to only interact with components that are co-located, or where there is spatial heterogeneity in the behaviour of agents.

The rich set of communication primitives is one of the distinctive features of CARMA. Specifically, CARMA supports both unicast and broadcast communication, and permits locally synchronous, but globally asynchronous communication. This richness is important to take into account the spatially distributed nature of CAS, where agents may have only local awareness of the system, yet the design objectives and adaptation goals are often expressed in terms of global behaviour. Representing these patterns of communication in classical process algebras or traditional stochastic process algebras would be difficult, and would require the introduction of additional model components to represent buffers, queues and other communication structures.

Another key feature of CARMA is its distinct treatment of the *environment*. It should be stressed that although this is an entity explicitly introduced within our models, it is intended to represent something more pervasive and diffusive of the real system, which is abstracted within the modelling to be an entity which exercises influence and imposes constraints on the different agents in the system. For example, in a model of a smart transport system, the environment may have responsibility for determining the rate at which entities (buses, bikes, taxis etc.) move through the city. However this should be recognised as an abstraction of the presence of other vehicles causing congestion which may impede the progress of the focus entities to a greater or lesser extent at different times of the day. The presence of an environment in the model does not imply the existence of centralised control in the system. The role of the environment is also related to the spatially distributed nature of CAS — we expect that the location *where* an agent is will have an effect on *what* an agent can do.

This view of the environment coincides with the view taken by many researchers within the situated multi-agent community e.g. [26]. Specifically, in [27] Weyns *et al.* argue about the importance of having a distinct environment within every multi-agent system. Whilst they are viewing such systems from the perspective of software engineers, many of their arguments are as valid when it comes to modelling a multi-agent or collective adaptive system. Thus our work can be viewed as broadly fitting within the same framework, albeit with a higher level of abstraction. Just as in the construction of a system, in the construction development of a model distinguishing clearly between the responsibilities of the agents and of the environment provides separation of concerns and assists in the management of complex systems.

In [27] the authors provide the following definition: "*The environment is a first-class abstraction that proves the surrounding conditions for agents to exist and that mediates both the interaction among agents and the access to resources.*" This is the role that the environment plays within CARMA models through the evolution rules. However, in contrast to the framework of Weyns *et al.*, the environment in a CARMA model is not an active entity in the same sense as the agents are active entities. In our case, the environment is constrained to work *through* the agents, by influencing their dynamic behaviour or by inducing changes in the number and types of agents making up the system.

In [24], Saunier *et al.* advocate the use of an *active environment* to mediate the interactions between agents; such an active environment is aware of the current context for each agent. The environment in CARMA also supports this view, as the evolution rules in the environment take into account the state of all the potentially participating components to determine both the rate and the probability of communications being successful, thus achieving a multicast communication not based on the address of the receiving agents, as suggested by Saunier *et al.* This is what we term "attribute-based communication" in CARMA. Moreover, when the application calls for a centralised information portal, the global store in CARMA can represent it. The higher level of abstraction offered by CARMA means that many implementation issues are ignored.

2.1   A Running Example

To describe basic features of CARMA a *running example* will be used. This is based on a *bike sharing system* (BSS) [10]. These systems are a recent, and increasingly popular, form of public transport in urban areas. As a resource-sharing system with large numbers of independent users altering their behaviour due to pricing and other incentives [14], they are a simple instance of a collective adaptive system, and hence a suitable case study to exemplify the CARMA language.

The idea in a bike sharing system is that bikes are made available in a number of stations that are placed in various areas of a city. Users that plan to use a bike for a short trip can pick up a bike at a suitable origin station and return it to any other station close to their planned destination. One of the major issues in bike sharing systems is the availability and distribution of resources, both in terms of available bikes at the stations and in terms of available empty parking places in the stations.

In our scenario we assume that the city is partitioned in homogeneous zones and that all the *stations* in the same zone can be equivalently used by any user in that zone. Below, we let $\{z_0, \ldots, z_n\}$ be the n zones in the city, each of which contains k parking stations.

2.2 A Gentle Introduction to CARMA

The bike sharing systems described in the previous section represent well typical scenarios that can be modelled with CARMA. Indeed, a CARMA system consists of a *collective* (N) operating in an *environment* ($\mathcal{E}$). The collective is a multiset of components that models the behavioural part of a system; it is used to describe a group of interacting *agents*. The environment models all those aspects which are intrinsic to the context where the agents under consideration are operating. The environment also mediates agent interactions.

Example 1. Bike Sharing System (1/7). In our running example the collective N will be used to model the behaviour of *parking stations* and *users*, while the environment will be used to model the city context where these agents operate like, for instance, the user arrival rate or the possible destinations of trips.   □

We let SYS be the set of CARMA *systems* S defined by the following syntax:

$$S ::= N \text{ in } \mathcal{E}$$

where is a collective and is an environment.

Collectives and Components. We let COL be the set of collectives N which are generated by the following grammar:

$$N ::= C \mid N \parallel N$$

A collective N is either a *component* C or the parallel composition of collectives $N_1 \parallel N_2$. The former identifies a multiset containing the single component C while the latter represents the union of the multisets denoted by N_1 and N_2, respectively. In the rest of this chapter we will sometimes use standard operations on multisets over a collective. We use $N(C)$ to indicate the multiplicity of

C in N, $C \in N$ to indicate that $N(C) > 0$ and $N - C$ to represent the collective obtained from N by removing component C.

The precise syntax of components is:

$$C ::= \mathbf{0} \mid (P, \gamma)$$

where we let COMP be the set of components C generated by the previous grammar.

A component C can be either the *inactive component*, which is denoted by $\mathbf{0}$, or a term of the form (P, γ), where P is a *process* and γ is a *store*. A term (P, γ) models an *agent* operating in the system under consideration: the process P represents the agent's behaviour whereas the store γ models its *knowledge*. A store is a function which maps *attribute names* to *basic values*. We let:

- ATTR be the set of *attribute names* a, a', $a_1, \ldots$, b, b', $b_1, \ldots$;
- VAL be the set of *basic values* v, v', $v_1, \ldots$;
- Γ be the set of *stores* $\gamma, \gamma_1, \gamma', \ldots$, i.e. functions from ATTR to VAL.

Example 2. Bike Sharing System (2/7). To model our *Bike Sharing System* in CARMA we need two kinds of components, one for each of the two groups of agents involved in the system, i.e. *parking stations* and *users*. Both kinds of component use the local store to publish the relevant data that will be used to represent the state of the agent. We can notice that, following this approach, bikes are not explicitly modelled in the system. This is because we are interested in modelling only the behaviour of the *active* components in the system. Under this perspective, bikes are just the resources exchanged by *parking stations* and *users*.

The local store of components associated with *parking stations* contains the following attributes:

- loc: identifying the zone where the parking station is located;
- capacity: describing the maximal number of parking slots available in the station;
- available: indicating the current number of bikes currently available in the parking station.

Similarly, the local store of components associated with *users* contains the following attributes:

- loc: indicating current user location;
- dest: indicating user destination.   □

Processes. The behaviour of a component is specified via a process P. We let PROC be the set of CARMA processes P, $Q, \ldots$ defined by the following grammar:

$$P, Q ::= \mathbf{nil} \qquad\qquad act ::= \alpha^\star[\pi_s]\langle\overrightarrow{e}\rangle\sigma$$
$$\mid\ act.P \qquad\qquad\quad \mid\ \alpha[\pi_r]\langle\overrightarrow{e}\rangle\sigma$$
$$\mid\ P + Q \qquad\qquad\quad \mid\ \alpha^\star[\pi_s](\overrightarrow{x})\sigma$$
$$\mid\ P \mid Q \qquad\qquad\quad \mid\ \alpha[\pi_r](\overrightarrow{x})\sigma$$
$$\mid\ [\pi]P$$
$$\mid\ \mathbf{kill} \qquad\qquad e ::= a \mid \mathsf{my}.a \mid x \mid v \mid \mathsf{now} \mid \cdots$$
$$\mid\ A \qquad (A \stackrel{\triangle}{=} P) \quad \pi_s, \pi_r, \pi ::= \top \mid \bot \mid e_1 \bowtie e_2 \mid \neg\pi \mid \pi \wedge \pi \mid \cdots$$

Above, the following notation is used:

- α is an *action type* in the set ACTTYPE;
- π is a *predicate*;
- x is a *variable* in the set of variables VAR;
- e is an expression in the set of expressions EXP[1];
- $\overrightarrow{\cdot}$ indicates a sequence of elements;
- σ is an *update*, i.e. a function from Γ to $Dist(\Gamma)$ in the set of *updates* Σ; where $Dist(\Gamma)$ is the set of probability distributions over Γ.

CARMA processes are built by using standard operators of process algebras. Basic processes can be either **nil** or **kill**. The former represents the *inactive process* while the latter is used, when activated, to *destroy* a component. We assume that the term **kill** always occurs under the scope of an action prefix.

Choice $(\cdot + \cdot)$ and parallel composition $(\cdot|\cdot)$ are the usual process algebra operators: $P_1 + P_2$ indicates a process that can behave either like P_1 or like P_2; while the behaviour of $P_1|P_2$ is the result of the interleaving between P_1 and P_2. In the next section, when the stochastic operational semantics of CARMA will be presented, we will show how possible alternative computations of a process P are probabilistically selected.

Process behaviour can be influenced by the store γ of the hosting component. This is the case of the *guard* operator $[\pi]P$ where the process P is activated when the predicate π, i.e. a boolean expression over *attribute names*, is satisfied (otherwise it is inactive). This operator can be used to enable a given behaviour only when some conditions are satisfied. In the case of our *Bike Sharing System*, if P_c is the behaviour modelling *bike retrieval*, a prediate of the form available > 0 can be used to enable P_c only when there are bikes available.

CARMA processes located in different components interact while performing four types of actions: *broadcast output* $(\alpha^\star[\pi]\langle\overrightarrow{e}\rangle\sigma)$, *broadcast input* $(\alpha^\star[\pi](\overrightarrow{x})\sigma)$, *output* $(\alpha[\pi]\langle\overrightarrow{e}\rangle\sigma)$, and *input* $(\alpha[\pi](\overrightarrow{x})\sigma)$.

The admissible communication partners of each of these actions are identified by the predicate π. Note that, in a component (P, γ) the store γ regulates the behaviour of P. Primarily, γ is used to evaluate the predicate associated with

---

[1] The precise syntax of expressions e has been deliberately omitted. We only assume that expressions are built using the appropriate combinations of *values*, *attributes* (sometime prefixed with my), variables and the special term now. The latter is used to refer to current time unit.

an action in order to filter the possible synchronisations involving process P. In addition, γ is also used as one of the parameters for computing the actual rate of actions performed by P. The process P can change γ immediately after the execution of an action. This change is brought about by the *update* σ. The update is a function that when given a store γ returns a probability distribution over Γ which expresses the possible evolutions of the store after the action execution.

The *broadcast output* $\alpha^\star[\pi]\langle \overrightarrow{e} \rangle \sigma$ models the execution of an action α that spreads the values resulting from the evaluation of expressions $\overrightarrow{e}$ in the local store γ. This message can be potentially received by any process located at components whose store satisfies predicate π. This predicate may contain references to attribute names that have to be evaluated under the local store. For instance, if loc is the attribute used to store the position of a component, action

$$\alpha^\star[\text{my.loc} == \text{loc}]\langle \overrightarrow{v} \rangle \sigma$$

potentially involves all the components located at the same location. The *broadcast output* is non-blocking. The action is executed even if no process is able to receive the values which are sent. Immediately after the execution of an action, the update σ is used to compute the (possible) *effects* of the performed action on the store of the hosting component where the output is performed.

To receive a broadcast message, a process executes a *broadcast input* of the form $\alpha^\star[\pi](\overrightarrow{x})\sigma$. This action is used to receive a tuple of values $\overrightarrow{v}$ sent with an action α from a component whose store satisfies the predicate $\pi[\overrightarrow{v}/\overrightarrow{x}]$. The transmitted values can be part of the predicate π. For instance, $\alpha^\star[x > 5](x)\sigma$ can be used to receive a value that is greater than 5.

The other two kinds of action, namely *output* and *input*, are similar. However, differently from broadcasts described above, these actions realise a *point-to-point* interaction. The *output* operation is blocking, in contrast with the non-blocking broadcast output.

Example 3. Bike Sharing System (3/7). We are now ready to describe the behaviour of *parking stations* and *users* components.

Each parking station is modelled in CARMA via a component of the form:

$$(\ G|R \ , \{\text{loc} = \ell, \text{capacity} = i, \text{available} = j\})$$

where loc is the attribute that identifies the zone where the parking station is located; capacity indicates the number of parking slots available in the station; available is the number of available bikes.

Processes G and R, which model the procedure to *get* and *return* a bike in the parking station, respectively, are defined as follows:

$$G \triangleq [\text{available} > 0] \ \text{get}[\text{my.loc} == \text{loc}]\langle \bullet \rangle \{\text{available} \leftarrow \text{available} - 1\}.G$$
$$R \triangleq [\text{available} < \text{capacity}] \ \text{ret}[\text{my.loc} == \text{loc}]\langle \bullet \rangle \{\text{available} \leftarrow \text{available} + 1\}.R$$

When the value of attribute available is greater than 0, process G executes the *unicast output* with action type get that potentially involves components satisfying the predicate my.loc $==$ loc, i.e. the ones that are located in the same zone[2]. When the output is executed the value of the attribute available is decreased by one to model the fact that one bike has been retrieved from the parking station.

Process R is similar. It executes the *unicast output* with action type ret that potentially involves components satisfying predicate my.loc $==$ loc. This action can be executed only when there is at least one parking slot available, i.e. when the value of attribute available is less than the value of attribute capacity. When the output considered above is executed, the value of attribute available is increased by one to model the fact that one bike has been returned in the parking station.

Users, who can be either *bikers* or *pedestrians*, are modelled via components of the form:

$$(Q, \{\mathsf{loc} = \ell_1, \mathsf{dest} = \ell_2\})$$

where loc is the attribute indicating where the user is located, while dest indicates the user destination. Process Q models the current state of the user and can be one of the following processes:

$$P \overset{\triangle}{=} \mathsf{get}[\mathsf{my.loc} == \mathsf{loc}](\bullet).B$$

$$B \overset{\triangle}{=} \mathsf{move}^\star[\bot]\langle\bullet\rangle\{\mathsf{loc} \leftarrow \mathsf{dest}\}.W$$

$$W \overset{\triangle}{=} \mathsf{ret}[\mathsf{my.loc} == \mathsf{loc}](\bullet).\mathbf{kill}$$

Process P represents a *pedestrian*, i.e. a user that is waiting for a bike. To *get a bike* a *pedestrian* executes a *unicast input* over activity get while selecting only parking stations that are located in his/her current location (my.loc $==$ loc). When this action is executed, a *pedestrian* becomes a *biker* B.

A biker can *move* from the current zone to the destination. This activity is modelled with the execution of a broadcast output via action type move. Note that, the predicate used to identify the target of the actions is $\bot$, denoting the value *false*. This means that this action actually does not synchronise with any component (since $\bot$ is never satisfied). This kind of *pattern* is used in CARMA to model *spontaneous actions*, i.e. actions that render the execution of an activity and that do not require synchronisation. After the broadcast move* the value of attribute loc is updated to dest and process W is activated. We will see in the next section that the actual rate of this action is determined by the environment and may also depend on the current time.

Process W represents a user who is waiting for a parking slot. This process executes an input over ret. This models the fact that the user has found a parking station with an available parking slot in their zone. After the execution of this input the user *disappears* from the system since the process **kill** is activated.

To model the arrival of new users, the following component is used:

$$(A, \{\mathsf{loc} = \ell\})$$

---

[2] Here we use $\bullet$ to denote the unit value.

where attribute loc indicates the location where users arrive, while process A is:

$$A \stackrel{\triangle}{=} \mathsf{arrival}^\star[\bot]\langle\bullet\rangle\{\}.A$$

This process only performs the spontaneous action arrival. The precise role of this process will be clear in a few paragraphs when the environment will be described.   □

Environment. An environment consists of two elements: a *global store* γ_g, that models the overall state of the system, and an *evolution rule* ρ.

Example 4. Bike Sharing System (4/7). The global store can be used to describe global information that may affect the system behaviour. In our *Bike Sharing System* we use the attribute user to record the number of active users.

The *evolution rule* ρ is a function which, depending on the *current time*, on the global store and on the current state of the collective (i.e., on the configurations of each component in the collective) returns a tuple of functions $\varepsilon = \langle \mu_p, \mu_w, \mu_r, \mu_u \rangle$ known as the *evaluation context* where $\mathrm{ACT} = \mathrm{ACTTYPE} \cup \{\alpha^\star | \alpha \in \mathrm{ACTTYPE}\}$ and:

- $\mu_p : \Gamma \times \Gamma \times \mathrm{ACT} \rightarrow [0,1]$, $\mu_p(\gamma_s, \gamma_r, \alpha)$ expresses the probability that a component with store γ_r can receive a broadcast message from a component with store γ_s when α is executed;
- $\mu_w : \Gamma \times \Gamma \times \mathrm{ACT} \rightarrow [0,1]$, $\mu_w(\gamma_s, \gamma_r, \alpha)$ yields the weight will be used to compute the probability that a component with store γ_r can receive a unicast message from a component with store γ_s when α is executed;
- $\mu_r : \Gamma \times \mathrm{ACT} \rightarrow \mathbb{R}_{\geq 0}$, $\mu_r(\gamma_s, \alpha)$ computes the execution rate of action α executed at a component with store γ_s;
- $\mu_u : \Gamma \times \mathrm{ACT} \rightarrow \Sigma \times \mathrm{COL}$, $\mu_u(\gamma_s, \alpha)$ determines the updates on the environment (global store and collective) induced by the execution of action α at a component with store γ_s.

For instance, the probability to receive a given message may depend on the *number or faction* of components in a given state. Similarly, the actual rate of an action may be a function of the number of components whose store satisfies a given property.

Functions μ_p and μ_w play a similar role. However, while the former computes the probability that a component receives a broadcast message, the latter associates to each unicast interaction with a weight, i.e. a non negative real number. This weight will be used to compute the probability that a given component with store γ_r receives a unicast message over activity α from a component with store γ_r. This probability is obtained by dividing the weight $\mu_w(\gamma_s, \gamma_r, \alpha)$ by the *total weights* of all possible receivers.

Example 5. Bike Sharing System (5/7). In our scenario, function μ_w can have the following form:

$$\mu_w(\gamma_s, \gamma_r, \alpha) = \begin{cases} 1 & \alpha = \mathsf{get} \wedge \gamma_s(\mathsf{loc}) = \gamma_r(\mathsf{loc}) \\ 1 & \alpha = \mathsf{ret} \wedge \gamma_s(\mathsf{loc}) = \gamma_r(\mathsf{loc}) \\ 0 & \text{otherwise} \end{cases}$$

where γ_s is the store of the sender, γ_r is the store of the receiver. The above function imposes that all the users in the same zone have the same weight, that is 1 when a user is located in the same zone of the parking station and 0 otherwise. This means that each user in the same zone have the same probability to be selected for getting a bike or for using a parking slot at a station. The weight associated to all the other interactions is 0.   □

Function μ_r computes the rate of a unicast/broadcast output. This function takes as parameter the local store of the component performing the action and the action on which the interaction is based. Note that the environment can disable the execution of a given action. This happens when the function μ_r (resp. μ_p or μ_w) returns the value 0.

Example 6. Bike Sharing System (6/7). In our example μ_r can be defined as follows:

$$\mu_r(\gamma_s, \alpha) = \begin{cases} \lambda_g & \alpha = \mathsf{get} \\ \lambda_r & \alpha = \mathsf{ret} \\ mtime(\mathsf{now}, \gamma_s(\mathsf{loc}), \gamma_s(\mathsf{dest})) & \alpha = \mathsf{move}^* \\ atime(\mathsf{now}, \gamma_s(\mathsf{loc}), \gamma_g(\mathsf{users})) & \alpha = \mathsf{arrival}^* \\ 0 & \text{otherwise} \end{cases}$$

We say that actions get and ret are executed at a constant rate; the rate of movement is a function (*mtime*) of actual time (now) and of starting location and final destination. Rate of user arrivals (computed by function *atime*) depends on current time now on location loc and on the number of users that are currently active in the system[3]. All the other interactions occurs with rate 0.   □

Finally, the function μ_u is used to update the global store and to activate a new collective in the system. The function μ_u takes as parameters the store of the component performing the action together with the action type and returns a pair (σ, N). Within this pair, σ identifies the update on the global store whereas N is a new collective installed in the system. This function is particularly useful for modelling the arrival of new agents into a system.

---

[3] Here we assume that functions *mtime* and *atime* are obtained after some observations on real systems.

Example 7. Bike Sharing System (7/7). In our scenario function update is used to model the arrival of new users and it is defined as follows:

$$\mu_u(\gamma_s, \alpha) = \begin{cases} \{\text{users} \leftarrow \gamma_g(\text{users}) + 1\}, & \\ \quad (W, \{\text{loc} = \gamma_s(\text{loc}), \text{dest} = destLoc(\text{now}, \gamma_s(\text{loc}))\}) & \alpha = \text{arrival}^\star \\ \{\text{users} \leftarrow \gamma_g(\text{users}) - 1\}, 0 & \alpha = \text{ret} \\ \{\}, 0 & \text{otherwise} \end{cases}$$

When action arrival* is performed a component associated with a new user is created in the same location as the sender (see Example 3). The destination of the new user will be determined by function *destLoc* that takes the current system time and starting location and returns a probability distribution over locations. Moreover, the global store records that a new user entered in the system. The number of active users is decremented by 1 each time action ret is performed. All the other actions do not trigger any update on the environment.   □

3   CARMA Semantics

The operational semantics of CARMA specifications is defined in terms of three functions that compute the possible *next states* of a *component*, a *collective* and a *system*:

1. the function $\mathbb{C}$ that describes the behaviour of a single component;
2. the function $\mathbb{N}_\varepsilon$ builds on $\mathbb{C}$ to describe the behaviour of collectives;
3. the function $\mathbb{S}_t$ that shows how CARMA systems evolve.

Note that, classically behaviour of (stochastic) process algebras is represented via *transition relations*. These relations, defined following a Plotkin-style, are used to infer possible computations of a process. Note that, due to *nondeterminism*, starting from the same process, different evolutions can be inferred. However, in CARMA, there is not any form of nonterminism while the selection of possible next state is governed by a probability distribution.

In this chapter we use an approach based on FuTS style [11]. Using this approach, the behaviour of a term is described using a function that, given a *term* and a *transition label*, yields a function associating each component, collective, or system with a non-negative number. The meaning of this value depends on the context. It can be the rate of the exponential distribution characterising the time needed for the execution of the action represented by ℓ; the probability of receiving a given broadcast message or the weight used to compute the probability that a given component is selected for the synchronisation. In all the cases the zero value is associated with unreachable terms.

We use the FuTS style semantics because it makes explicit an underlying (time-inhomogeneous) Action Labelled Markov Chain, which can be simulated with standard algorithms [16] but is nevertheless more compact than Plotkin-style semantics, as the functional form allows different possible outcomes to be treated within a single rule. A complete description of FuTS and their use can be found in [11].

Table 1. Operational semantics of components (Part 1)

$$\overline{\mathbb{C}[(\mathbf{nil},\gamma),\ell] = \emptyset} \ \ \mathbf{Nil} \qquad\qquad \overline{\mathbb{C}[0,\ell] = \emptyset} \ \ \mathbf{Zero}$$

$$\frac{[\![\pi_s]\!]_\gamma = \pi_s' \quad [\![\overrightarrow{e}\,]\!]_\gamma = \overrightarrow{v} \quad \mathbf{p} = \sigma(\gamma)}{\mathbb{C}[(\alpha^\star[\pi_s]\langle\overrightarrow{e}\,\rangle\sigma.P,\gamma),\alpha^\star[\pi_s']\langle\overrightarrow{v}\,\rangle,\gamma] = (P,\mathbf{p})} \ \ \mathbf{B\text{-}Out}$$

$$\frac{[\![\pi_s]\!]_\gamma = \pi_s' \quad [\![\overrightarrow{e}\,]\!]_\gamma = \overrightarrow{v} \quad \ell \neq \alpha^\star[\pi_s']\langle\overrightarrow{v}\,\rangle,\gamma}{\mathbb{C}[(\alpha^\star[\pi_s]\langle\overrightarrow{e}\,\rangle\sigma.P,\gamma),\ell] = \emptyset} \ \ \mathbf{B\text{-}Out\text{-}F1}$$

$$\frac{\gamma_r \models \pi_s \quad \gamma_s \models \pi_r[\overrightarrow{v}/\overrightarrow{x}] \quad \mathbf{p} = \sigma[\overrightarrow{v}/\overrightarrow{x}](\gamma_2)}{\mathbb{C}[(\alpha^\star[\pi_r](\overrightarrow{x})\sigma.P,\gamma_r),\alpha^\star[\pi_s](\overrightarrow{v}),\gamma_s] = (P[\overrightarrow{v}/\overrightarrow{x}],\mathbf{p})} \ \ \mathbf{B\text{-}In}$$

$$\frac{\gamma_r \not\models \pi_s \vee \gamma_s \not\models \pi_r[\overrightarrow{v}/\overrightarrow{x}]}{\mathbb{C}[(\alpha^\star[\pi_r](\overrightarrow{x})\sigma.P,\gamma_r),\alpha^\star[\pi_s](\overrightarrow{v}),\gamma_s] = \emptyset} \ \ \mathbf{B\text{-}In\text{-}F1}$$

$$\frac{\ell \neq \alpha^\star[\pi_s](\overrightarrow{v}),\gamma_s}{\mathbb{C}[(\alpha^\star[\pi_r](\overrightarrow{x})\sigma.P,\gamma_r),\ell] = \emptyset} \ \ \mathbf{B\text{-}In\text{-}F2}$$

3.1 Operational Semantics of Components

The behaviour of a single component is defined by a function

$$\mathbb{C} : \textsc{Comp} \times \textsc{Lab} \to [\textsc{Comp} \to \mathbb{R}_{\geq 0}]$$

Function $\mathbb{C}$ takes a component and a transition label, and yields a function in $[\textsc{Comp} \to \mathbb{R}_{\geq 0}]$. $\textsc{Lab}$ is the set of transition labels ℓ which are generated by the following grammar, where π_s is defined in Sect. 2.2:

$$
\begin{aligned}
\ell ::= \ &\alpha^\star[\pi_s]\langle\overrightarrow{v}\,\rangle,\gamma \ \ \text{Broadcast Output} \\
\mid \ &\alpha^\star[\pi_s](\overrightarrow{v}),\gamma \ \ \text{Broadcast Input} \\
\mid \ &\alpha[\pi_s]\langle\overrightarrow{v}\,\rangle,\gamma \ \ \text{Unicast Output} \\
\mid \ &\alpha[\pi_s](\overrightarrow{v}),\gamma \ \ \text{Unicast Input}
\end{aligned}
$$

These labels are associated with the four CARMA input-output actions and contain a reference to the action which is performed (α or $\alpha^\star$), the predicate π_s used to identify the target of the actions, and the value which is transmitted or received.

Function $\mathbb{C}$ is formally defined in Tables 1 and 2 and shows how a single component evolves when a *input/output* action is executed. For any component C and transition label ℓ, $\mathbb{C}[C,\ell]$ indicates the possible next states of C after the transition ℓ. These states are weighted. If $\mathbb{C}[C,\ell] = \mathscr{C}$ and $\mathscr{C}(C') = p$ then C evolves to C' with a weight p when ℓ is executed.

The process **nil** denotes the process that cannot perform any action. The behaviour associated to this process at the level of components can be derived via the rule **Nil**. This rule states that the inactive process cannot perform any action. This is derived from the fact that function $\mathbb{C}$ maps any label to function $\emptyset$ (rule **Nil**), where $\emptyset$ denotes the 0 constant function.

The behaviour of a *broadcast output* $(\alpha^\star[\pi_s]\langle\overrightarrow{e}\rangle\sigma.P, \gamma)$ is described by rules **B-Out** and **B-Out-F1**. Rule **B-Out** states that a broadcast output $\alpha^\star[\pi_s]\langle\overrightarrow{e}\rangle\sigma$ sends message $[\![\overrightarrow{e}]\!]_\gamma{}^4$ to all components that satisfy $[\![\pi_s]\!]_\gamma = \pi'_s$. The possible next local stores after the execution of an action are determined by the update σ. This takes the store γ and yields a probability distribution $\mathbf{p} = \sigma(\gamma) \in Dist(\Gamma)$. In rule **B-Out**, and in the rest of the chapter, the following notations are used:

- let $P \in$ Proc and $\mathbf{p} \in Dist(\Gamma)$, $(P, \mathbf{p})$ is a probability distribution in $Dist(\text{Comp})$ such that:

$$(P, \mathbf{p})(C) = \begin{cases} 1 & P \equiv Q|\mathbf{kill} \wedge C \equiv \mathbf{0} \\ \mathbf{p}(\gamma) & C \equiv (P, \gamma) \wedge P \not\equiv Q|\mathbf{kill} \\ 0 & \text{otherwise} \end{cases}$$

- let $\mathbf{c} \in Dist(\text{Comp})$ and $r \in \mathbb{R}_{\geq 0}$, $r \cdot \mathbf{c}$ denotes the function $\mathscr{C} : \text{Comp} \rightarrow \mathbb{R}_{\geq 0}$ such that: $\mathscr{C}(C) = r \cdot \mathbf{c}(C)$

Note that, after the execution of an action a component can be destroyed. This happens when the continuation process after the action prefix contains the term **kill**. For instance, by applying rule **B-Out** we have that:

$$\mathbb{C}[(\alpha^\star[\pi_s]\langle v\rangle\sigma.(\mathbf{kill}|Q), \gamma), \alpha^\star[\pi_s]\langle v\rangle, \gamma] = [\mathbf{0} \mapsto r]$$

Rule **B-Out-F1** states that a *broadcast output* can be only involved in labels of the form $\alpha^\star[\pi_s]\langle\overrightarrow{v}\rangle, \gamma$.

Computations related to a broadcast input are labelled with $\alpha^\star[\pi_s](\overrightarrow{v}), \gamma_1$. There, π_s is the predicate used by the sender to identify the target components while $\overrightarrow{v}$ is the sequence of transmitted values. Rule **B-In** states that a component $(\alpha^\star[\pi_r](\overrightarrow{x})\sigma.P, \gamma_r)$ can evolve with this label when its store γ_r (the store of the receiver) satisfies the sender predicate, i.e. $\gamma_r \models \pi_s$, while the store of the sender, i.e. γ_s satisfies the predicate of the receiver $\pi_r[\overrightarrow{v}/\overrightarrow{x}]$.

Rule **B-In-F1** models the fact that if a component is not in the set of possible receivers ($\gamma_r \not\models \pi_s$) or the received values do not satisfy the expected requirements then the component cannot receive a broadcast message. Finally, the rule **B-In-F2** models the fact that $(\alpha^\star[\pi_r](\overrightarrow{x})\sigma.P, \gamma_r)$ can only perform a broadcast input on action α and that it always refuses input on any other action type $\beta \neq \alpha$.

The behaviour of *unicast output* and *unicast input* is defined by the first five rules of Table 2. These rules are similar to the ones already presented for broadcast output and broadcast input.

---

[4] We let $[\![\cdot]\!]_\gamma$ denote the evaluation function of an expression/predicate with respect to the store γ.

Table 2. Operational semantics of components (Part 2)

$$\frac{[\![\pi_s]\!]_\gamma = \pi'_s \quad [\![\vec{e}\,]\!]^t_\gamma = \vec{v} \quad \mathbf{p} = \sigma(\gamma)}{\mathbb{C}[(\alpha[\pi_s]\langle\vec{e}\,\rangle\sigma.P,\gamma),\alpha[\pi'_s]\langle\vec{v}\,\rangle,\gamma] = (P,\mathbf{p})}\ \text{Out}$$

$$\frac{[\![\pi_s]\!]_\gamma = \pi'_s \quad [\![\vec{e}\,]\!]^t_\gamma = \vec{v} \quad \ell \neq \alpha[\pi'_s]\langle\vec{v}\,\rangle,\gamma}{\mathbb{C}[(\alpha[\pi_s]\langle\vec{e}\,\rangle\sigma.P,\gamma),\ell] = \emptyset}\ \text{Out-F}$$

$$\frac{\gamma_r \models \pi_s \quad \gamma_s \models \pi_r[\vec{v}/\vec{x}] \quad \mathbf{p} = \sigma[\vec{v}/\vec{x}](\gamma_2)}{\mathbb{C}[(\alpha[\pi_r](\vec{x})\sigma.P,\gamma_r),\alpha[\pi_s](\vec{v}),\gamma_s] = (P[\vec{v}/\vec{x}],\mathbf{p})}\ \text{In}$$

$$\frac{\gamma_r \not\models \pi_s \vee \gamma_s \not\models \pi_r[\vec{v}/\vec{x}]}{\mathbb{C}[(\alpha[\pi_r](\vec{x})\sigma.P,\gamma_r),\alpha[\pi_r](\vec{v}),\gamma_r] = \emptyset}\ \text{In-F1} \qquad \frac{\ell \neq \alpha[\pi_s](\vec{v}),\gamma_s}{\mathbb{C}[(\alpha[\pi_r](\vec{x})\sigma.P,\gamma_r),\ell] = \emptyset}\ \text{In-F2}$$

$$\frac{\mathbb{C}[(P,\gamma),\ell] = \mathscr{C}_1 \quad \mathbb{C}[(Q,\gamma),\ell] = \mathscr{C}_2}{\mathbb{C}[(P+Q,\gamma),\ell] = \mathscr{C}_1 \oplus \mathscr{C}_2}\ \text{Plus}$$

$$\frac{\gamma \models \pi \quad \mathbb{C}[(P,\gamma),\ell] = \mathscr{C}}{\mathbb{C}[([\pi]P,\gamma),\ell] = \mathscr{C}}\ \text{Guard} \qquad \frac{\gamma \not\models \pi}{\mathbb{C}[([\pi]P,\gamma),\ell] = \emptyset}\ \text{Guard-F}$$

$$\frac{\mathbb{C}[(P,\gamma),\ell] = \mathscr{C}_1 \quad \mathbb{C}[(Q,\gamma),\ell] = \mathscr{C}_2}{\mathbb{C}[(P|Q,\gamma),\ell] = \mathscr{C}_1|Q \oplus P|\mathscr{C}_2}\ \text{Par} \qquad \frac{A \stackrel{\triangle}{=} P \quad \mathbb{C}[(P,\gamma),\ell] = \mathscr{C}}{\mathbb{C}[(A,\gamma),\ell] = \mathscr{C}}\ \text{Rec}$$

The other rules of Table 2 describe the behaviour of other process operators, namely *choice* $P+Q$, *parallel composition* $P|Q$, *guard* and *recursion*. The term $P+Q$ identifies a process that can behave either as P or as Q. The rule **Plus** states that the components that are reachable by $(P+Q,\gamma)$ are the ones that can be reached either by (P,γ) or by (Q,γ). In this rule we use $\mathscr{C}_1 \oplus \mathscr{C}_2$ to denote the function that maps each term C to $\mathscr{C}_1(C) + \mathscr{C}_2(C)$, for any $\mathscr{C}_1, \mathscr{C}_2 \in$ [COMP $\rightarrow \mathbb{R}_{\geq 0}$].

In $P|Q$ the two composed processes interleave for all the transition labels. In the rule the following notations are used:

– for each component C and process Q we let:

$$C|Q = \begin{cases} 0 & C \equiv 0 \\ (P|Q,\gamma) & C \equiv (P,\gamma) \end{cases}$$

$Q|C$ is symmetrically defined.
– for each $\mathscr{C} :$ COMP $\rightarrow \mathbb{R}_{\geq 0}$ and process Q, $\mathscr{C}|Q$ (resp. $Q|\mathscr{C}$) denotes the function that maps each term of the form $C|Q$ (resp. $Q|C$) to $\mathscr{C}(C)$, while the others are mapped to 0;

Rule **Rec** is standard. The behaviour of $([\pi]P,\gamma)$ is regulated by rules **Guard** and **Guard-F**. The first rule states that $([\pi]P,\gamma)$ behaves exactly like (P,γ)

when γ satisfies predicate π. However, in the first case the *guard* is removed when a transition is performed. In contrast, no component is reachable when the *guard* is not satisfied (rule **Guard-F**).

The following lemma guarantees that for any C and for any ℓ $\mathbb{C}[C, \ell]$ is either a probability distribution or the 0 constant function $\emptyset$.

3.2   Operational Semantics of Collectives

The operational semantics of a *collective* is defined via the function

$$\mathbb{N}_\varepsilon : \text{Col} \times \text{Lab}_I \rightarrow [\text{Col} \rightarrow \mathbb{R}_{\geq 0}]$$

that is formally defined in Table 3, where we use a straightforward adaptation of the notations introduced in the previous section. This function shows how a collective reacts when a broadcast/unicast message is received. Indeed, Lab_I denotes the subset of Lab with only input labels:

$$\ell ::= \alpha^\star[\pi_s](\overrightarrow{v}), \gamma \qquad \text{Broadcast Input}$$
$$| \quad \alpha[\pi_s](\overrightarrow{v}), \gamma \qquad \text{Unicast Input}$$

Given a collective N and an input label $\ell \in \text{Lab}_I$, function $\mathbb{N}_\varepsilon[N, \ell]$ returns a function $\mathscr{N}$ that associates each collective N' reachable from N via ℓ with a value in $\mathbb{R}_{\geq 0}$. If ℓ is a broadcast input $(\alpha^\star[\pi_s](\overrightarrow{v}), \gamma)$ this value represents the probability that the collective is reachable after ℓ. When ℓ is a unicast input $\alpha[\pi_s](\overrightarrow{v}), \gamma$, $\mathscr{N}(N')$ is the weight that will be used, at the level of systems, to compute the probability that N' is selected after ℓ. Note that this difference is due from the fact that while the probability to receive a *broadcast input* can be computed *locally* (each component identifies its own probability), to compute the probability to receive a *unicast input* the complete collective is needed. Function $\mathbb{N}_\varepsilon$ is also parametrised with respect to the *evaluation function* ε, obtained from the environment where the collective operates, that is used to compute the above mentioned *weights*.

The first four rules in Table 3 describe the behaviour of the single component at the level of collective. Rule **Zero** is similar to rule **Nil** of Table 1 and states that inactive component **0** cannot perform any action. Rule **Comp-B-In** states that if (P, γ) can receive a message sent via a broadcast with activity α $(\mathbb{C}[(P, \gamma), \alpha^\star[\pi_s](\overrightarrow{v}), \gamma] = \mathscr{N} \neq \emptyset)$ then the component receives the message with probability $\mu_p(\gamma, \alpha^\star)$ while the message is not received with probability $1 - \mu_p(\gamma, \alpha^\star)$. In the first case, the resulting function is renormalised by $\oplus \mathscr{N}$ to indicate that each element in P receives the message with the same probability. There we use $\oplus \mathscr{N}$ to denote $\sum_{N \in \text{Col}} \mathscr{N}(N)$. On the contrary, rule **Comp-B-In-F** states that if (P, γ) is not able to receive a broadcast message, $(\mathbb{C}[(P, \gamma), \alpha^\star[\pi_s](\overrightarrow{v}), \gamma] = \emptyset)$, with probability 1 the message is received while the component remains unchanged.

Rule **Comp-In** is similar to **Comp-B-In**. It simply lifts the transition at the level of component to the level of collective while the resulting function is

Table 3. Operational semantics of collective

$$\overline{\mathbb{N}_\varepsilon[0,\ell]=0}\ \ \textbf{Zero}$$

$$\frac{\mathbb{C}[(P,\gamma),\alpha^\star[\pi_s](\overrightarrow{v}),\gamma]=\mathcal{N}\quad \mathcal{N}\neq 0\quad \varepsilon=\langle\mu_p,\mu_w,\mu_r,\mu_u\rangle}{\mathbb{N}_\varepsilon[(P,\gamma),\alpha^\star[\pi_s](\overrightarrow{v}),\gamma]=\frac{\mu_p(\gamma,\alpha^\star)}{\oplus\mathcal{N}}\cdot\mathcal{N}+[(P,\gamma)\mapsto(1-\mu_p(\gamma,\alpha^\star))]}\ \ \textbf{Comp-B-In}$$

$$\frac{\mathbb{C}[(P,\gamma),\alpha^\star[\pi_s](\overrightarrow{v}),\gamma]=0}{\mathbb{N}_\varepsilon[(P,\gamma),\alpha^\star[\pi_s](\overrightarrow{v}),\gamma]=[(P,\gamma)\mapsto 1]}\ \ \textbf{Comp-B-In-F}$$

$$\frac{\mathbb{C}[(P,\gamma_2),\alpha[\pi_s](\overrightarrow{v}),\gamma_1]=\mathcal{N}\quad \mathcal{N}\neq 0\quad \varepsilon=\langle\mu_p,\mu_w,\mu_r,\mu_u\rangle}{\mathbb{N}_\varepsilon[(P,\gamma_2),\alpha[\pi_s](\overrightarrow{v}),\gamma_1]=\mu_w(\gamma_1,\gamma_2,\alpha)\cdot\frac{\mathcal{N}}{\oplus\mathcal{N}}}\ \ \textbf{Comp-In}$$

$$\frac{\mathbb{C}[(P,\gamma_2),\alpha[\pi_s](\overrightarrow{v}),\gamma_1]=0}{\mathbb{N}_\varepsilon[(P,\gamma_2),\alpha[\pi_s](\overrightarrow{v}),\gamma_1]=0}\ \ \textbf{Comp-In-F}$$

$$\frac{\mathbb{N}_\varepsilon[N_1,\alpha^\star[\pi_s](\overrightarrow{v}),\gamma]=\mathcal{N}_1\quad \mathbb{N}_\varepsilon[N_2,\alpha^\star[\pi_s](\overrightarrow{v}),\gamma]=\mathcal{N}_2}{\mathbb{N}_\varepsilon[N_1\parallel N_2,\alpha^\star[\pi_s](\overrightarrow{v}),\gamma]=\mathcal{N}_1\parallel\mathcal{N}_2}\ \ \textbf{B-In-Sync}$$

$$\frac{\mathbb{N}_\varepsilon[N_1,\alpha[\pi_s](\overrightarrow{v}),\gamma]=\mathcal{N}_1\quad \mathbb{N}_\varepsilon[N_2,\alpha[\pi_s](\overrightarrow{v}),\gamma]=\mathcal{N}_2}{\mathbb{N}_\varepsilon[N_1\parallel N_2,\alpha[\pi_s](\overrightarrow{v}),\gamma]=\mathcal{N}_1\parallel N_2\oplus N_1\parallel\mathcal{N}_2}\ \ \textbf{In-Sync}$$

multiplied by the weight $\mu_p(\gamma_1,\gamma_2,\alpha)$. The latter is the probability that this component is selected for the synchronisation. As in **Comp-B-In**, function $\mathcal{N}$ is divided by $\oplus\mathcal{N}$ to indicate that any possible receiver in P is selected with the same probability. Rule **Comp-In-F** is applied when a component is not involved in a synchronisation.

Rule **B-In-Sync** states that that two collectives N_1 and N_2 that operate in parallel synchronise while performing a broadcast input. This models the fact that the input can be potentially received by both of the collectives. In this rule we let $\mathcal{N}_1\parallel\mathcal{N}_2$ denote the function associating the value $\mathcal{N}_1(N_1)\cdot\mathcal{N}_2(N_2)$ with each term of the form $N_1\parallel N_2$ and 0 with all the other terms. We can observe that if

$$\mathbb{N}_\varepsilon[N,\alpha^\star[\pi_s](\overrightarrow{v}),\gamma]=\mathcal{N}$$

then, as we have already observed for rule **Comp-B-In**, $\oplus\mathcal{N}=1$ and $\mathcal{N}$ is in fact a probability distribution over COL.

Rule **In-Sync** controls the behaviour associated with unicast input and it states that a collective of the form $N_1\parallel N_2$ performs a *unicast input* if this is performed either in N_1 or in N_2. This is rendered in the semantics as an interleaving rule, where for each $\mathcal{N}:\text{COL}\to\mathbb{R}_{\geq 0}$, $\mathcal{N}\parallel N_2$ denotes the function associating $\mathcal{N}(N_1)$ with each collective of the form $N_1\parallel N_2$ and 0 with all other collectives.

3.3   Operational Semantics of Systems

The operational semantics of systems is defined via the function

$$\mathbb{S}_t : \text{SYS} \times \text{LAB}_S \rightarrow [\text{SYS} \rightarrow \mathbb{R}_{\geq 0}]$$

that is formally defined in Table 4. This function only considers synchronisation labels LAB_S:

$$\ell ::= \alpha^\star[\pi_s]\langle\overrightarrow{v}\rangle, \gamma \qquad \text{Broadcast Output}$$
$$| \quad \tau[\alpha[\pi_s]\langle\overrightarrow{v}\rangle, \gamma] \qquad \text{Unicast Synchronization}$$

The behaviour of a CARMA system is defined in terms of a *time-inhomogeneous Action Labelled Markov Chain* whose transition matrix is defined by function $\mathbb{S}_t$. For any system S and for any label $\ell \in \text{LAB}_S$, if $\mathbb{S}_t[S, \ell] = \mathscr{S}$ then $\mathscr{S}(S')$ is the rate of the transition from S to S'. When $\mathscr{S}(S') = 0$ then S' is not reachable from S via ℓ.

The first rule is **Sys-B**. This rule states that, when $\varepsilon = \langle\mu_p, \mu_w, \mu_r, \mu_u\rangle = \rho(t, \gamma_g, N)$, a system of the form N **in** (γ_g, ρ) at time t can perform a *broadcast output* when there is a component $C \in N$ that performs the output while the remaining part of the collective $(N - C)$ performs the complementary input. The outcome of this synchronisation is computed by the function $\mathsf{bSync}_\varepsilon$ defined below:

$$\frac{\varepsilon = \langle\mu_p, \mu_w, \mu_r, \mu_u\rangle \quad \mathbb{C}[C, \alpha^\star[\pi_s]\langle\overrightarrow{v}\rangle, \gamma] = \mathscr{C} \quad \mathbb{N}_\varepsilon[N, \alpha^\star[\pi_s]\langle\overrightarrow{v}\rangle, \gamma] = \mathscr{N}}{\mathsf{bSync}_\varepsilon(C, N, \alpha^\star[\pi_s]\langle\overrightarrow{v}\rangle, \gamma) = \mu_r(\gamma, \alpha^\star[\pi_s]\langle\overrightarrow{v}\rangle, \gamma) \cdot \mathscr{C} \parallel \mathscr{N}}$$

This function combines the outcome of the broadcast output performed by C, ($\mathscr{C}$) with the complementary input performed by N ($\mathscr{N}$), the result is then multiplied by the rate of the action induced by the environment $\mu_r(\gamma_C, \alpha^\star[\pi_s]\langle\overrightarrow{v}\rangle, \gamma)$. Note that, since both $\mathscr{C}$ and $\mathscr{N}$ are probability distributions, the same is true for $\mathscr{C} \parallel \mathscr{N}$.

To compute the total rate of a synchronisation we have to sum the outcome above for all the possible senders $C \in N$ multiplied by the multiplicity of C component in N ($N(C)$). After the synchronisation, the global store is updated and a new collective can be created according to function μ_u. In rule **Sys-B** the following notations are used. For each collective N_2, $\mathscr{N}$: COL $\rightarrow \mathbb{R}_{\geq 0}$, $\mathscr{S}$: SYS $\rightarrow \mathbb{R}_{\geq 0}$ and $\mathbf{p} \in Dist(\Gamma)$ we let $\mathscr{N}$ **in** $(\mathbf{p}, \rho)$ denote the function mapping each system N **in** (γ, ρ) to $\mathscr{N}(N) \cdot \mathbf{p}(\gamma)$.

The second rule is **Sys** that regulates unicast synchronisations, which is similar to **Sys-B**. However, there function $\mathsf{uSync}_\varepsilon$ is used. This function is defined below:

$$\frac{\varepsilon = \langle\mu_p, \mu_w, \mu_r, \mu_u\rangle \quad \mathbb{C}[C, \alpha^\star[\pi_s]\langle\overrightarrow{v}\rangle, \gamma] = \mathscr{C} \quad \mathbb{N}_\varepsilon[N, \alpha^\star[\pi_s]\langle\overrightarrow{v}\rangle, \gamma] = \mathscr{N} \neq \emptyset}{\mathsf{uSync}_\varepsilon(C, N, \alpha^\star[\pi_s]\langle\overrightarrow{v}\rangle, \gamma) = \mu_r(\gamma_C, \alpha^\star[\pi_s]\langle\overrightarrow{v}\rangle, \gamma) \cdot \mathscr{C} \parallel \frac{\mathscr{N}}{\oplus \mathscr{N}}}$$

$$\frac{\mathbb{N}_\varepsilon[N, \alpha^\star[\pi_s]\langle\overrightarrow{v}\rangle, \gamma] = \emptyset}{\mathsf{uSync}_\varepsilon(C, N, \alpha^\star[\pi_s]\langle\overrightarrow{v}\rangle, \gamma) = \emptyset}$$

Similarly to $\mathsf{bSync}_\varepsilon$, this function combines the outcome of a unicast output performed by C, $(\mathscr{C})$ with the complementary input performed by N $(\mathscr{N})$. The result is then multiplied by the rate of the action induced by the environment $\mu_r(\gamma_C, \alpha^\star[\pi_s]\langle\overrightarrow{v}\rangle, \gamma)$. However, in $\mathsf{uSync}_\varepsilon$ we have to renormalise $\mathscr{N}$ by the value $\oplus\mathscr{N}$. This guarantees that the total synchronisation rate does not exceeds the capacity of the sender. Note that, $\mathscr{N}$ is not a probability distribution while $\frac{\mathscr{N}}{\oplus\mathscr{N}}$ is.

4   Carma Implementation

To support simulation of Carma models, a prototype simulator has been developed. This simulator, which has been implemented in Java, can be used to perform stochastic simulation and will be the basis for the implementation of other analysis techniques. An Eclipse plug-in for supporting specification and analysis of CAS in Carma has also been developed. In this plug-in, Carma systems are specified by using an appropriate high-level language for designers of CAS, named the Carma *Specification Language*. This is mapped to the process algebra, and hence will enable qualitative and quantitive analysis of CAS during system development by enabling a design workflow and analysis pathway. The intention of this high-level language is not to add to the expressiveness of Carma, which we believe to be well-suited to capturing the behaviour of CAS, but rather to ease the task of modelling for users who are unfamiliar with process algebra and similar formal notations. Both the simulator and the Eclipse plug-in are available at https://quanticol.sourceforge.net/.

In the rest of this section, we first describe the Carma *Specification Language* then an overview of the Carma Eclipse Plug-in is provided. In Sect. 5 we will show how the *Bike Sharing System* considered in Sect. 2 can be modelled, simulated and analysed with the Carma tools.

4.1   Carma Specification Language

In this section we present the language that supports the design of CAS in Carma. To describe the main features of this language, following the same approach used in Sect. 2, we will use the *Bike Sharing System*.

Each Carma specification, also called a Carma *model*, provides definitions for:

- structured *data types* and the relative *functions*;
- prototypes of *components*;
- *systems* composed of collective and environment;
- *measures*, that identify the relevant data to *measure* during simulation runs.

Data Types. Three basic types are natively supported in our specification language. These are: `bool`, for booleans, `int`, for integers, and `real`, for real values. However, to model complex structures, it is often useful to introduce custom

Table 4. Operational Semantics of Systems.

$$\rho(t, \gamma_g, N) = \varepsilon = \langle \mu_p, \mu_w, \mu_r, \mu_u \rangle \quad \mu_u(\gamma_g, \alpha^\star) = (\sigma, N')$$
$$\frac{\sum_{C \in N} N(C) \cdot \mathsf{bSync}(C, N - C, \alpha^\star[\pi_s]\langle \overrightarrow{v} \rangle, \gamma) = \mathscr{N}}{\mathbb{S}_t[N \text{ in } (\gamma_g, \rho), \alpha^\star[\pi_s]\langle \overrightarrow{v} \rangle, \gamma] = \mathscr{N} \parallel N' \text{ in } (\sigma(\gamma_g), \rho)} \quad \textbf{Sys-B}$$

$$\rho(t, \gamma_g, N) = \varepsilon = \langle \mu_p, \mu_w, \mu_r, \mu_u \rangle \quad \mu_u(\gamma_g, \alpha^\star) = (\sigma, N')$$
$$\frac{\sum_{C \in N} N(C) \cdot \mathsf{uSync}(C, N - C, \tau[\alpha[\pi_s]\langle \overrightarrow{v} \rangle, \gamma]) = \mathscr{N}}{\mathbb{S}_t[N \text{ in } (\gamma_g, \rho), \tau[\alpha[\pi_s]\langle \overrightarrow{v} \rangle, \gamma] = \mathscr{N} \parallel N' \text{ in } (\sigma(\gamma_g), \rho)} \quad \textbf{Sys}$$

types. In a CARMA specification two kind of custom types can be declared: *enumerations* and *records*.

Like in many other programming languages, an *enumeration* is a data type consisting of a set of *named values*. The enumerator names are identifiers that behave as constants in the language. An attribute (or variable) that has been declared as having an enumerated type can be assigned any of the enumerators as its value. In other words, an enumerated type has values that are different from each other, and that can be compared and assigned, but which are not specified by the programmer as having any particular concrete representation. The syntax to declare a new *enumeration* is:

enum *name* = *elem*$_1$,..., *elem*$_n$;

where *name* is the name of the declared enumeration while *elem*$_i$ are its value names. Enumeration names start with a capitalised letter while the enumeration values are composed by only capitalised letters.

Example 8. Enumerations can be used to define predefined set of values that can be used in the specification. For instance one can introduce an enumeration to identify the possible four directions of movement:

enum Direction = NORTH, SOUTH, EAST, WEST;

To declare aggregated data structures, a CAS designer can use *records*. A record consists of a sequence of a set of typed fields:

record *name* = [*type*$_1$ *field*$_1$,..., *type*$_n$ *field*$_n$];

Each field has a type *type*$_i$ and a name *field*$_i$: *type*$_i$ can be either a built-in type or one of the new declared types in the specification; *field*$_i$ can be any valid identifier.

Example 9. Record can be used to model structured elements. For instance, a position over a grid can be rendered as follows:

record Position = [int x, int y];

A record can be created by assigning a value to each field, within square brackets:

$$[\ field_1 := expression_1\ ,\dots,\ \ field_n := expression_n\ ]$$

Example 10. The instantiation of a location referring to the point located at $(0,0)$ has the following form:

```
[ x:=0 , y:=0 ]
```

Given a variable (or attribute) having a record type, each field can be accessed using the *dot* notation:

$$variable\,.\,field_i$$

Constants and Functions. A CARMA specification can also contain *constants* and *functions* declarations having the following syntax:

```
const name = expression;

fun type name( type₁ arg₁ ,..., typeₖ argₖ ) {
    ...
}
```

where the body of an expression consists of standard statements in a high-level programming language. The type of a constant is not declared but inferred directly from the assigned expression.

Example 11. A constant can be used to represent the number of *zones* in the *Bike Sharing System*:

```
const ZONES =   5;
```

Moreover, functions can be used to perform complex computations that cannot be done in a single expression:

```
fun real ReceivingProb( int size ) {
  if (size != 0) {
    return 1.0/real(size);
  } else {
    return 0.0;
  }
}
```

Components Prototype. A *component prototype* provides the general structure of a component that can be later instantiated in a CARMA system. Each prototype is parameterised with a set of typed parameters and defines: the store; the component's behaviour and the initial configuration. The syntax of a *component prototype* is:

```
component name( type₁ arg₁ ,..., typeₙ argₙ) {
    store { ···
        attr_kind anameᵢ := expressionᵢ; ···
    }
    behaviour { ···
        procᵢ = pdefᵢ; ···
    }
    init { P₁|···|Pw }
}
```

Each component prototype has a possibly empty list of arguments. Each argument arg_i has a type $type_i$ that can be one of the built-in types (bool, int and real), a custom type (an enumeration or record), or the type process that indicates a component behaviour. These arguments can be used in the body of the component. The latter consists of three (optional) blocks: store, behaviour and init.

The block store defines the list of attributes (and their initial values) exposed by a component. Each attribute definition consists of an attribute kind $attr\_kind$ (that can be either attrib or const), a $name$ and an expression identifying the initial attribute value. When an attribute is declared as const, it cannot be changed. The actual type of an attribute is not declared but inferred from the expression providing its initialisation value.

The block behaviour is used to define the processes that are specific to the considered components and consists of a sequence of definitions of the form

$$proc_i = pdef_i;$$

where $proc_i$ is the process name while $pdef_i$ is its definition having the following syntax[5]:

pdef ::= *pdef+pdef*

   | [*expr*] *pdef*

   | *act . proc*

act ::= *act_name[expr]<expr₁,..., exprₙ>{aname₁:= expr′₁,...,anameₖ:=expr′ₖ}*

   | *act_name*[expr]<expr₁,..., exprₙ>{aname₁:=expr′₁,...,anameₖ:=expr′ₖ}*

   | *act_name[expr](var₁,..., varₙ){aname₁:=expr′₁,...,anameₖ:=expr′ₖ}*

   | *act_name*[expr](var₁,..., varₙ){aname₁:=expr′₁,...,anameₖ:=expr′ₖ}*

Finally, block init is used to specify the initial behaviour of a component. It consists of a sequence of terms P_i separated by the symbol |. Each P_i can be a process defined in the block behaviour, kill or nil.

Example 12. The prototypes for Station, Users and Arrival components, already described in Example 2, can be defined as follows:

---

[5] All the operators are right associative and presented in the order of priority.

```
component Station ( int loc , int capacity , int available
      )
{
   store {
      attrib loc := loc ;
      attrib available := available ;
      attrib capacity := capacity ;
   }
   behaviour {
      G = [my. available >0]
                        get<>{ my. available := my. available −1 }.G;
      R = [my. available <my. capacity ]
                        ret <>{ my. available := my. available+1 }.R;
   }
   init {
      G|R
   }
}

component User ( int loc , int dest ) {
   store {
      attrib loc := loc ;
      attrib dest := dest ;
   }
   behaviour {
      P = get [ my. loc == loc ]() .B;
      B = move*[ false ]<>{ my. loc := my. dest }.W;
      W = ret [ my. loc == loc ]() . kill ;
   }
   init {
      P
   }
}

component Arrival ( int loc ) {
   store {
      attrib loc := loc ;
   }
   behaviour {
      A = arrival *[false]<>.A;
   }
   init {
      A
   }
}
```

System Definitions. A system definition consists of two blocks, namely `collective` and `environment`, that are used to declare the collective in the system and its environment, respectively:

```
system name {
   collective {
      inist_stmt
   }
   environment { ··· 
   }
}
```

Above, *inist_stmt* indicates a sequence of commands that are used to instantiate components. The basic command to create a new component is:

$$\text{new } name(\ expr_1, \ldots, expr_n\ )$$

where *name* is the name of a component prototype. However, in a system a large number of collectives can occur. For this reason, our specification language provides specific constructs for the instantiation of multiple copies of a component. A first construct is the *range operator*. This operator is of the form:

$$[\ expr_1\ :\ expr_2\ :\ expr_3\ ]$$

and can be used as an argument of type integer. It is equivalent to a sequence of integer values starting from $expr_1$, ending at $expr_2$. The element $expr_3$ (that is optional) indicates the step between two elements in the sequence. When $expr_3$ is omitted, value 1 is assumed. The *range operator* can be used where an integer parameter is expected. This is equivalent to having multiple copies of the same instantiation command where each element in the sequence replaces the command.

For instance, assuming ZONES to be the constant identifying the number of zones in the city, while CAPACITY and INITIAL_AVAILABILITY refer to the station capacity and to the initial availability, respectively, the instantiation of the stations can be modelled as:

$$\text{new Station(}\ [0\!:\!\text{ZONES}\!-\!1]\ ,\ \text{CAPACITY, INITIAL\_AVAILABILITY}$$
$$)\,;$$

The command above is equivalent to:

$$\text{new Station(}\ 0\ ,\ \text{CAPACITY, INITIAL\_AVAILABILITY}\ )\,;$$
$$\vdots$$
$$\text{new Station(}\ \text{ZONES}\!-\!1\ ,\ \text{CAPACITY, INITIAL\_AVAILABILITY}\ )\,;$$

Two other commands are used to control instantiation of components. These are:

```
for ( var_name = expr1 ; expr2 ; expr3 ) {
   inist_stmt
}

if ( expr ) {
   inist_stmt
} else {
```

```
        inist_stmt
   }
```

The former is used to iterate an instantiation block for a given number of times while the latter can be used to differentiate the instantiation depending on a given condition.

Example 13. The following block can be used to instantiate SITES copies of component Station at each zone. The same block instantiates a component Arrival at each zone:

```
collective {
   for ( i ; i<ZONES ; 1 ) {
      for ( j ; j<SITES ; 1 ) {
         new Station( i , CAPACITY, INITIAL_AVAILABILITY );
      }
      new Arrival(i);
   }
}
```

The syntax of a block environment is the following:

```
environment {
   store { ··· }
   prob { ··· }
   weight { ··· }
   rate { ··· }
   update { ··· }
}
```

The block store defines the *global store* and has the same syntax as the similar block already considered in the component prototypes.

Example 14. In the *Bike Sharing System* we use a global attribute to count the amount of *active users* in the system:

```
store {
   attrib users := 0;
}
```

Blocks prob and weight are used to compute the probability to receive a message. Syntax of prob is the following:

```
prob { ···
      [guard_i] act_i : expr_i; ···
      default : expr;
}

weight { ···
      [guard_i] act_i : expr_i; ···
      default : expr;
}
```

In the above, each $guard_i$ is a boolean expression over the global store and the stores of the two interacting components, i.e. the sender and the receiver, while act_i denotes the action used to interact. In $guard_i$ attributes of sender and receiver are referred to using sender.a and receiver.a, while the values published in the global store are referenced by using global.a. This probability value may depend on the number of components in a given state. To compute this value, expressions of the following form can be used:

$$\#\{\ \Pi\ |\ expr\ \}$$

This expression denotes the number of components in the system satisfying boolean expression $expr$ where a process of the form Π is executed. In turn, Π is a pattern of the following form:

$$\Pi ::= *\ |\ *[proc]\ |\ comp[\ *\ ]\ |\ comp[proc]$$

Example 15. In our example the block weight can be instantiated as follows:

```
weight{
    [receiver.loc==sender.loc]  get:  1;
    [receiver.loc==sender.loc]  ret:  1;
    default:  0;
}
```

Above, we say that each user in a zone receives a bike/parking slot with the same probability. All the other interactions are disabled having the associated weight equal to 0.

Block rate is similar and it is used to compute the rate of an unicast/broadcast output. This represents a function taking as parameter the local store of the component performing the action and the action type used. Note that the environment can disable the execution of a given action. This happens when evaluation of block rate (resp. prob) is 0. Syntax of rate is the following:

```
rate {  ···
    [guard_i]  act_i  :  expr_i;  ···
    default  :  expr;
}
```

Differently from prob, in rate guards $guard_i$ are evaluated by considering only the attributes defined in the store of the component performing the action, referenced as sender.a, or in the global store, accessed via global.a.

Example 16. In our example rate can be defined as follow:

```
rate{
    [true]  get:  get_rate;
    [true]  ret:  ret_rate;
    [true]  move*:  move_rate;
    [true]  arrival*:
        (global.users<TOTAL_USERS?arrival_rate:0.0);
    [true]  default:  1;
}
```

Above we say that actions move*, get and ret are executed at a constant rate. Rate of user arrivals depends on the number of active users. Action arrival* is executed with rate arrival_rate when the total number of users active in the system is less than TOTAL_USERS. Otherwise, the same action is disabled (i.e. executed with rate 0.0).

Finally, the block update is used to update the global store and to install a new collective in the system. Syntax of update is:

$$update \{ \cdots$$
$$[guard_i] \ act_i \ : \ attr\_updt_i \ ; inst\_cmd_i \ ; \ \cdots$$
$$\}$$

As for rate, guards in the update block are evaluated on the store of the component performing the action and on the global store. However, the result is a sequence of attribute assignments followed by an instantiation command (above considered in the collective instatiation). If none of the guards are satisfied, or the performed action is not listed, the global store is not changed and no new collective is instantiated. In both cases, the collective generating the transition remains in operation. This function is particularly useful for modelling the arrival of new agents into a system.

Example 17. In our scenario block update is used to model the arrival of new users and the exit of existing ones. It is defined as follows:

```
update {
    [true]  arrival*: users := global.users+1 , new User(
        sender.loc , U[0:ZONES−1] );
    [true]  ret: users := global.users−1;
}
```

When action arrival* is performed a component associated with a new user is created in the same location as the sender (see Example 3). The destination of the new user is probabilistically selected. Indeed, above we use U[0:ZONES-1] to indicate the uniform probability over the integer values between 0 and ZONES-1 (included). When a bike is returned, the user exits from the system (process kill is enabled) and the global attribute users is updated accordingly.

Measure Definitions. To extract observations from a model, a CARMA specification also contains a set of *measures*. Each measure is defined as:

$$measure \ m\_name[\ var_1=range_1, \ \ldots, \ var_n=range_n \ ] \ = \ expr;$$

Expression *expr* can be used to count, by using expressions of the form by using expressions of the form #{ Π | *expr* } already described above, or to compute statistics about attribute values of components operating in the system: min{ *expr* | *guard* }, max{ *expr* | *guard* } and avg{ *expr* | *guard* }. These expressions are used to compute the minimum/maximum/average value of expression *expr* evaluated in the store of all the components satisfying boolean expression *guard*, respectively.

Example 18. In our scenario, we are interested in measuring the number of available bikes in a zone. For this reason, the following measures are used:

```
measure AverageBikes[l:=0:4] =
    avg{ my.available | my.loc == l };
measure MinBikes[l:=0:4] =
    min{ my.available | my.loc == l };
measure MaxBikes[l:=0:4] =
    max{ my.available | my.loc == l };
```

4.2 CARMA Eclipse Plug-In

The CARMA specification language is implemented as an Eclipse plug-in using the Xtext framework. It can be downloaded using the standard procedure in Eclipse by pointing to the update site at http://quanticol.sourceforge. net/updates/[6]. After the installation, the CARMA editor will open any file in the workspace with the `carma` extension.

Given a CARMA specification, the CARMA Eclipse Plug-In automatically generates the Java classes providing the machinery to simulate the model. This generation procedure can be specialised to enable the use of different kind of simulators. Currently, a simple ad-hoc simulator, is used. The simulator provides generic classes for representing simulated systems (named here *models*). To perform the simulation each *model* provides a collection of *activities* each of which has its own *execution rate*. The simulation environment applies a standard *kinetic Monte-Carlo* algorithm to select the next activity to be executed and to compute the execution time. The execution of an *activity* triggers an update in the simulation model and the simulation process continues until a given simulation time is reached. In the classes generated from a CARMA specification, these activities correspond to the *actions* that can be executed by processes located in the system components. Each of these activities in fact mimics the execution of a transition of the CARMA operational semantics. Specific *measure functions* can be passed to the simulation environment to collect simulation data at given intervals. To perform statistical analysis of collected data the *Statistics package* of *Apache Commons Math Library* is used[7].

To access the simulation features, a user can select the menu *Carma→Simulation*. When this menu is selected, a dialogue box pops up to choose the simulation parameters (see Fig. 2). This dialogue box is automatically populated with appropriate values from the model. When the selection of the simulation parameters is completed, the simulation is started. The results are reported within the *Experiment Results View* (see Fig. 3). Two possible representation are available. The former, on the left side of Fig. 3, provides a graphical representation of collected data; the latter, on the right side of Fig. 3, shows average and standard deviation of the collected values, which correspond to the *measures* selected during the simulation set-up, are reported in a tabular form.

---

[6] Detailed installation instructions can be found at http://quanticol.sourceforge.net.
[7] http://commons.apache.org.

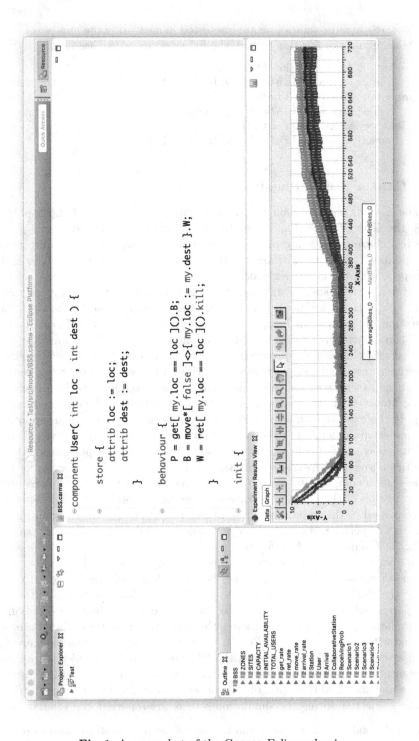

Fig. 1. A screenshot of the CARMA Eclipse plug-in.

Fig. 2. CARMA Eclipse Plug-In: Simulation Wizard.

These values can then be exported in CSV format and used to build suitable plots in the preferred application.

5   Carma Tools in Action

In this section we present the *Bike Sharing System* in its entirety and demonstrate the quantitative analysis which can be undertaken on a CARMA model. One of the main advantages of the fact that we structure a CARMA system specification in two parts – a collective and an environment – is that we can evaluate the same collective in different enclosing environments.

We now consider a scenario with 5 zones and instantiate the environment of the *Bike Sharing Systems* with respect to two different specifications for the environment:

Scenario 1: Users always arrive in the system at the same rate;
Scenario 2: User arrival rate is higher at the beginning (modelling the fact that bikes are mainly used in the morning) and then decreases.

The first scenario is the one presented in Sect. 4 and reported below for completeness:

```
system Scenario1 {
```

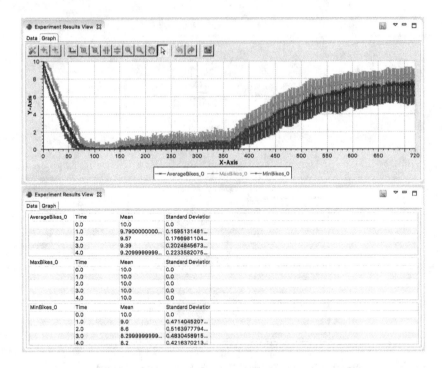

Fig. 3. CARMA Eclipse Plug-In: Experiment Results View.

```
collective {
  for(i;i<ZONES;1) {
    for(j;j<SITES;1) {
      new Station(i,CAPACITY,INITIAL_AVAILABILITY);
    }
    new Arrival(i);
  }
}
environment {
  store {
    attrib users := 0;
  }
  prob {
    default: 1;
  }
  weight{
    [receiver.loc==sender.loc] get: 1;
    [receiver.loc==sender.loc] ret: 1;
    default: 0;
  }
  rate {
    get: get_rate;
```

```
      ret :  ret_rate ;
      move *:  move_rate ;
      arrival *:  (global . users <TOTAL_USERS? arrival_rate
          :0.0) ;
      default :  1;
   }
   update {
      arrival *:
        users:=global . users+1,
        new  User(sender . loc ,U[0:ZONES−1]) ;
      ret :
        users:=global . users −1;
   }
  }
 }
```

The second scenario can be simply obtained by changing the rate block as
follows:

```
   rate {
      get :  get_rate ;
      ret :  ret_rate ;
      move *:  move_rate ;
      arrival *:
        (global . users <TOTAL_USERS?
          (now<360?4*arrival_rate : arrival_rate /2):0.0) ;
      default :  1;
   }
```

The results of the simulation of the two CARMA models are reported in Fig. 4
where we report max/average/min number of bikes available at zone 0. Due to
the symmetry of the considered model, any other location in the border presents
similar results.

We can notice that, in both the scenarios the use of stations is not well
balanced. Indeed, when the system is not overloaded, there are stations that are
almost empty while others are full. This is due to the fact that stations do not
collaborate and *concur* to attract users. To overcome this problem we change the
behaviour of stations to let them exchange information about their availability.
The new prototype is the following:

```
   component  CollaborativeStation ( int  loc  , int  capacity  ,
      int  available ) {
    store {
      attrib  loc  :=  loc ;
      attrib  available  :=  available ;
      attrib  capacity  :=  capacity ;
      attrib  get_enabled  :=  true;
      attrib  ret_enabled  :=  true;
   }

   behaviour {
```

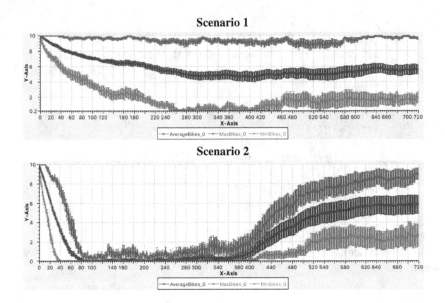

Fig. 4. Bike Sharing System: Simulation Results — 10 *simulation runs*

```
G = [my.available >0 && my.get_enabled]
      get<>{ my.available := my.available -1 }.G;
R = [my.available <my.capacity && my.ret_enabled]
      ret<>{ my.available := my.available+1 }.R;
C =
   [my.get_enabled || my.ret_enabled] spread*< my.
      available >.C
   +
   spread*[true]( x )
         { my.get_enabled := my.available >= x , my.
            ret_enabled := my.available <= x }.C;
}

init {
  G|R|C
}

}
```

CollaborativeStations use action spread* to communicate to components in the same zone the number of bikes locally available. Actions get and ret, used by users to get and return a bike, are enabled only when no other components with an higher number of bikes/parking slots is present in the zone. The simulation of these collectives in the two scenarios is reported in Fig. 5. We can notice that in both the scenarios the average number of available bikes is the same as in Fig. 4. However, differently from in Fig. 4, the use of bikes in the stations is more balanced.

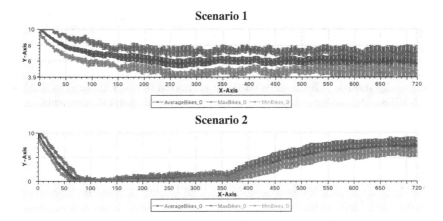

Fig. 5. Bike Sharing System (Collaborative Stations): Simulation Results — 10 *simulation runs*

6 Concluding Remarks

In this paper we have presented CARMA, a novel modelling language which aims to represent collectives of agents working in a specified environment and support the analysis of quantitative aspects of their behaviour such as performance, availability and dependability. CARMA is a stochastic process algebra-based language combining several innovative features such as the separation of behaviour and knowledge, locally synchronous and globally asynchronous communication, attribute-defined interaction and a distinct environment which can be changed independently of the agents. We have demonstrated the use of CARMA on a simple example, showing the ease with which the same system can be studied under different contexts or environments.

Together with the modelling language presented as a stochastic process algebra, we have also described a high level language (named the CARMA Specification Language) that can be used as a front-end to support the design of CARMA models and to support quantitative analyses that, currently, are performed via simulation. To support simulation of CARMA models a prototype simulator has been also developed. This simulator, which has been implemented in Java, can be used to perform stochastic simulation and can be used as the basis for implementing other analysis techniques. These tools are available in an Eclipse plug-in that has been used to specify and verify a simple scenario.

One of the main issues related with CAS is scalability. For this reason is strongly desirable to develop alternative semantics that, abstract on the precise identities of components in a system and when appropriate offer mean-field approximation [6]. We envisage providing CARMA with a fluid semantics and in general the exploitation of scalable specification and analysis techniques [25] to provide a key focus for on-going work. In this direction we refer also here to [21] where the process language ODELINDA has been proposed which provides an *asynchronous*, tuple-based, interaction paradigm for CAS. The language is equipped both with an individual-based Markovian semantics and with a population-based Markovian semantics. The latter forms the basis for a continuous, fluid-flow, semantics definition, in a way similar to [13].

Acknowledgements. This work is partially supported by the EU project QUANTI-COL, 600708. The authors thank Stephen Gilmore for his helpful comments on the chapter.

References

1. Alrahman, Y.A., De Nicola, R., Loreti, M.: On the power of attribute-based communication. CoRR, abs/1602.05635 (2016)
2. Alrahman, Y.A., De Nicola, R., Loreti, M., Tiezzi, F., Vigo, R.: A calculus for attribute-based communication. In: Proceedings of SAC, pp. 1840–1845. ACM (2015)
3. Bergstra, J.A., Ponse, A., Smolka, S.A.: Handbook of Process Algebra. Elsevier, Amsterdam (2001)
4. Bernardo, M., Gorrieri, R.: A tutorial on EMPA: a theory of concurrent processes with nondeterminism, priorities, probabilities and time. Theoret. Comput. Sci. **202**(1–2), 1–54 (1998)
5. Bohnenkamp, H.C., D'Argenio, P.R., Hermanns, H., Katoen, J.-P.: MODEST: a compositional modeling formalism for hard and softly timed systems. IEEE Trans. Software Eng. **32**(10), 812–830 (2006)
6. Bortolussi, L., Gast, N.: Mean-field limits beyond ordinary differential equations. Springer. In: SFM (2016)
7. Bortolussi, L., De Nicola, R., Galpin, V., Gilmore, S., Hillston, J., Latella, D., Loreti, M., Massink, M.: Collective adaptive resource-sharing Markovian agents. In: Proceedings of the Workshop on Quantitative Analysis of Programming Languages, vol. 194, EPTCS, pp. 16–31 (2015)
8. Bortolussi, L., Policriti, A.: Hybrid dynamics of stochastic programs. Theor. Comput. Sci. **411**(20), 2052–2077 (2010)
9. Ciocchetta, F., Hillston, J.: Bio-PEPA: a framework for the modelling and analysis of biological systems. Theoret. Comput. Sci. **410**(33), 3065–3084 (2009)
10. De Maio, P.: Bike-sharing: Its history, impacts, models of provision, and future. J. Public Transp. **12**(4), 41–56 (2009)
11. De Nicola, R., Latella, D., Loreti, M., Massink, M.: A uniform definition of stochastic process calculi. ACM Comput. Surv. **46**(1), 5 (2013)
12. De Nicola, R., Loreti, M., Pugliese, R., Tiezzi, F.: A formal approach to autonomic systems programming: The SCEL language. TAAS **9**(2), 7 (2014)
13. Feng, C., Hillston, J.: PALOMA: a process algebra for located Markovian agents. In: Norman, G., Sanders, W. (eds.) QEST 2014. LNCS, vol. 8657, pp. 265–280. Springer, Heidelberg (2014)
14. Fricker, C., Gast, N.: Incentives and redistribution in bike-sharing systems (2013). Accessed 17 Sept 2013
15. Galpin, V.: Spatial representations and analysis techniques. In: Bernardo, M., De Nicola, R., Hillston, J. (eds.) SFM 2016. LNCS, vol. 9700, pp. 120–155. Springer, Switzerland (2016)

16. Daniel, T.: Gillespie. A general method for numerically simulating the stochastic time evolution of coupled chemical reactions. J. Comput. Phys. **22**(4), 403–434 (1976)
17. Hermanns, H., Herzog, U., Katoen, J.-P.: Process algebra for performance evaluation. Theor. Comput. Sci. **274**(1–2), 43–87 (2002)
18. Hermanns, H., Rettelbach, M.: Syntax, semantics, equivalences and axioms for MTIPP. In: Herzog, U., Rettelbach, M., (eds.), Proceedings of 2nd Process Algebra and Performance Modelling Workshop (1994)
19. Hillston, J.: A compositional approach to performance modelling. Cambridge University Press, New York (1996). ISBN:0-521-57189-8
20. Hillston, J., Loreti, M.: Specification and analysis of open-ended systems with CARMA. In: Weyns, D., et al. (eds.) E4MAS 2014 - 10 years later. LNCS, vol. 9068, pp. 95–116. Springer, Heidelberg (2015). doi:10.1007/978-3-319-23850-0_7
21. Latella, D., Loreti, M., Massink, M.: Investigating fluid-flow semantics of asynchronous tuple-based process languages for collective adaptive systems. In: Holvoet, T., Viroli, M. (eds.) Coordination Models and Languages. LNCS, vol. 9037, pp. 19–34. Springer, Heidelberg (2015)
22. Latella, D., Loreti, M., Massink, M., Senni, V.: Stochastically timed predicate-based communication primitives for autonomic computing. In: Bertrand, N., Bortolussi, L., (eds.) Proceedings Twelfth International Workshop on Quantitative Aspects of Programming Languages and Systems, QApPL 2014, Grenoble, France, 12–13 , vol. 154, EPTCS, pp. 1–16, April 2014
23. Priami, C.: Stochastic π-calculus. Comput. J. **38**(7), 578–589 (1995)
24. Saunier, J., Balbo, F., Pinson, S.: A formal model of communication and context awareness in multiagent systems. J. Logic Lang. Inform. **23**(2), 219–247 (2014)
25. Vandin, A., Tribastone, M.: Quantitative abstractions for collective adaptive systems. Springer. In: SFM (2016)
26. Weyns, D., Holvoet, T.: A formal model for situated multi-agent systems. Fundam. Inform. **63**(2–3), 125–158 (2004)
27. Weyns, D., Omicini, A., Odell, J.: Environment as a first class abstraction in multiagent systems. Auton. Agent. Multi-Agent Syst. **14**(1), 5–30 (2007)

Spatial Representations and Analysis Techniques

Vashti Galpin[⊠]

Laboratory for Foundations of Computer Science,
School of Informatics, University of Edinburgh, Edinburgh, UK
Vashti.Galpin@ed.ac.uk

Abstract. Space plays an important role in the dynamics of collective adaptive systems (CAS). There are choices between representations to be made when we model these systems with space included explicitly, rather than being abstracted away. Since CAS often involve a large number of agents or components, we focus on *scalable* modelling and analysis of these models, which may involve approximation techniques. Discrete and continuous space are considered, for both models of individuals and models of populations. The aim of this tutorial is to provide an overview that supports decisions in modelling systems that involve space.

1 Introduction

Collective adaptive systems (CAS) are systems which consist of a number of components which interact (directly or indirectly) to achieve goals, by collaboration, and in some instances, by competition. These components may be static or mobile, as in the case of a robot swarm. Various smart transport systems provide examples of CAS; for example, bike-sharing schemes and ride sharing. Because movement is fundamental in these systems, space and spatial aspects are important characteristics and influence the behaviour that these systems demonstrate. Therefore, we wish to understand the dynamics of these systems and how these may vary with changes in the implementation of the system, and changes in use of the system. In the bike-sharing example, incentives can be offered to users to influence their behaviour in terms of the station a bike is returned to, or alternatively a system may suddenly show very poor performance when the user base grows beyond a certain size. Alterations to timetables of other public transport such as trains, could also impact the effectiveness of a bike-sharing scheme. Furthermore, roadworks or new lane markings can modify the space that the bike users travel through, affecting performance.

We model these systems to understand their behaviour because it is frequently not possible to experiment with the actual systems, either because of the disruption this will cause, or because the systems have not yet been constructed. In this chapter, we focus here on modelling dynamic systems (which we also refer to as time-based) that involve some notion of space. These aresystems

© Springer International Publishing Switzerland 2016
M. Bernardo et al. (Eds.): SFM 2016, LNCS 9700, pp. 120–155, 2016.
DOI: 10.1007/978-3-319-34096-8_5

where the behaviour of the system is observed as time passes[1]. When trying to understand the behaviour of a collective adaptive system by developing a model of the system, it can be moderately straightforward to programmatically construct an agent-based model where the agents move in a representation of real space. But often, for a realistic number of agents, it is not computationally feasible to simulate this model a sufficient number of times to understand its overall behaviour through the use of descriptive statistics. Additionally, an agent-based model is likely to have a very large states space because it considers individuals separately. The computational costs of many other analysis techniques are often dependent on the number of distinct states that the system can take on, and hence cannot be applied to these individual-based models.

Thus, detailed agent-based models may lead to more precision but at the cost of choices for analysis. Typically in modelling, one wishes to retain the details that the model is designed to answer, and to abstract from everything else. Therefore, carefully chosen abstractions are crucial, and this tutorial provides details about a particular type of abstraction and associated approximation of results, that of population-based modelling, rather than solely modelling individuals. These abstractions contribute to *scalable* analysis. By this, we mean that when modelling large systems with many components, our analysis can be computed in a reasonable time (with reasonable memory requirements), and as the system becomes larger, this analysis remains feasible. Concomitant with the scalability is a requirement that any analysis technique that involves approximation remains within reasonable distance from the true value. Obviously, there will be a system size at which the analysis becomes infeasible. In that case, possible solutions are then to consider whether size can be reduced by working with a more abstract model, or to consider a different approximation technique which is more scalable.

Furthermore, we focus on stochastic models. Stochasticity allows model behaviour to vary, and hence captures the variation we observe in the systems we wish to model. Specifically, we use random length durations drawn from exponential distributions. The exponential distribution is suitable and convenient for modelling because it has a single parameter (which is the inverse of the average duration), it is memoryless (which means that what happens next is only dependent on the current state, as opposed to any previous states, and this negates the need when simulating to keep track of prior states or amount of time elapsed), and other distributions can be approximated by combinations of exponential distributions. In their most basic form, our models are continuous-time Markov chains (CTMCs) and their discrete version, where probabilities are used to determine the next state, discrete-time Markov chains (DTMCs). We also consider extensions and variations of these models, but in general, any stochasticity in our models occurs because of exponentially-distributed durations

---

[1] Another approach to space is to consider it topologically, that is to consider the relationships between points in space. This can be applied to both discrete and continuous space. Details can be found elsewhere in this volume [19] in the context of spatial and spatio-temporal logics.

or probabilistic choices. One extension that we may use in some cases is allowing the exponential rate (and probabilities) to be functional and depend on time or other aspects of the model. This introduces time inhomogeneity into our models, and this is often important to capture variations in behaviour at different times of day, for example. The disadvantage of allowing time inhomogeneity is that it can reduce the number of analysis techniques that are applicable.

This presentation does not consider any languages for specifying models but instead focusses on mathematical representations of systems (which we will refer to as models) to which analysis techniques can be applied. The choice of representation for a model is often influenced by the type of analysis and approximation techniques that are available, and the aim of this tutorial to support such decisions when modelling space. This chapter starts with a discussion of the type of mathematical representations and analysis techniques that can be used if space is not considered explicitly, and then moves onto consider these with the addition of space. Techniques for discrete space are considered in detail in Sect. 3, followed by those for continuous space in Sect. 4. In these two sections, general concepts are introduced for the type of space, followed by a high-level discussion of the basic model and analysis techniques. Details are given of techniques that have relevance to CAS, followed by a brief review of how they have been used in different disciplines. Finally in Sect. 5, techniques that can be applied to both types of space, or to models containing both types of space are considered.

2   Representations for Dynamic Modelling

Before considering the role of space, we introduce a number of dimensions that we consider germane to our modelling, so that we can develop a classification of dynamic modelling techniques relating to the modelling context described in the introduction. Even without considering space, there are already a number of choices that lead to different ways in which to model dynamic systems in a quantified manner. We consider the dimensions and the choices on each dimension. For example, the time dimension considers how time is treated in different types of Markov chains. There are other aspects of time such as non-determinism and causality, but these are not a strong focus of our general modelling approach, and so are not included in the classification.

Time: Time is non-negative, strictly increasing and infinite, and can either be a non-negative real or integer. In some models, a finite end-point may be used to delimit the period of interest.

   discrete: In the context of this tutorial, discrete time is used in those modelling approaches where choices are probabilistic. At each clock tick (which can be associated with an integer if useful for the specific model), the next state is chosen probabilistically from all possible next states. For example, discrete time Markov chains (DTMCs) use this approach [53,70].

continuous: In this case, time is represented by the non-negative real numbers. Actions such as changing state have a duration associated with them. In the case of continuous time Markov chains (CTMCs), stochasticity is introduced by having random durations that are drawn from exponential distributions [70].

State: States can be viewed as capturing a quality or attribute of an individual. An individual is assumed to be in a single state at each point in time[2].

discrete: Usually when the states associated with an individual are discrete, there are a finite number of them. However, in the case of an attribute like *year-of-birth*, there may be a countably infinite number of values.

continuous: A continuous-valued state can be interpreted as measurement of some quantity associated with the individual. An example of this would be *temperature* or *height*.

Aggregation: Individuals can be considered separately, or the focus can be on the number of individuals in each state. This is more relevant to discrete state approaches than continuous state. In the continuous case, aggregation can be described by a function, or discretisation can be applied to obtain frequency data.

none: Behaviour of each individual is considered separately. This is often referred to as *agent-based* or *individual-based*.

state-based: The behaviour of groups of individuals is considered by counting the number of individuals in each state over time (giving a non-negative integer value), or by having a non-negative real-valued approximation to this number. This approach appears under a number of different names in the literature including *population-based, state frequency data, numerical vector form,* and *counting abstraction*. The term *occupancy measure* is used when counts are normalised by the population size.

These dimensions can be expressed in a table, which can then be populated with mathematical modelling techniques from the literature. Figure 1 illustrates this and describes the modelling techniques that fit each possible combination of elements for each dimension. There is the possibility of hybrid approaches for the state and aggregation dimensions and we discuss these briefly in Sect. 5.2.

2.1   Scalable Modelling and Analysis Techniques

As mentioned in the introduction, we focus on Markov chain models. Basic definitions can be found in the appendix. An important aspect of our modelling approach is the application of the mean-field technique where the analysis of a population CTMC or DTMC can be approximated by an analysis using ordinary

---

[2] An individual could have more than one attribute, and then the individual's state is multidimensional with a value for each attribute. In this case, the individual's state is a tuple of values.

TIME	discrete			
AGGR	none (individuals)		state (populations)	
STATE	discrete	continuous	discrete	continuous

DTMC [70]   LMP [73]   population   difference equations,
                                  DTMC [10]   ODEs [10, 67]

TIME	continuous			
AGGR	none (individuals)		state (populations)	
STATE	discrete	continuous	discrete	continuous

CTMC [70]   CTMP [24]   population   population
                                  CTMC [10, 58]   ODEs [10, 58]

Fig. 1. Classification of mathematical models in terms of time, aggregation and state (DTMC: discrete time Markov chain, LMP: labelled Markov process, ODE: ordinary differential equation, CTMC: continuous time Markov chain, CTMP: continuous-time Markov process

differential equations (ODEs) [10, 58]. As the number of states of a Markov chain increases (the "state-space explosion" problem), the analysis of the Markov chain becomes intractable. Modelling a large number of individuals can lead to a very large Markov chain. This can be mitigated by using a population Markov chain where behaviour is considered at a population level rather than at an individual level. The choice of a population Markov chain means we are interested in how many individuals from a population P_A are in each local state A_i, given by N_{A_i}, and the states in the Markov chain have the form $(N_{A_1}, \ldots, N_{A_n})$. However, for large systems this may still not be sufficient to obtain reasonable analysis times, and an approximation using ODEs obtained from the population Markov chain can be used. This gives a system of ODEs for the variables $(X_{A_1}, \ldots, X_{A_n})$. The population Markov chain considers non-negative integer-valued population counts whereas the ODEs take a fluid approach and population quantities are non-negative real values X_{A_i}. Considering the modelling techniques in Fig. 1 for both discrete time and continuous time, the Markov chain obtained by considering many individuals (in the first column) can be transformed into a smaller Markov chain (in the third column) which can then be approximated by ODEs (in the fourth column).

This transformation uses the mean-field approximation technique which comes from physics, where it refers to the approach where movement of an individual particle is considered in the field generated by other particles rather than trying to solve the more complex problem of many particles interacting [68]. In modelling of systems, it has come to mean an approach where it is assumed that when the number of individuals in a stochastic system becomes very large, the population-level behaviour of the system can be expressed as ODEs which provide an "average" behaviour. Results such as those proved by Kurtz [58] demonstrate that under certain conditions, convergence occurs, namely as the number of individuals tends to infinity, the difference between the stochastic trajectories of the subpopulation sizes and the deterministic trajectories of the subpopulation sizes tends to zero. Practically, in many cases, good approximations using the ODE approach over the stochastic approach can be achieved at relatively low numbers of individuals [85] and there are error bounds on the approximations [21]. The mean-field approach is discussed in more detail elsewhere in this volume [9].

Additionally, we will consider moment closure approaches to approximation. For a PCTMC, it is possible to obtain ODEs that describe how the moments (expected values) of variables and products of variables vary over time. Typically, this results in an infinite system of ODEs, because the ODE for each moment is dependent on higher moments. For example, the ODE for $E[X]$ may involve not only $E[X]$ and $E[Y]$ but also $E[X^2]$, $E[Y^2]$ and $E[XY]$. Likewise, the ODE for $E[XY]$ may involve expectations of the product of three variables. Moment closure techniques provide approximations for these higher-order moments through a number of techniques that will be described later in this tutorial, thus providing ODEs that give an approximation for the moments. The mean-field approach described above can be seen as a specific instance of moment closure where second order moments are replaced by the products of expectations ($E[XY]$ is approximated by $E[X]E[Y]$, for example) under certain conditions relating to mass actions and pairwise interactions. This is equivalent to assuming that variances and covariances are zero, and is a reasonable assumption to make if they are likely to be small enough to be safely abstracted from. Typically, in the spatial case, we wish to consider covariances and other higher moments to ensure that spatial variation is included and not abstracted from.

Returning to Fig. 1, Markov processes (in the second column of the figure) do not fit into this work flow (of transforming an individual-based model to a population-based model and then using an ODE approximation) and seem different from the other modelling techniques, as they are characterised by a continuous state space which can also be interpreted as any continuous aspect of a model, including space. We do not consider labelled Markov processes (LMPs) further in this chapter, but we will comment further on continuous-time Markov processes (CTMPs) in Sect. 4.1.

The research surveyed in this chapter involves transformation and analysis techniques. Transformations of models may be necessary for a different analysis to be applied. The counting abstraction as described above is an aggregation technique, and treating population sizes as being real-valued rather than integral, is fluidisation. Another form of aggregation is when multiple locations are considered as a single location. Finally, discretisation happens when some continuous value is transformed to a discrete value, such as transforming real space to discrete space. Hybridisation which can involve fluidisation to make some parts of a discrete model continuous, or discretisation to make parts of a continuous model discrete, is discussed in Sect. 5.2.

2.2   Introducing Space

In this tutorial, **Space** will be considered in two different ways.

continuous: Here, space is represented by real values in the case of one-dimensional space, pairs of real values in the two-dimensional case and triples of real values in the three-dimensional case. It is always (uncountably) infinite but may be bounded in extent. Continuous space used in this way can be seen as an exact representation of actual physical space.

discrete: Approaches that use discrete space assume a number (usually finite) of distinct locations where connectivity between locations is described by an adjacency relation[3]. At each location, there may be multiple individuals, although in some cases, such as cellular automata [49], this may be restricted to a single individual. A location may be an abstraction or aggregation of actual space.

The table in Fig. 2 shows the mathematical models for the different combinations of time, aggregation, state and space. Here, we have chosen to focus on continuous time models; however there are discrete time models of various approaches, for example, some variants of interacting particle systems (IPSs) use probabilities [29]. We now consider each entry of the table in Fig. 2 briefly together with illustrative diagrams.

TIME	continuous			
AGGR	none (individuals)		state (populations)	
STATE	discrete	continuous	discrete	continuous
SPACE				
discrete	CTMC, IPS [29]	TDSHA [12] PDMP [22]	patch population CTMC [17]	patch population ODEs [17]
continuous	molecular dynamics [20] agents	CTMP [24]	spatio-temporal point processes [78]	PDEs [46]

Fig. 2. Classification of mathematical models in terms of time, aggregation, state and space (CTMC: continuous time Markov chain, IPS: interacting particle systems, TDSHA: transition-driven stochastic hybrid automata, PDMP: piecewise deterministic Markov process, ODE: ordinary differential equation, CTMP: continuous-time Markov process, PDE: partial differential equation)

---

[3] For CAS, we are usually interested in the adjacency of different regions of space, and as we will see later, we use graphs to describe this relationship. Another approach is where space has a nested arrangement, as seen in biological modelling. This containment relationship can be represented graphically by trees, but we do not focus on this arrangement of space further.

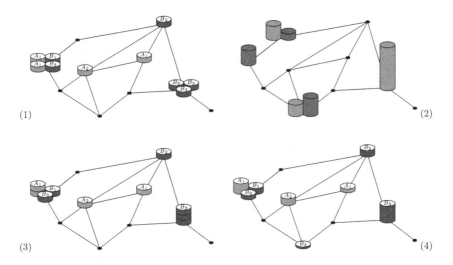

Fig. 3. Discrete space: (1) no aggregation, discrete state; (2) no aggregation, continuous state; (3) aggregation of state, possible aggregation of space, discrete state (4) aggregation of state, possible aggregation of space, continuous state

2.3   Discrete Space Illustrated

The approaches in the discrete-space category consider space to consist of a (usually) finite number of locations that have connections between them. The most straightforward way is to consider these models as graphs with the locations as nodes and the links as edges. Discrete space is illustrated in Figs. 3 and 4, showing the general case of an arbitrary graph, and the case of a more regular graph structure, respectively. Regular space models are those that have a regular pattern of locations [28, 29]. For example, the locations could be laid out in the rectangular grid, or a hexagonal tiling. The locations that represent space can be situated at the nodes of the regular graphs or in the spaces (faces) created by the regular graph as shown in Fig. 4. Regular space will be more formally defined in Sect. 3.

In the diagrams, we assume individuals are from two populations. The first, P_A consists of red and white tokens, and has states A_1 and A_2. The second, P_B consists of blue and white tokens with states B_1, B_2 and B_3. The current state of an individual is indicated on the top of the token. The four diagrams in each figure represent four single points in time and do not show change over time[4].

Figures 3(1) and 4(1) show discrete-space models of individuals with discrete states, hence there is no aggregation into populations. Some models only allow one individual in each location, such as interacting particle systems (IPSs) [29] and cellular automata (CA) [49], but others may allow multiple individuals.

---

[4] For two-dimensional and three-dimensional space, the best visualisation method for change over time is video. For one-dimensional space, a graph with two axes can be used.

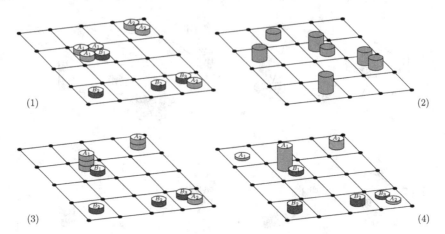

Fig. 4. Regular discrete space: (1) no aggregation, discrete state; (2) no aggregation, continuous state; (3) aggregation of state, possible aggregation of space, discrete state (4) aggregation of state, possible aggregation of space, continuous state

In the case of single individuals at a node, this can be indicated by a flat token as illustrated in Fig. 6.

Models of discrete space without aggregation and with continuous state are shown in Figs. 3(2) and 4(2). The continuous state is indicated by a solid token where the height indicates the value of a single continuous state. This is an inherently continuous value rather than the notion of population size approximation by continuous values described earlier in this section, and could be a measurement such as strength of radio signal or length of battery life. Different colours have been used in the diagram to make it clear that the values are continuous but not a population approximation. In Figs. 3(2) and 4(2), there is an assumption of at most one individual per node and face, respectively, and two values associated with that individual.

Next we consider discrete-state aggregation in the context of discrete space, as illustrated in Figs. 3(3) and 4(3) by the fact that individual tokens are grouped into stacks at nodes in the network, and it is the size of the stack that is relevant rather than the location of each individual. Finally, in the case of continuous state aggregation in discrete space, each region or point is associated with approximations to the discrete population shown in Figs. 3(3) and 4(3). These are illustrated in Figs. 3(4) and 4(4). At each node, for each state in each population, there is a real number that approximates the number of individuals in that state. This is illustrated by a token with a real-valued height for each state in each population. Note that in Fig. 3(4), the lowest node has a non-zero value for blue tokens in state B_3 although there were none in the CTMC model in Fig. 3(3), illustrating that approximation can occur.

2.4   Continuous Space Illustrated

We first consider continuous space with no aggregation and discrete state. This covers approaches where each individual's location and state are modelled separately from those of other individuals. An example of this type of model is where the movement and interaction of each molecule is modelled individually, as in molecular dynamics [7]. Agent-based models take a similar approach. Figure 5(1) illustrates this. The continuous space is indicated by a bounded area and each individual is shown at its own location. These models are typically computationally expensive to simulate.

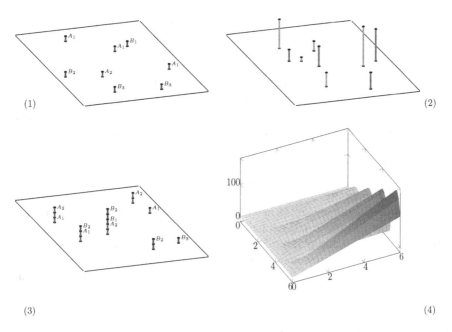

Fig. 5. Continuous space: (1) no aggregation, discrete state; (2) no aggregation, continuous state; (3) aggregation, discrete state; (4) aggregation, continuous state.

Moving on to state that is continuous rather than discrete, leads to continuous-time Markov processes (CTMPs) [24], if we assume some of the continuous dimensions relate to space and the others to state. Since there is no state-based aggregation, this approach models individuals rather than populations. The continuous space is indicated by a bounded area and each individual is shown at its own location. The continuous state is indicated by the varying heights of the tokens, and in Fig. 5(2), it is assumed that there is only one (non-spatial) measurement per individual, although multiple different measurements are possible.

For the case of aggregation with discrete state, each point in space can be filled by zero, one or more individuals [78]. Hence for each point in space, it is possible to aggregate the number of individuals in each state. Figure 5(3) shows a fairly sparse number of individuals but much denser arrangements are also possible. Finally, when aggregation is continuous in nature, then at each point in space, there is a real value describing an approximation to the number of individuals at that point [20,71]. In the case of two-dimensional space, the population of each state can be represented in three-dimensions by surfaces as defined by partial differential equations (PDEs). Figure 5(4) illustrates a surface describing the number of individuals at each point for state A_1. In contrast to Fig. 5(3), this figure illustrates a very dense situation.

2.5   Summary

As is the case with techniques that do not include space, presented in Fig. 1, the techniques using continuous state without aggregation (the second column of models in Fig. 2) seem distinctly different to the other approaches. The techniques that can be applied to models without space described earlier in this chapter (approximation by ODEs of a population DTMC or CTMC) can be applied to discrete space since the Markov chain involved is a population Markov chain that takes location into account. Furthermore, taking the hydrodynamic limit of IPS (which are discrete space models without aggregation) models provides PDEs [23].

In all of the models described in the previous section, there may be interaction between individuals (even if this interaction is expressed at the population level). Opportunity for interaction is often related to colocation or proximity (which requires some notion of neighbourhood or distance). Many models capture movement of individuals explicitly and then use colocation or proximity to determine the possibility of interaction, although there are some models that only use proximity without movement such as IPSs and CA. We discuss movement in more detail when we consider the analysis techniques for the two different kinds of space.

3   Discrete-Space Modelling Techniques

We now consider discrete space in more detail and formality, so we introduce both notation and concepts relevant to discrete space. We will focus here on the continuous-time models, with pointers to the discrete-time models where appropriate.

In the most general case, we assume a finite (or at most countably infinite) set of points or locations L with some naming convention [41]. Most generally, the set of locations L can be taken as the vertices of an undirected graph, and the connections between locations (the adjacency relation) can be defined as edges in that graph. The edges of the graph E_L are drawn from the subsets of size two of the location set $\mathscr{P}_2(L)$, so $E_L \subseteq \mathscr{P}_2(L)$. Each edge has the form $\{l_1, l_2\}$, and edges of the form $\{l, l\}$ are permitted. We have chosen to use an undirected graph which is to be understood as allowing movement or interaction in at least one direction between the two locations. The absence of an edge means that movement and interaction can never take place, in either direction. Parameters associated with an edge express (possibly in a time-varying manner) the propensity for movement or interaction in either direction. If a parameter is zero at a particular time for a particular direction, it means that no active interaction or movement can take place at that time point. Hence, the graph of locations provides a skeleton for describing what movement or interaction is possible.

Locations in discrete space models can have two main sources, either they are essentially locations on a map, such as bike-stations or bus stops, or alternatively each location represents a region on a two-dimensional map, and space is aggregated. These are called patch-based models. The edges of the graph can be determined by various factors. Adjacency of regions is an obvious choice, but there may be other context-specific elements, for example, presence of connections between regions such as railway lines or similar. A topic whose exploration is beyond the scope of this chapter is that of how to divide a map in regions. A simple approach is to base it on a tiling of the plane using triangles, quadrilaterals or hexagons. More complex approaches involve taking local information into account and creating irregular patches. Computer networks can be seen as being located in discrete space, either physically or logically.

An issue for discrete space (and continuous space) is determining what happens at the boundaries of the space. One approach is to ensure there are none by working with infinite structures such as infinite graphs, or alternatively boundaryless structures such as tori. A rectangular region can be transformed into a torus by joining the top and bottom edges (to form a cylinder) and then joining the left and right ends (by curving the tube). Other approaches work with boundaries and either choose to keep individuals inside the region (by reflection or other techniques) or to treat boundary locations as sources and/or sinks.

The discrete space approach as described above is very general as it allows arbitrary graphs over locations, as well as heterogeneity for parameters. In the literature there are modelling techniques that are defined for specific graph subclasses and we will discuss some of these below.

3.1   Spatial Parameters and Regularity

A modelling technique with discrete space will have parameters that depend on locations, or links between locations. We can consider two groups of parameters; those that are associated with locations, namely with vertices of the graph and those that are associated with interaction or movement, namely the edges of the graph, and we define two functions to describe these parameter sets as follows

- $\lambda(l)$ for $l \in L$, and
- $\eta(l_1, l_2)$ and $\eta(l_2, l_1)$ for $\{l_1, l_2\} \in E_L$.

The range of these functions will remain abstract for the purposes of this discussion. Note that although the edges of the graph are not directed, the function η is sensitive to direction. Movement is obviously directional but interaction can be undirected when considering an abstract view of effect or communication. Alternatively, it can be directed if one party is the sender and the other the recipient. Our choice of an undirected graph allows these details to be expressed in parameters. In the rest of this chapter, the term *transfer* will be used to refer to both movement and interaction.

We present the following definitions, leading to a definition of spatial homogeneity (a term which is used in the literature but not formally defined), by considering the location-related parameters. A spatial model is

- *location homogeneous* if $\lambda(l_i) = \lambda(l_j)$ for all locations $l_i, l_j \in L$.
- *transfer homogeneous* if $\eta(l_i, l_j) = \eta(l_j, l_i) = \eta(l_{i'}, l_{j'}) = \eta(l_{j'}, l_{i'})$ for all edges $\{l_i, l_j\}, \{l_{i'}, l_{j'}\} \in E_L$.
- *(spatially) parameter homogeneous* if it is both location and transfer homogeneous.
- *spatially homogeneous* if it is parameter homogeneous, and its location graph is complete[5]. Regular connections between locations which do not give total connectivity are discussed below.

Models with spatial homogeneity have a symmetry that can allow for analyses that are not possible for more complex models. Examples are the bike-sharing system considered in [39] where the metrics of interest are the number of empty and full bike stations.

Spatial inhomogeneity/heterogeneity can be introduced in two ways: the first involves connectivity where equal accessibility is no longer assumed, and the second where all locations are still accessible from all locations, but parameters vary between locations. Note that if a parameter $\rho_{i,j} \in \eta(l_i, l_j)$ is constant for all i and j but other parameters vary by locations, then the model is spatially inhomogeneous.

Regular discrete space covers those discrete space models where the organisation of space is regular (rather than an arbitrary graph where each vertex may have an arbitrary number of edges). By contrast to spatial homogeneity, regularity of space is more difficult to define formally when starting from a graph (and we do not give details here), although it is very straightforward to identify visually [72]. Terms such as lattice, grid or mesh are frequently used to describe a graph based on a square or rectangular tiling of the plane. The other two regular tiling possibilities are equilateral triangles and regular hexagons. Alternatively, a graph with regular structure can be constructed by identifying points in $\mathbb{Z} \times \mathbb{Z}$ or $\mathbb{R} \times \mathbb{R}$, and adding links. We will not attempt that level of generality for discrete space beyond saying that regular space should have the property that at each location (except possibly at boundary locations) there is a uniform way to determine the immediate neighbours[6]. One-dimensional regular space can be represented simply as an undirected path. We do not tackle a formal definition of three-dimensional regular space.

---

[5] A complete undirected graph has an edge $\{l, l'\}$ between each pair of vertices l and l'.
[6] We exclude from this definition n-hop neighbours in an arbitrary graph (see definition of n-hop in the next subsection).

3.2   Neighbours and Neighbourhoods

In an undirected graph of locations representing discrete space, the links between locations are used to define neighbours. Given a location l, its *immediate neighbours* are those vertices l' such that $\{l, l'\}$ is an edge in the graph. Its *n-hop neighbours* are those that can be reached through a path in the location graph of at most n steps (but usually excluding the location l itself). In the case of a regular grid graph, the immediate neighbours (west, north, east and south) are referred to as the Von Neumann neighbourhood. The larger neighbourhood that includes the northwest, northeast, southeast and southwest points as well as the immediate neighbourhood is known as the Moore neighbourhood. Both types of neighbourhoods can be extended to n-hop neighbours and also applied to hexagonal and triangular regular location graphs, with obvious adaptations.

This is a purely spatial approach to defining neighbourhoods. However, in some cases, it can be the entity or process itself that defines its neighbourhood depending on its capabilities. Other approaches use a (perception) function that determines the *de facto* neighbours of an individual by specifying the other individuals with which it can interact.

3.3   Techniques for Individual Discrete-Space Models

We now consider the different modelling techniques that have been applied to discrete space starting with those that do not involve aggregation of state. When there is no aggregation and state is discrete, the focus is on individuals and an example is an agent-based system over discrete space. Each individual has some state and is located at exactly one location. There may be a restriction to one individual per location. To describe these models in their most general form, we assume that each individual I (where I is a unique name for the individual) has associated time-based information:

- $\mathsf{loc}(I, t) \in L$ which is its location at time t
- $\mathsf{state}(I, t) \in \{A_1, \ldots, A_n\}$ which is its state[7] at time t

Additionally there are rules that describe how an individual can change location or change state. Since this is a continuous time model, these rules specify rates to describe how long it takes on average for the changes to occur. Each rate defines an exponential distribution, and may be constant or the rates may be functions that depend on the presence of others at that location, the characteristics of the location or the current time (thus introducing time-inhomogeneity). The behaviour of the agents in this modelling technique is thus described as they individually change state and/or location. Assuming a fixed population size, we can model this system as a CTMC, where each state in the CTMC is a tuple consisting of information about each individual in the system. If we assume N individuals then a state has the following form

$$\big((\mathsf{loc}(I_1, t), \mathsf{state}(I_1, t)), \ldots, (\mathsf{loc}(I_N, t), \mathsf{state}(I_N, t))\big)$$

---

[7] If the population P_A has multiple attributes $A^{[1]}, \ldots, A^{[p]}$, then $\mathsf{state}(I, t) = (A_{i_1}^{[1]}, \ldots, A_{i_p}^{[p]})$ represents a tuple of attributes.

There are $(L \times n)^N$ states in this Markov chain if there are L locations, n states and every combination of state and location is possible for all individuals.

Simulation suits this type of model, and techniques for simulating systems where behaviour is based on functional exponential rates are well understood [43]. They can also be analysed using standard numerical CTMC techniques for steady state and transient behaviour. However, a large number of individuals can make this computationally infeasible.

Next, we consider discrete space modelling techniques without aggregation but where the state is continuous. Therefore, instead of having a rule describing how (discrete) state change can happen, there needs to be a rule describing how the continuous state changes over time. A good candidate for this type of rule is an ODE. These techniques are hybrid in that they exhibit both continuous behaviour with respect to state and stochastic behaviour with respect to space. Transition-driven stochastic hybrid automata (TDSHAs) [12] and piecewise deterministic Markov chains (PDMPs) [22] are suitable modelling techniques. Both of these also introduce the possibility of instantaneous behaviour.

3.4   Pair Approximation: Spatial Moment Closure Based on Structure

The technique called pair approximation, which we will refer to as structure-based moment closure, provides ODEs which describe the changes over time in the probabilities of certain pairs (adjacent locations) in the model [66, 91]. From these ODEs, the proportion of locations in a particular state can be determined.

It is applied to a specific class of discrete-space models of individuals with discrete state, namely graph-transformation models. In these models, each node either represents a single individual or a single position in space or location which can take on exactly one of a small number of states. Whether the node itself is modelled or an individual at the node is modelled, the node is the agent in the model. Hence there is no distinction between location and agent, unlike in population discrete-space models.

The dynamics of the model are defined in terms of graph transformation rules with associated exponential rates (when using continuous time). A graph transformation rule describes how a small subgraph or pattern can be transformed in another pattern. There are two possible types of transformation: those that change the state of the nodes in the graph and those that modify the graph

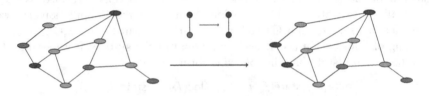

Fig. 6. A graph-transformation rule applied to an individual discrete-space model with discrete state

by removing or adding nodes or edges. Here, we investigate a static model of space and so we only consider the first type of transformation in this chapter. An example of such a rule is given in Fig. 6. The lack of distinction between location and agent is indicated by the fact that the disks are flat rather than raised tokens, as mentioned earlier.

As an example, consider a graph-based SIR model[8] where each node is an individual who can be in one of a number of states (susceptible, infected, recovered, hence the abbreviation SIR) and the edges of the graph link individuals that can affect each other. The graph-transformation rules include a linked pair consisting of one susceptible and one infected being modified to a linked pair consisting of two infected nodes (as illustrated in Fig. 6), and a infected node being modified to a recovered node. In ecological modelling, nodes may represent a patch of ground which can be in a number of states including filled by a plant of a specific species, empty but suitable for growth or infertile. Often the nodes are laid out in a grid pattern, and the transformation rules describe how plants spread, and how nodes become fertile or infertile.

The stochastic graph transformation model is used to obtain ODEs which describe the change in how often each pattern appears over time. By patterns, we mean small graphs consisting of nodes with states of interest. The reason this technique is called pair approximation is because one can consider the patterns of interest to be a graph consisting of two linked vertices, with the two vertices having specific states, and one wants to know how many times this pattern appears in the graph of the model. Much of the existing research assumes a finite grid/lattice [66,91], but one can also consider the more general case of arbitrary graphs rather than regular ones.

Deriving the ODE for a particular pattern may involve understanding how often a different pattern occurs (because the one pattern is transformed into the other by the stochastic process). Typically, to understand the various pair patterns that can occur, the number of certain triplet patterns must be known, and at the next step of obtaining ODEs for triplet patterns, the number of specific quadruplets must be known. This generation of ODEs is similar to that of the moment ODEs described in Sect. 2.1 and leads to an infinite system of ODEs. This system of ODEs can be closed using certain closure techniques (which will be discussed in more detail in Sect. 3.6) and thereby give an approximation to the true value. Structure-based moment closure has also been considered as a multi-scale technique [31]. In this case, different sizes of neighbourhood are used for different types of interaction.

3.5   Techniques for Population Discrete-Space Models

We now move on to consider discrete space when aggregation of state occurs resulting in populations, whose sizes are either integral or real-valued. It is

---

[8] This is different to the population SIR model that appears in another chapter in this volume [9] because there is at most one individual at each node in the graph, and that individual has an associated state, rather than subpopulations in each state.

assumed that we have many individuals to whom the same set of rules apply with the same parameters, and we choose to view them as a population and to reason about them as a population. These models are population CTMCs where subpopulations in different locations are viewed as separate subpopulations. These are also called patch-based models and there are various examples in the literature [17,93].

We consider a population P_A. At each point in time, each individual in P_A is in exactly one of its local *states* $A_1, \ldots, A_n$. Let $N_{A_i}(t)$ refer to the number of individuals in population P_A that are in state A_i at time t. These are called *subpopulations*. The total number of individuals in the population at time t can be expressed as $N_A(t) = \sum_{i=1}^{n} N_{A_i}(t)$. Furthermore, if no births or deaths are assumed, and an individual must be in one of the available states[9], then $N_A(t_1) = N_A(t_2)$ for all times t_1 and t_2 and the size of P_A is a constant N_A. We use $X_{A_i}(t) \in \mathbb{R}_{\geq 0}$ to represent a non-negative real-valued description of the population P_A which is an approximation to $N_{A_i}(t)$.

If we assume that we have a fixed number of locations, $l_1, \ldots, l_L$, we can now obtain the counts of subpopulations at each location. So for P_A, we have a value $N_{A_i}^{(k)}$ which is the number of individuals at location k in state i. Additionally

$$N_{A_i} = \sum_{k=1}^{L} N_{A_i}^{(k)} \quad \text{and} \quad N_A^{(k)} = \sum_{i=1}^{n} N_{A_i}^{(k)} \quad \text{and} \quad N_A = \sum_{i=1}^{n} N_{A_i} = \sum_{k=1}^{L} N_A^{(k)}$$

We can create a continuous time Markov chain smaller than that of the previous section consisting of at most $(N_A + 1)^{L \times n}$ states where each state has the form

$$\left(N_{A_1}^{(1)}, \ldots, N_{A_n}^{(1)}, \ldots, N_{A_1}^{(k)}, \ldots, N_{A_n}^{(k)}, \ldots, N_{A_1}^{(L)}, \ldots, N_{A_n}^{(L)} \right)$$

This provides a discrete aggregated representation of individuals in space where for each location, we know how many individuals are in each state without knowing exactly which individual at that location is in which state. An example of behaviour in such a model is illustrated in Fig. 7 where an individual in state B_3 moves from one location to another and the population sizes at those locations change as a result of this movement.

In the case of continuous state aggregation, the notation $X_{A_i}^{(k)}$ is used for the real value that describes the quantity of individuals in state i at location k. Since this can be a non-integer value, it is an approximation to the actual count $N_{A_i}^{(k)}$. Since the subpopulation sizes are treated as continuous values, a standard modelling technique is to express the change in this quantity in terms of an ODE.

$$\frac{dX_{A_i}^{(k)}}{dt} = F_{i,k}\left(\left(X_{A_1}^{(1)}, \ldots, X_{A_n}^{(1)}, \ldots, X_{A_1}^{(k)}, \ldots, X_{A_n}^{(k)}, \ldots, X_{A_1}^{(L)}, \ldots, X_{A_n}^{(L)} \right), t \right)$$

---

[9] In some models, births and deaths can be included for a fixed size population by introducing a "dead" state. However, this requires that there is a finite maximum population size.

Fig. 7. Behaviour in a population discrete-space model with discrete state

This is a population ODE because it tracks the changes in subpopulation sizes over time. There are $L \times n$ variables in total; one for each combination of state and location. The inclusion of t as an argument to $F_{i,j}$ indicates that it can be a time-inhomogeneous ODE. This ODE often has the following form

$$\frac{dX_{A_i}^{(k)}}{dt} = f_{i,k}\left(X_{A_1}^{(k)},\ldots,X_{A_n}^{(k)}\right) +$$

$$\sum_{j=1, j\neq k}^{L}\left(g_{i,k,j}\left(X_{A_1}^{(k)},\ldots,X_{A_n}^{(k)},X_{A_1}^{(j)},\ldots,X_{A_n}^{(j)}\right) - h_{i,k,j}\left(X_{A_1}^{(k)},\ldots,X_{A_n}^{(k)},X_{A_1}^{(j)},\ldots,X_{A_n}^{(j)}\right)\right)$$

where $f_{i,k}$ captures the local behaviour which only depends on the subpopulation sizes locally, $g_{i,k,j}$ describes the inflow of population from location j to location k, $h_{i,k,j}$ describes the outflow of population from location k to location j, and these flows depend only on the subpopulation sizes in location k and location j. This is a time-homogeneous ODE since change over time is only dependent on subpopulation sizes (that are dependent on time) rather than on time directly. For both the general and regular space cases and assuming only movement/interaction between 1-hop neighbours, then a term $X_{A_i}^{(j)}$ should only appear in the right hand side of the ODE if $\{l_k, l_j\}$ is an edge in the location graph.

In both models, discrete population and continuous population, rates are functional and there is no specific requirement for them to be continuous, although discontinuities in rate functions may affect the applicability of certain analysis techniques.

Since PCTMCs with locations are PCTMCs then the usual linear algebra numerical techniques that can be applied to PCTMCs to understand the probability of being in a specific state at steady state, or at a particular time during transient behaviour, can be applied. The computational feasibility is limited by the size of the state space.

Simulation is also applicable to simulate individual trajectories of behaviour using an algorithm such as that proposed by Gillespie [44]. A basic assumption is that the model has the property of being well-mixed, that is the entities in the model are evenly distributed throughout space and hence there is no spatial heterogeneity. If sufficient trajectories are simulated, statistical measures can be calculated across all trajectories. In the case of PCTMCs with locations, the assumption of well-mixedness must be made for each location.

Finally, the techniques based on Kurtz's result [58] that express the average behaviour of a PCTMC as ODEs also apply to the fluidisation of a PCTMC with locations. The assumption of well-mixedness also applies, as with Gillespie simulation. Although the ODEs provide an approximation to the true values, this is achieved much faster as it is easier to calculate the trajectory of a set of coupled ODEs than it is to do multiple simulations for statistical analysis. Techniques such as exact fluid lumpability and related approximation techniques [87,88] identify when it is possible to apply an aggregation when dealing with ODEs and these techniques are discussed further elsewhere in this volume [90].

We can also consider homogeneity of parameters. In the case of spatial homogeneity, the fact that parameters are identical may make the model amenable to an analytic approach, rather than requiring simulation [39]. However, variations in parameters and rates do not affect the speed of analysis, although it may make the description of the PCTMC more complex. This is because these analyses consider each possible transition (or term in the ODEs) individually and have no way to speed up analysis by considering transitions with the same rate (or identical terms in the ODEs) together (either as a group or to reduce calculation).

Another issue to consider that relates to spatial heterogeneity is that of dynamic space where nodes can leave and join a network and links can be added or removed. Although we do consider that time-homogeneity may be a feature of our PCTMCs and associated ODEs because rates are dependent on time, we do not consider dynamic location graphs here, because of the complexity introduced by this additional change in behaviour over time.

3.6   Aggregate Moment Closure: Spatial Moment Closure Based on Averages

We now consider existing techniques from the literature referred to as *spatial moment closure* that can abstract from the details of space but still provide a spatially based approach. We will use the term *aggregate moment closure* for the techniques that are applicable to population discrete-space models because it is more descriptive. Aggregate moment closure requires fluidisation of the population model, derivation of moment ODEs, and application of an approximation technique to close the moment ODEs.

In this approach, moment ODEs (see the appendix for a definition) are obtained for averages over all locations (or values for a specific attribute) for various subpopulations. When applied to spatial models, it is a spatial abstraction technique because information about what happens in individual locations is lost. The basic approach is to obtain an ODE for each subpopulation for the ensemble[10] of the average over all locations for that subpopulation. This will then (in most cases) be expressed in terms of the expectation of the product of two variables (a higher order moment). The ODE for this can then be derived and this again is likely to contain even higher order moments. In most cases, the

---

[10] The mean (at time t) over all stochastic realisations (at time t).

system of ODEs is not closed (or it is not reasonable to determine whether it is closed), and it can be closed by approximating higher order moments after a certain level. Earlier it was mentioned that the mean-field approximation (in the sense of Kurtz) is given by the first moment ODEs with approximations for variances and covariances based on an assumption that these were zero or negligible (see also [9] in this volume). Because the covariance captures spatial variation, we must have ODEs for at least second moments but third and higher moments can be approximated. There are four ways to approach this approximation.

- Assume that the higher order moments above this level provide negligible contributions and ignore them by approximating them with zero. A related approach is to assume that higher order cumulants are zero [65].
- Use the technique of stochastic linearisation which approximates the expectations of products with the product of expectations for higher order moments above this level. It is not sufficient to express second order terms as the product of first order terms as mentioned above, hence this technique can only be applied to third and higher order moments [61]. The modified mean-field approach from ecology takes a similar approach by approximating higher moments with powers of first order moments [74].
- Assume that the data has a particular distribution and use that distribution to determine the values of the higher order moments above this level. The log normal distribution is frequently used because of its positive support which makes it suitable for population modelling [61,62].
- Apply a Taylor expansion of moments, as used in scale transition theory [18] which formalises how local dynamics relate to global dynamics, particularly in the case of nonlinearity.

Most applications of this technique assume a complete graph, or alternatively when neighbourhood is used in an incomplete graph, approximate the results with those obtained from a complete graph [62].

Another approach to moment closure is language-based where information from the model specification language is used to determine which moments are likely to be negligible [36]. A neighbourhood relation is derived from the (language-based) model to determine when it is appropriate to approximate the expectation of a product with the product of expectations. This relation could also use spatial information to determine approximation.

3.7 Multi-scale Techniques Based on Differences in Rates

As mentioned previously, rates can vary, and it may be possible to exploit this variation in the analysis techniques. There are well-known techniques that use differences in interaction rates between entities, such as the Quasi-Steady-State Assumption (QSSA) which assumes an equilibrium for the parts of the system that have fast interaction rates and then derives expressions for the slower parts of the system [45,79]. This can be done both within a stochastic approach and a deterministic approach using ODEs. Another technique is timescale decomposition applied to CMTCs which have the characteristic that its states can be partitioned into groups such that transitions between group members are fast, and transitions between groups are slow. This permits an approximation technique that allows for the CTMC represented by each group of states to be solved separately and then combined into a solution for the whole CTMC [80].

In ecological modelling, spatial aggregation methods consider the combination of different time scales that are location-based [1]. Starting with an assumption that interactions that occur at a location are slow and movement between locations is fast, the usual ODEs for a population model can be derived, consisting of terms for migration and terms for local interaction. It is assumed that the terms for migration are multiplied by the inverse of the scale parameter, a value much smaller than 1. This expresses the difference between the fast migration and slow local interaction. Through a change of variables from subpopulation size at a location to a pair consisting of density at a location and total subpopulation over all locations, with a related change in the time variable that divides time by the scale parameter, a slow-fast system can be obtained to which either the quasi-steady-state assumption or Fenichel's theorem [37,92] can be applied to obtain a reduced system. This technique can perform much better than the spatial moment technique when there is substantial demographic variation across patches but it does require differences in rates.

In other models outside of ecology, particularly those involving computer systems, it is likely to be the case that the pattern will be the opposite as movement between locations is typically physical, whereas interaction within locations may be computer-based and much faster than physical movement and then techniques based on QSSA is more appropriate.

3.8   Applications of Discrete Space Models

In this section, some applications of the discrete space models that have been presented are now discussed briefly. For a detailed survey of the applications of discrete space models, the reader is referred to [41].

Ecology: Space plays a crucial role in many ecological models and ecologists are interested in global qualities of the whole space such as whether species persist or can co-exist, as well as dynamic patterns such as stationarity, oscillatory behaviour, chaos or multistability [68]. Berec [5] provides a classification of spatial models where he considers the time, space and population as different dimensions. Reaction-dispersal networks (also called metapopulation models) are continuous-time, discrete-space, continuous-population models that describe change over time by a system of ODEs over species in locations. They are the same as ODE patch models in our terminology. Coupled-map lattices are a discrete-time model defined systems of difference equations [51] and regular discrete space, and allow continuous population sizes. Morozov and Poggiale [68] highlight that the term "mean-field" can be used in the ecology literature to both describe the non-spatial Kurtz-based approximation technique as well spatial approaches.

Biology: Bittig and Uhrmacher [7] identify five distinct methods for spatial modelling in cell biology that offer different granularities in their approximation of physical reality. Two of these are continuous space approaches and are discussed in Sect. 4.5. The discrete-space models are those that use compartments as a nested arrangement of space, discrete-space lattice approaches with a single molecule at each face of the lattice, and discrete-space lattice approaches where multiple molecules are permitted at each face. For an overview of techniques to model diffusion, both stochastically and continuously, see [33]. Patch models are also used to model biochemical reaction systems [61]. Pattern formation is also important in biology and Turing's paper gave an initial insight into this process [89]. Pattern formation is considered in [19] in this volume.

Epidemiology: Riley [75] identifies four distinct approaches to disease spread modelling that considers different levels of interaction: patch-based, distance, multigroup and individual. Patch-based or *metapopulation* models[11] are used extensively in modelling of epidemics [27]. These models often focus on the calculation of the basic reproduction number, R_0, which determines whether a disease will die out or spread to the whole population. Individual discrete space models have also been used for disease modelling [59], as illustrated in Sect. 3.4.

Networking: Computer networks, in particular ad hoc networks and mobile networks, often require spatial modelling for evaluation. For example, computer and mobile phone virus spread modelling involves spatial aspects and much of this research draws on epidemiological approaches [48,55]. Routing protocols may have spatial aspects that can be discrete or continuous [95]. Patch models have been used to model information transmission between mobile nodes [17,35,94].

Forest fires: Propagation of forest fires is investigated using Multi-class Multi-type Markovian Agent Model (M²MAM) [16]. The approach models individual agents in discrete space and from this, a patch ODE model is derived. Forest fires have also been modelled using stochastic cellular automata in a climate model [60].

Robotics: A robotics case study consists of a swarm of robots that have to collectively identify a shortest path [63]. The division of a path into separate sections which are considered as discrete locations provides a way to approximate the traversal time by real robots and the convergence on the shortest path.

Emergency Egress: The modelling of evacuation from a multi-story building [64] involves a multi-story building with building elements such as rooms, corridors and stairwells, doors and exits. To model the movement of people and the time to evacuate the building, a discrete-space model using patches was developed.

---

[11] The basic epidemiological SIR model is called the compartment model [13] and this consists of a single population with no spatial aspects. It should not be confused with the compartment models in biology which are patch-based models.

Crowd Behaviour: Spontaneous drinking parties are a common phenomenon in cities in the south of Spain [76]. A model shows that the introduction of small variations that break symmetry, both in space and in the degree of connectivity between locations and in the behaviour of the individuals can lead to new behaviour [11]. This example is considered elsewhere in this volume [90].

Bike Sharing: Bike sharing systems have been modelled with homogeneous discrete space using a population CTMC approach with an associated mean-field model [39]. When space is not homogeneous, a clustering approach has been used to group similar locations together [40]. This example is also considered in this volume [90].

A number of the above examples are CAS. Other CAS examples where discrete-space techniques are applicable include smart transport and smart grids. The next section considers modelling with continuous space.

4   Continuous-Space Modelling Techniques

Continuous space is more straightforward to define than discrete space. In this section, we will focus on two-dimensional space; however, both one- and three-dimensional space may be useful in various contexts. Continuous space can either be the Euclidean plane extending infinitely in all directions, $\mathbb{R} \times \mathbb{R}$, or it can be a bounded connected (contiguous) subset of this plane. Points in the plane can be referred to by their coordinates $(x, y) \in \mathbb{R} \times \mathbb{R}$. As with discrete space, we can consider two cases, depending on whether we focus on individuals or populations.

This section starts with considering individual-based continuous-space models. Next, population continuous-space models are presented, followed by two techniques that are relevant for population discrete-space modelling, but involve continuous-space models or techniques as well. The section ends with examples of the application of continuous-space techniques in various disciplines.

4.1   Techniques for Individual Continuous-Space Models

In these models, we consider identifiable individuals. There are many different models of the movement of individuals through two-dimensional space, such as models of animal movement and models for ad hoc and opportunistic networks [14]. These are often stochastic and capture the probability of movement in a particular direction at a certain speed. Additionally, it may be necessary to determine what happens at the boundary of the space. Often, it is assumed that the space is the surface of a torus and hence has no boundaries – this is more common than assuming the surface of a sphere, as it is hard to map subsets of $\mathbb{R} \times \mathbb{R}$ to the sphere. There are also models to describe the movement of a related group of individuals through the space [14]. Connectivity models on the other hand, describe interaction (for example, contact duration and time between contacts) rather than location [52] so they are implicit movement models. Interaction can be interpreted as dynamic graphs with the individuals as the nodes.

Next, we consider the form that these models can take. If I is an individual, then it has associated information, similar to the discrete state case.

- $\mathsf{loc}(I,t) \in \mathbb{R} \times \mathbb{R}$ which is its location at time t, and
- $\mathsf{state}(I,t) \in \{A_1, \dots, A_n\}$ which is its state[12] at time t.

There are rules which describe how the individual changes state and these may take into account the individual's current location, and rules that describe an individual's movement through space which may take into account the individual's state. As with discrete space, the rates for state change are exponential and can be functional. Unlike with discrete space, it is not useful to construct a Markov chain whose states are obtained from the locations and states of each individual. Discrete event simulation can be used to explore the behaviour of these systems [38].

In the case that the state is continuous, then

$\mathsf{state}(I,t) \in \mathbb{R}^n$ for $n \geq 1$, which is continuous and represents its state at time t.

As with the discrete space case, some way is required that describes the change of state over time, and an ODE can be used for this. Some models require both discrete and continuous non-aggregated states and this requires a hybrid solution. Agent-based models in continuous space are examples of an individual continuous-space model where individuals can take on discrete states or continuous values.

A different approach to modelling continuous state with continuous time is that of continuous time Markov processes (CTMP) [24]. A CTMP is a tuple (S, Σ, R, L) where (S, Σ) forms a specific type of topological manifold and $R : S \times \Sigma \to \mathbb{R}_{\geq 0}$ is a rate function which is measurable in its first coordinate and a measure on its second coordinate. L is a state labelling function. Applying this in the context of space, the manifold is $(\mathbb{R} \times \mathbb{R}, \Sigma)$ where Σ consists of the open sets of $\mathbb{R} \times \mathbb{R}$, hence defining a σ-algebra. A notion of path through this space can be defined describing the behaviour of an individual. Furthermore, if there are additional continuous quantities associated with the individual then additional dimensions of $\mathbb{R}$ can be used.

4.2   Techniques for Population Continuous-Space Models

When individuals are aggregated into populations, there is no need to keep track of them individually and *densities* become more important. In spatio-temporal point processes[13], each point in space (x, y) has an associated integral count for a state in a population at a specific point in time t. We can denote this as

---

[12] As with discrete space, if the population P_A has multiple attributes $A^{[1]}, \dots, A^{[p]}$, then $\mathsf{state}(I,t) = (A_{i_1}^{[1]}, \dots, A_{i_p}^{[p]})$ representing a tuple of attributes.

[13] In contrast to spatio-temporal point processes, spatial point processes describe distributions in space, and do not include a notion of change over time [3] and hence are not relevant in this context.

$N_{A_i}((x,y),t)$ and its behaviour is described by a function $\lambda((x,y),t)$. In general, λ can depend on all preceding events, but in the case of a Poisson process, it only depends on (x,y) and t [78]. If λ is a constant, then there is no spatial heterogeneity. If the equation defining λ includes comparison with other points, then either clustering or inhibitory behaviour can be defined. If time and space are independent then λ can be defined by $\lambda((x,y),t) = \lambda_1(x,y)\lambda_2(t)$. The form of λ may also describe a reduction in the population at a specific point (x,y) and dispersal of that population to other points, thus capturing movement.

For continuous aggregation of populations, we now consider the classical model of movement in continuous space, that of partial differential equations. For populations described by $X_{A_i}((x,y),t)$, the general form is

$$F_i\left(x,y,t,X_{A_1},\ldots,X_{A_n},\frac{\partial X_{A_i}}{\partial x},\frac{\partial X_{A_i}}{\partial y},\frac{\partial X_{A_i}}{\partial t},\frac{\partial^2 X_{A_i}}{\partial x^2},\frac{\partial^2 X_{A_i}}{\partial xy},\frac{\partial^2 X_{A_i}}{\partial y^2}\right) = 0$$

if we assume that we are interested in second order partial derivatives over space only for the population $X_{A_i}((x,y),t)$. Note that writing the PDE in this form simply allows it to be described as a function over all the derivatives of interest rather than as a single partial derivative being equal to a function of other derivatives. When interactions between populations are to be modelled, diffusion-reaction PDEs are used since they can express movement as diffusion and interaction as reactions [20,89]. The diffusion terms can also capture drift which accounts for obstacles or external stimuli such as wind, the likelihood of continuing in the same direction, the effect of the density of other individuals, and the impact of environmental characteristics. The reaction term describes interactions between individuals. Examples are given in the following sections. There are various techniques for solving PDEs which we will not consider here, many of which involve discretising the plane into a mesh [81].

We now consider two approaches to modelling discrete space where continuous space plays an important role, in the sense that transformation from one type of space to another is involved.

4.3   PDE-Based Analysis of Discrete-Space Models

Tschaikowski and Tribastone [88] have considered an approach which involves taking a discrete space model with random walks to continuous space through spatial fluidisation and then using PDE analysis techniques to get good approximation results.

They studied population-based CTMCs where agents are subject to a random walk on the uniform lattice $\mathscr{R} := \{(i\Delta s, j\Delta s) \mid 0 \leq i,j \leq K\}$ in the unit square $[0;1]^2$ with $\Delta s := 1/K$ and $K \geq 1$. Each agent may attain one of the local states $A_1,\ldots,A_L$ while being at any point in $\mathscr{R}$, meaning that the CTMC state

$$A := (A_1^{(x,y)},\ldots,A_L^{(x,y)})_{(x,y)\in\mathscr{R}}$$

provides the agent populations in each local state at each region. Agents in the same region may cooperate with each other by performing local interactions from a rich class of functions. The spatial domain is assumed to have absorbing or reflective boundary conditions. The former can be used to model a hostile environment, while the latter account for closed environments. It can be shown that the CTMC of size $\mathscr{O}(N^{L \cdot K^2})$ converges to the solution of an ODE system of size $\mathscr{O}(L \cdot K^2)$ as $N \to \infty$. While this is a major improvement because the complexity drops from exponential to polynomial, the ODE system may be hard to solve if K is large.

Fortunately, it is possible to identify a finite difference scheme [42] which solves the ODE system of size $\mathscr{O}(L \cdot K^2)$ and that can be also interpreted as a finite difference scheme [84] of a PDE system of size L. By combining this with the former result, one then proves that the solutions of the ODE system of size $\mathscr{O}(L \cdot K^2)$ converge, as $K \to \infty$, to the solution of a PDE system of size L. This is not a purely theoretical result because one solves PDE systems by discretising them to large ODE systems and the discretization induced by a PDE solver is purely dependent on the PDE system itself and thus may be substantially coarser than the one induced by the spatial domain $\mathscr{R}$ which can be arbitrarily fine. Indeed, substantial speed-ups have been reported in [86,88], thus showing that a characterization of mobile systems in terms of PDEs gives rise to shorter calculation times.

4.4   Fluid Approximation and Spatial Discretisation Applied to Agent-Based Continuous Space Models

The use of fluid approximation of population and spatial discretisation has been applied in an ad hoc manner to a 2-dimensional space model of delay-tolerant networks [35]. A general approach based on Markovian agents has been proposed for 1-dimensional space which aggregates and fluidises individuals and discretises space.

Feng developed a continuous-space model with individual agents (using the process algebra stochastic HYPE) for a delay-tolerant network which used wild animals as nodes. Due to computational limitations, the analysis was restricted in terms of how many nodes could be modelled. The model was then transformed to a discrete-space model by dividing up space according to waterhole locations, and using the continuous space model to derive parameters for movement [35]. This enabled the population-based modelling of systems with many more nodes and still provided good approximations.

More recently, a proposal has been made to apply this process in a general way to 1-dimensional space. Specifically, it considers models which consist of Markovian agents (MAs) moving on a bounded one-dimensional continuous space. Markovian agents are a formalism that involves message-passing between agents, and whose overall behaviour can be expressed as a CTMC or a set of ODEs [15]. A detailed definition of Markovian agents is beyond the scope of this tutorial.

The analysis of interest is the transient evolution of the state density distribution of agents of class c in state i at position l and at time t. The change in this value over a small amount of time can be expressed in terms of those agents at location l who change state and those agents who move to l. The movement speed of MAs solely depends on the current state of the agents. A new term to describe the agents that move can be derived from the Taylor expansion of the movement term. The change in value can be then be expressed as a PDE in terms of both time and distance (in 1-dimension). Assuming upper and lower bounds, the upwind semi-discretisation technique [47] can be applied to discretise the distance aspect of the PDE leading to a set of ODEs expressing the change of state density at each discretised location.

4.5   Applications of Continuous Space

As with the case for discrete space, the aim of this section is to briefly consider various applications and a survey can be found in [41].

Ecology: Spatio-temporal point processes have been used to model plant growth and dispersal [8] and other applications [25]. Markov random graphs on continuous space over continuous time can also be considered as spatio-temporal point processes [50]. Holmes *et al* [46] review the use of PDEs in ecological applications, and consider the different forms of PDEs that are used for different models including Brownian (random) motion, drift and the *telegraph* equation.

Biology: Bittig and Uhrmacher [7] describe two continuous space approaches for cellular modelling: particle space and PDEs. In the former, each molecule is modelled separately and these models can be simulated more efficiently by assuming that each particle is only affected by nearby events. When using PDEs, often only simple diffusion based on Brownian motion is required. Fange *et al* [34] describe three different techniques for spatially heterogeneous stochastic kinetics as *microscopic* when each individual particle is considered in terms of its position (continuous-space), as *mesoscopic* when the Reaction Diffusion Master Equation (RDME) is used (discrete space) and as *macroscopic* when PDEs are used. PDEs can also be obtained by taking the hydrodynamic limit of IPSs, namely as the number of particles tends to infinity [23,30]. Pattern formation is important in biology and an important PDE in this context is the Swift-Hohenberg equation [82].

Epidemiology: Spatial point processes have been used to model the spread of foot and mouth disease [26]. Kendall [54] proposed the first spatial epidemic PDE model based on the Kermack-McKendrick nonspatial compartment model, and this has been extended to the Diekmann-Thieme model where traits of individuals affect both their susceptibility to infection and their infectiveness to other individuals [77].

Networking: There is a substantial amount of work on mobility models, both at the analytical level and experimentally through traces in the domain of networking [14,69]. Connectivity models provide an abstraction of mobility models in that they provide information about intercontact time [52]. Stochastic geometry has been applied to wireless networks [2] and epidemiological approaches using PDEs have been used for routing in networks [57].

Continuous-space techniques can be applied to CAS modelling when individual movement is to be tracked, or when it is possible to aggregate movement using PDEs because of the large subpopulation sizes. However, any techniques that tracks individuals is unlikely to be scalable. In the next section, hybrid approaches are considered that can be used to mitigate this problem.

5   Other Approaches to Modelling Space

The techniques discussed in this section are not specific to whether a model is an individual or a population model and may also apply to models that have characteristics of both. Using logic-based approaches, spatial and spatio-temporal model checking can be applied to either sort of model and are addressed in another chapter in this volume [19].

5.1   Crowding

In biological modelling of cells, crowding (occupation of space) is an important issue, because cells have limited volume and it can be important to consider how much space various molecules take up, and how this may affect reactions, as well as the health of the cell. Models range from those that model continuous space in which each entity has a volume and collision between molecules are explicitly modelled, to grid-based approaches where there is space for only one entity in each location [56,83]. The lattice-based approaches can be similar to individual discrete-space models but use regular graph rather than arbitrary graphs.

For population discrete-space models, crowding can be modelled by imposing maximum quantities on locations. Functional rates for movement into a location can be defined to be zero when the maximum population count for a location has been reached This can lead to discontinuous rate functions. Crowding can be important in CAS, as we may want to impose occupation limits, such as the number of people in a shared taxi, or the capacity of a bike station in a bike-sharing scheme.

5.2   Hybrid Approaches

Hybridness is a ubiquitous feature in many models of real systems. As far as space is concerned, there are many ways in which one can construct hybrid models. Here we list some possibilities for future research, with CAS examples from smart transport.

– Space may be seen or modelled differently depending on which kind of agent we are considering in the model. An example taken from biology is in the description of large and small molecules. The former are often modelled as individual objects having a precise position in continuous space. The latter are described as populations, and hence represented by counting variables, in subregions of space [6]. This produces a model combining individual objects moving in space with discretised stochastic diffusion process. If we consider models of interaction of pedestrians with public transportation, we can investigate a scenario in which buses are modelled as individual entities moving in continuous space, while pedestrians or bus users are modelled as populations moving from one discrete location in the city to another, or on and off a bus. Alternatively, buses outside the city centre could be modelled as moving in continuous space, whereas those within the city centre are modelled as a population with movement rates that are determined by the number of buses.

- Another source of hybridness in spatial modelling can be related to different representations of space at different scales or in different locations. The simplest scenario to consider is a high level representation of space in terms of locations, and a low level description of space inside each location in terms of a grid or continuous space. In this case, one has to define appropriate interfaces between the dynamics at the two scales, in terms of abstraction and concretisation functions mapping the low level into the high level and vice versa. By contrast to the previous example, one may wish to model details of the bus movement within the city centre but represent the flow of buses in and out of the centre to different suburbs in a discrete-space style.

- A similar situation to the previous one is a scenario in which one special location of interest is treated in detail, while the rest of the system is approximated in a coarser manner as a single component. The detailed model of a region may be either continuous or grid-based, while the rest of the system can be abstracted as a location-based model, possibly homogeneous, hence resorting to some kind of aggregate moment closure technique. An example of collective adaptive system of this kind may be a crowd movement scenario, in which different squares of a city are described in detail, and the flow of people in and out of each square is represented in a location-based style.

- Similarly, there may be situations in which different locations require a different level of detail in their treatment. For instance, in a crowd movement scenario, we may be interested in tracking the density of people on bikes in the streets or in a square, which calls for a continuous space representation and a PDE dynamics, but coupling this model with a model describing the number of people at bike stations, in order to keep track of the inflow and outflow of people from the streets or the square.

- From a more classical perspective, we can imagine hybrid models in space where small and large populations are both present [9]. This may be location specific, and change as the system evolves. Then, we can construct hybrid models in which some populations are kept discrete in some locations, but are approximated continuously in other ones.

Analysing hybrid spatial models can be challenging, but also opens new ways of using locally different forms of spatial abstraction techniques. As an example, consider a multi-scale scenario where the local space is described as a fine grid, while globally space is represented by a collection of locations. In such a situation, we may use structure-based moment closure approximation locally (if that is accurate enough), de facto reducing the model to a standard location population ODE. In the case of the hybrid treatment of populations, simulation of TDSHA (transition-driven stochastic hybrid automata) [12] or PDMPs (piecewise deterministic Markov processes) [22] can be used.

6 Conclusion

To conclude, this tutorial has provided information about the choices than can be made when modelling space in a quantified manner, focussing on the modelling of CAS. Scalability of techniques have been considered, with specific references to moving away from individual-based modelling to population modelling, using both exact and approximate techniques. There has been an exploration of techniques for both discrete and continuous space, as well a review of how techniques have been applied in the literature, and specific details of techniques that have been considered for CAS.

Acknowledgements. This work is supported by the EU project QUANTICOL, 600708. The author thanks Jane Hillston and Mieke Massink for their useful comments.

Appendix: Discrete and continuous time Markov Chains

This section briefly introduces these concepts, as they would be used in stochastic modelling both without aggregation of state and with aggregation of state (population-based Markov chains) [4,10].

Definition 1. *A* discrete time Markov chain (DTMC) *is a tuple* $\mathcal{M}_D = (\mathcal{S}, \mathbf{P})$ *where*

- $\mathcal{S}$ *is a finite set of* states, *and*
- $\mathbf{P} : \mathcal{S} \times \mathcal{S} \to [0,1]$ *is a* probability matrix *satisfying* $\sum_{S' \in \mathcal{S}} \mathbf{P}(S, S') = 1$ *for all* $S \in \mathcal{S}$.

A DTMC is *time-abstract* [4] in the sense that time is viewed as a sequence of discrete steps or clock ticks. It describes behaviour as follows: if an entity or individual is currently in state $S \in \mathcal{S}$ then the probability of the entity being in state S' at the next time step is defined by $\mathbf{P}(S, S')$. Under certain conditions, the steady state of the DTMC can be determined and this describes when the DTMC is at equilibrium and gives the (unchanging) probability of being in any of the states of $\mathcal{S}$. By contrast, transient state probabilities can be determined at each point in time before steady state is achieved.

Definition 2. *A* continuous time Markov chain (CTMC) *is a tuple* $\mathcal{M}_C = (\mathcal{S}, \mathbf{R})$ *where*

- $\mathcal{S}$ *is a finite set of* states, *and*
- $\mathbf{R} : \mathcal{S} \times \mathcal{S} \to \mathbb{R}_{\geq 0}$ *is a* rate matrix.

CTMCs are *time-aware* [4] since they use continuous time. If an entity is currently in state S, then $\mathbf{R}(S, S')$ is a non-negative number that defines an exponential distribution from which the duration of the time taken to transition from state S to state S' can be drawn. As with DTMCs and under certain conditions, transient and steady state probabilities can be calculated which describe the probability of being in each state at a particular time t or in the long run, respectively.

Let $E(S) = \sum_{S' \in \mathscr{S}} \mathbf{R}(S, S')$ be the exit rate of state S'. Then the embedded DTMC of a CTMC has entries in its probability matrix of the form $\mathbf{P}(S, S') = \mathbf{R}(S, S')/E(S)$ if $E(S) > 0$ and $\mathbf{P}(S, S') = 0$ otherwise. DTMCs and CTMCs can be state-labelled (usually with propositions) or transition-labelled (usually with actions). The research in QUANTICOL focusses on transition-labelled Markov chains. We next consider population Markov chains, both discrete time and continuous time. Instead of considering an entity with states, we now consider a vector of counts $\mathbf{X}$ that describes how many entities are in each state; thus it is a population view rather than an individual view. Our definition in the continuous-time case is slightly simpler than that appearing in another chapter in this volume [9] since transitions do not have guards and we do not parameterise the Markov chain with the population size.

Definition 3. *A population discrete time Markov chain (PDTMC) is a tuple $\mathscr{X}_D = (\mathbf{X}, \mathscr{D}, \mathscr{T})$ where*

- *$\mathbf{X} = (X_1, \ldots, X_n)$ is a vector of variables*
- *$\mathscr{D}$ is a countable set of states defined as $\mathscr{D} = \mathscr{D}_1 \times \ldots \times \mathscr{D}_n$ where each $\mathscr{D}_i \subseteq \mathbb{N}$ represents the domain of X_i*
- *$\mathscr{T} = \{\tau_1, \ldots \tau_m\}$ is the set of transitions of the form $\tau_j = (\mathbf{v}, p)$ where*
  - *$\mathbf{v} = (v_1, \ldots, v_n) \in \mathbb{N}^n$ is the state change or update vector where v_i describes the change in number of units of X_i caused by transition τ_j*
  - *$p : \mathscr{D} \to \mathbb{R}_{\geq 0}$ is the probability function of transition τ_j that defines a sub-probability distribution, namely $\sum_{\tau \in \mathscr{T}} p_\tau(\mathbf{d}) \leq 1$ for all $\mathbf{d} \in \mathscr{D}$, such that $p(\mathbf{d}) = 0$ whenever $\mathbf{d} + \mathbf{v} \notin \mathscr{D}$*

Definition 4. *A population continuous time Markov chain (PCTMC) is a tuple $\mathscr{X}_C = (\mathbf{X}, \mathscr{D}, \mathscr{T})$ where*

- *$\mathbf{X}$ and $\mathscr{D}$ are defined as in the previous definition,*
- *$\mathscr{T} = \{\tau_1, \ldots \tau_m\}$ is the set of transitions of the form $\tau_j = (\mathbf{v}, r)$ where*
  - *$\mathbf{v}$ is defined as in the previous definition,*
  - *$r : \mathscr{D} \to \mathbb{R}_{\geq 0}$ is the rate function of transition τ_j with $r(\mathbf{d}) = 0$ whenever $\mathbf{d} + \mathbf{v} \notin \mathscr{D}$.*

In both types of population Markov chain, the associated Markov chain can be obtained. In both cases, $\mathscr{D}$ is the state space $\mathscr{S}$. For the population DTMC, the probability matrix of its associated DTMC is defined as

$$\mathbf{P}(\mathbf{d}, \mathbf{d}') = \sum_{\tau \in \mathscr{T}, \mathbf{v}_\tau = \mathbf{d}' - \mathbf{d}} p_\tau(\mathbf{d}) \text{ whenever } \mathbf{d} \neq \mathbf{d}'$$

and since probability functions define sub-probabilities then the rest of the probability mass must be accounted for by defining

$$\mathbf{P}(\mathbf{d}, \mathbf{d}) = 1 - \sum_{\tau \in \mathcal{T}, \mathbf{v}_\tau \neq 0} p_\tau(\mathbf{d}).$$

For the population CTMC, the rate matrix of its associated CTMC is

$$\mathbf{R}(\mathbf{d}, \mathbf{d'}) = \sum_{\tau \in \mathcal{T}, \mathbf{v}_\tau = \mathbf{d'} - \mathbf{d}} r_\tau(\mathbf{d}) \text{ whenever } \mathbf{d} \neq \mathbf{d'}$$

and if the summation is empty, then $\mathbf{R}(\mathbf{d}, \mathbf{d'}) = 0$.

As the size of the population increases, it has been shown [58] under specific conditions that cover a large range of models that the behaviour of an (appropriately normalised) population CTMC at time t is very close to the solution of a set of ODEs, expressed in the form $\mathbf{X}(t) = (X_1(t), \ldots, X_n(t))$ defining a trajectory over time. The ODEs can be expressed in terms of a single vector ODE as

$$\dot{\mathbf{X}} = \frac{d\mathbf{X}}{dt} = \mathbf{f}(\mathbf{X})$$

where $\mathbf{f}(\mathbf{X})$ is a function derived from the specifics of the PCTMC (see [9] in this volume for details). It is also possible to approximate the moments of a PCTMC using the ODEs [32]

$$\frac{\mathrm{d}}{\mathrm{d}t} E[M(\mathbf{X}(t))] = \sum_{\tau \in \mathcal{T}} E[(M(\mathbf{X}(t) + \mathbf{v}_\tau) - M(\mathbf{X}(t)))r_\tau(\mathbf{X}(t))]$$

where $M(\mathbf{X})$ denotes the moment to be calculated, $\mathbf{v}_\tau$ and $r_\tau(\mathbf{X}(t))$ represents the update vector and the rate of a transition τ, respectively.

References

1. Auger, P., Poggiale, J., Sánchez, E.: A review on spatial aggregation methods involving several time scales. Ecol. Complex. **10**, 12–25 (2012)
2. Baccelli, F., Błaszczyszyn, B.: Stochastic Geometry and Wireless Networks: Volume I and II. NOW Publishers, Hanover (2009)
3. Baddeley, A., Bárány, I., Schneider, R.: Spatial point processes and their applications. Stochastic Geometry. Lecture Notes in Mathematics, vol. 1892, pp. 1–75. Springer, Heidelberg (2007)
4. Baier, C., Katoen, J.P., Hermanns, H., Wolf, V.: Comparative branching-time semantics for Markov chains. Inf. Comput. **200**, 149–214 (2005)
5. Berec, L.: Techniques of spatially explicit individual-based models: construction, simulation, and mean-field analysis. Ecol. Model. **150**, 55–81 (2002)
6. Bittig, A., Haack, F., Maus, C., Uhrmacher, A.: Adapting rule-based model descriptions for simulating in continuous and hybrid space. In: Proceedings of CMSB 2011, pp. 161–170. ACM (2011)
7. Bittig, A., Uhrmacher, A.: Spatial modeling in cell biology at multiple levels. In: Winter Simulation Conference (WSC 2010), pp. 608–619. IEEE (2010)

8. Bolker, B., Pacala, S.: Using moment equations to understand stochastically driven spatial pattern formation in ecological systems. Theor. Popul. Biol. **52**, 179–197 (1997)

9. Bortolussi, L., Gast, N.: Mean-field limits beyond ordinary differential equations. In: Bernardo, M., De Nicola, R., Hillston, J. (eds.) SFM 2016. LNCS, vol. 9700, pp. 61–82. Springer, Switzerland (2016)

10. Bortolussi, L., Hillston, J., Latella, D., Massink, M.: Continuous approximation of collective systems behaviour: a tutorial. Perform. Eval. **70**, 317–349 (2013)

11. Bortolussi, L., Latella, D., Massink, M.: Stochastic process algebra and stability analysis of collective systems. In: De Nicola, R., Julien, C. (eds.) COORDINATION 2013. LNCS, vol. 7890, pp. 1–15. Springer, Heidelberg (2013)

12. Bortolussi, L., Policriti, A.: Hybrid dynamics of stochastic programs. Theor. Comput. Sci. **411**, 2052–2077 (2010)

13. Brauer, F.: Compartmental models in epidemiology. In: Allen, L., Brauer, F., van den Driessche, P., Wu, J. (eds.) Mathematical Epidemiology, pp. 19–80. Springer, Heidelberg (2008)

14. Camp, T., Boleng, J., Davies, V.: A survey of mobility models for ad hoc network research. Wirel. Commun. Mob. Comput. **2**, 483–502 (2002)

15. Cerotti, D., Gribaudo, M., Bobbio, A.: Markovian agents models for wireless sensor networks deployed in environmental protection. Reliab. Eng. Syst. Saf. **130**, 149–158 (2014)

16. Cerotti, D., Gribaudo, M., Bobbio, A., Calafate, C.T., Manzoni, P.: A markovian agent model for fire propagation in outdoor environments. In: Aldini, A., Bernardo, M., Bononi, L., Cortellessa, V. (eds.) EPEW 2010. LNCS, vol. 6342, pp. 131–146. Springer, Heidelberg (2010)

17. Chaintreau, A., Le Boudec, J.Y., Ristanovic, N.: The age of gossip: spatial mean field regime. In: Proceedings of SIGMETRICS/Performance 2009, pp. 109–120. ACM (2009)

18. Chesson, P.: Scale transition theory: its aims, motivations and predictions. Ecol. Complex. **10**, 52–68 (2012)

19. Ciancia, V., Latella, D., Loreti, M., Massink, M.: Spatial logic and spatial model checking for closure spaces. In: Bernardo, M., De Nicola, R., Hillston, J. (eds.) SFM 2016. LNCS, vol. 9700, pp. 156–201. Springer, Switzerland (2016)

20. Codling, E., Plank, M., Benhamou, S.: Random walk models in biology. J. Roy. Soc. Interface **5**, 813–834 (2008)

21. Darling, R., Norris, J.: Differential equation approximations for Markov chains. Probab. Surv. **5**, 37–79 (2008)

22. Davis, M.: Markov Models and Optimization. Chapman & Hall, Boca Raton (1993)

23. De Masi, A., Presutti, E.: Mathematical Methods for Hydrodynamic Limits. Lecture Notes in Mathematics. Springer, Berlin (1991)

24. Desharnais, J., Panangaden, P.: Continuous stochastic logic characterizes bisimulation of continuous-time Markov processes. J. Logic Algebraic Program. **56**, 99–115 (2003)

25. Diggle, P.: Spatio-temporal point processes: methods and applications. Working paper, Department of Biostatistics, Johns Hopkins University (2005)

26. Diggle, P.: Spatio-temporal point processes, partial likelihood, foot and mouth disease. Stat. Methods Med. Res. **15**, 325–336 (2006)

27. van den Driessche, P.: Spatial structure: patch models. In: Allen, L., Brauer, F., van den Driessche, P., Wu, J. (eds.) Mathematical Epidemiology, pp. 179–190. Springer, Heidelberg (2008)

28. Durrett, R., Levin, S.: The importance of being discrete (and spatial). Theor. Popul. Biol. **46**, 363–394 (1994)
29. Durrett, R., Levin, S.: Stochastic spatial models: a user's guide to ecological applications. Philos. Trans. Roy. Soc. B: Biol. Sci. **343**, 329–350 (1994)
30. Durrett, R., Neuhauser, C.: Particle systems and reaction-diffusion equations. The Ann. Probab. **22**, 289–333 (1994)
31. Ellner, S.: Pair approximation for lattice models with multiple interaction scales. J. Theor. Biol. **210**, 435–447 (2001)
32. Engblom, S.: Computing the moments of high dimensional solutions of the master equation. Appl. Math. Comput. **180**, 498–515 (2006)
33. Erban, R., Chapman, J., Maini, P.: A practical guide to stochastic simulations of reaction-diffusion processes (2007). arXiv preprint arXiv:0704.1908
34. Fange, D., Berg, O., Sjöberg, P., Elf, J.: Stochastic reaction-diffusion kinetics in the microscopic limit. Proc. Nat. Acad. Sci. **107**, 19820–19825 (2010)
35. Feng, C.: Patch-based hybrid modelling of spatially distributed systems by using stochastic HYPE - ZebraNet as an example. In: Proceedings of QApPL 2014 (2014)
36. Feng, C., Hillston, J., Galpin, V.: Automatic moment-closure approximation of spatially distributed collective adaptive systems. ACM TOMACS **26**, 26:1–26:22 (2016)
37. Fenichel, N.: Persistence and smoothness of invariant manifolds for flows. Indiana Univ. Math. J. **21**, 1972 (1971)
38. Fishman, G.: Discrete-Event Simulation. Springer, New York (2001)
39. Fricker, C., Gast, N.: Incentives and regulations in bike-sharing systems with stations of finite capacity (2012). arXiv preprint arXiv:1201.1178
40. Fricker, C., Gast, N., Mohamed, H.: Mean field analysis for inhomogeneous bike sharing systems. DMTCS Proc. **01**, 365–376 (2012)
41. Galpin, V., Feng, C., Hillston, J., Massink, M., Tribastone, M., Tschaikowski, M.: Review of time-based techniques for modelling space. Technical report TR-QC-05-2014, QUANTICOL (2014)
42. Gear, C.W.: Numerical Initial Value Problems in Ordinary Differential Equations. Prentice Hall, Upper Saddle River (1971)
43. Gillespie, D.: Exact stochastic simulation of coupled chemical reactions. J. Phys. Chem. **81**, 2340–2361 (1977)
44. Gillespie, D.: Stochastic simulation of chemical kinetics. Ann. Rev. Phys. Chem. **58**, 35–55 (2007)
45. Gorban, A., Radulescu, O., Zinovyev, A.: Asymptotology of chemical reaction networks. Chem. Eng. Sci. **65**, 2310–2324 (2010)
46. Holmes, E., Lewis, M., Banks, J., Veit, R.: Partial differential equations in ecology: spatial interactions and population dynamics. Ecology **75**, 17–29 (1994)
47. Horton, G., Kulkarni, V., Nicol, D., Trivedi, K.: Fluid stochastic Petri nets: Theory, applications, and solution techniques. Eur. J. Oper. Res. **105**, 184–201 (1998)
48. Hu, H., Myers, S., Colizza, V., Vespignani, A.: WiFi networks and malware epidemiology. Proc. Nat. Acad. Sci. **106**, 1318–1323 (2009)
49. Ilachinski, A.: Cellular Automata: A Discrete Universe. World Scientific, Singapore (2001)
50. Isham, V.: An introduction to spatial point processes and Markov random fields. Int. Stat. Rev./Rev. Int. Stat. **41**, 21–43 (1981)
51. Kaneko, K.: Diversity, stability, and metadynamics: remarks from coupled map studies. In: Bascompte, J., Solé, R. (eds.) Modeling Spatiotemporal Dynamics in Ecology, pp. 27–45. Springer, New York (1998)

52. Kathiravelu, T., Pears, A.: Reproducing opportunistic connectivity traces using connectivity models. In: 2007 ACM CoNEXT Conference, p. 34. ACM (2007)

53. Kemeny, J., Snell, J.: Finite Markov Chains. Springer, New York (1976)

54. Kendall, D.: Mathematical models of the spread of infection. In: Mathematics and Computer Science in Biology and Medicine, pp. 213–225. Medical Research Council London (1965)

55. Kephart, J., White, S.: Directed-graph epidemiological models of computer viruses. In: Proceedings of the IEEE Computer Society Symposium on Research in Security and Privacy, pp. 343–359. IEEE (1991)

56. Klann, M., Koeppl, H.: Spatial simulations in systems biology: from molecules to cells. Int. J. Mol. Sci. **13**, 7798–7827 (2012)

57. Klein, D., Hespanha, J., Madhow, U.: A reaction-diffusion model for epidemic routing in sparsely connected MANETs. In: Proceedings of INFOCOM 2010, pp. 1–9. IEEE (2010)

58. Kurtz, T.: Approximation Popul. Process. SIAM, Philadelphia (1981)

59. Levin, S., Durrett, R.: From individuals to epidemics. Philos. Trans. Roy. Soc. Lond. Ser. B: Biol. Sci. **351**, 1615–1621 (1996)

60. Lichtenegger, K., Schappacher, W.: A carbon-cycle-based stochastic cellular automata climate model. Int. J. Mod. Phys. C **22**, 607–621 (2011)

61. Marion, G., Mao, X., Renshaw, E., Liu, J.: Spatial heterogeneity and the stability of reaction states in autocatalysis. Phys. Rev. E **66**, 051915 (2002)

62. Marion, G., Swain, D., Hutchings, M.: Understanding foraging behaviour in spatially heterogeneous environments. J. Theor. Biol. **232**, 127–142 (2005)

63. Massink, M., Brambilla, M., Latella, D., Dorigo, M., Birattari, M.: On the use of Bio-PEPA for modelling and analysing collective behaviors in swarm intelligence. Swarm Intell. **7**, 201–228 (2013)

64. Massink, M., Latella, D., Bracciali, A., Harrison, M., Hillston, J.: Scalable context-dependent analysis of emergency egress models. Formal Aspects Comput. **24**, 267–302 (2012)

65. Matis, J., Kiffe, T.: Effects of immigration on some stochastic logistic models: a cumulant truncation analysis. Theor. Popul. Biol. **56**, 139–161 (1999)

66. Matsuda, H., Ogita, N., Sasaki, A., Satō, K.: Statistical mechanics of population: the lattice Lotka-Volterra model. Prog. Theor. Phys. **88**, 1035–1049 (1992)

67. McCaig, C., Norman, R., Shankland, C.: From individuals to populations: a mean field semantics for process algebra. Theor. Comput. Sci. **412**, 1557–1580 (2011)

68. Morozov, A., Poggiale, J.C.: From spatially explicit ecological models to mean-field dynamics: the state of the art and perspectives. Ecol. Complex. **10**, 1–11 (2012)

69. Musolesi, M., Mascolo, C.: Mobility models for systems evaluation. In: Garbinato, B., Miranda, H., Rodrigues, L. (eds.) Middleware Netw. Eccentric Mob. Appl., pp. 43–62. Springer, Heidelberg (2009)

70. Norris, J.: Markov Chains. Cambridge University Press, Cambridge (1998)

71. Okubo, A., Levin, S.A.: Diffusion and Ecological Problems: Modern Perspectives. Springer, New York (2001)

72. Othmer, H., Scriven, L.: Instability and dynamic pattern in cellular networks. J. Theor. Biol. **32**, 507–537 (1971)

73. Panangaden, P.: Labelled Markov Processes. Imperial College Press, London (2009)

74. Pascual, M., Roy, M., Laneri, K.: Simple models for complex systems: exploiting the relationship between local and global densities. Theor. Ecol. **4**, 211–222 (2011)

75. Riley, S.: Large-scale spatial-transmission models of infectious disease. Science **316**, 1298–1301 (2007)

76. Rowe, J.E., Gomez, R.: El Botellón: modeling the movement of crowds in a city. Complex Syst. **14**, 363–370 (2003)
77. Ruan, S.: Spatial-temporal dynamics in nonlocal epidemiological models. In: Takeuchi, Y., Iwasa, Y., Sato, K. (eds.) Mathematics for Life Science and Medicine. Biological and Medical Physics, Biomedical Engineering, pp. 97–122. Springer, Heidelberg (2007)
78. Schoenberg, F., Brillinger, D., Guttorp, P.: Point processes, spatial-temporal. In: El-Shaarawi, A., Piegorsch, W. (eds.) Encyclopedia of Environmetrics, pp. 1573–1577. Wiley Online Library, New York (2002)
79. Segel, L., Slemrod, M.: The quasi-steady-state assumption: a case study in perturbation. SIAM Rev. **31**, 446–477 (1989)
80. Simon, H.A., Ando, A.: Aggregation of variables in dynamic systems. Econometrica **29**, 111–138 (1961)
81. Slepchenko, B., Schaff, J., Macara, I., Loew, L.: Quantitative cell biology with the virtual cell. Trends Cell Biol. **13**, 570–576 (2003)
82. Swift, J., Hohenberg, P.: Hydrodynamic fluctuations at the convective instability. Phys. Rev. A **15**, 319 (1977)
83. Takahashi, K., Arjunan, S., Tomita, M.: Space in systems biology of signaling pathways-towards intracellular molecular crowding in silico. FEBS Lett. **579**, 1783–1788 (2005)
84. Thomas, J.W.: Numerical Partial Differential Equations: Finite Difference Methods. Springer, New York (1995)
85. Tribastone, M., Gilmore, S., Hillston, J.: Scalable differential analysis of process algebra models. IEEE Trans. Softw. Eng. **38**, 205–219 (2012)
86. Tschaikowski, M., Tribastone, M.: A Partial-differential Approximation for Spatial Stochastic Process Algebra. In: Proceedings of VALUETOOLS 2014 (2014)
87. Tschaikowski, M., Tribastone, M.: Exact fluid lumpability in Markovian process algebra. Theor. Comput. Sci. **538**, 140–166 (2014)
88. Tschaikowski, M., Tribastone, M.: Spatial fluid limits for stochastic mobile networks. Performance Evaluation (2015), under minor revision
89. Turing, A.: The chemical basis of morphogenesis. Philos. Trans. Roy. Soc. (Part B) **237**, 37–72 (1953)
90. Vandin, A., Tribastone, M.: Quantitative abstractions for collective adaptive systems. In: Bernardo, M., De Nicola, R., Hillston, J. (eds.) SFM 2016. LNCS, vol. 9700, pp. 202–232. Springer, Switzerland (2016)
91. Webb, S., Keeling, M., Boots, M.: Host-parasite interactions between the local and the mean-field: how and when does spatial population structure matter? J. Theor. Biol. **249**, 140–152 (2007)
92. Wiggins, S.: Normally Hyperbolic Invariant Manifolds in Dynamical Systems. Springer Science & Business Media, New York (1994)
93. Wu, J., Loucks, O.: From balance of nature to hierarchical patch dynamics: a paradigm shift in ecology. Q. Rev. Biol. **70**, 439–466 (1995)
94. Zhou, X., Ioannidis, S., Massoulié, L.: On the stability and optimality of universal swarms. ACM SIGMETRICS Perform. Eval. Rev. **39**, 301–312 (2011)
95. Zungeru, A., Ang, L.M., Seng, K.P.: Classical and swarm intelligence based routing protocols for wireless sensor networks: a survey and comparison. J. Netw. Comput. Appl. **35**, 1508–1536 (2012)

Spatial Logic and Spatial Model Checking for Closure Spaces

Vincenzo Ciancia[1], Diego Latella[1], Michele Loreti[2,3], and Mieke Massink[1(✉)]

[1] Istituto di Scienza e Tecnologie dell'Informazione 'A. Faedo', CNR, Pisa, Italy
mieke.massink@isti.cnr.it
[2] Università di Firenze, Florence, Italy
[3] IMT Alti Studi, Lucca, Italy

Abstract. Spatial aspects of computation are increasingly relevant in Computer Science, especially in the field of *collective adaptive systems* and when dealing with systems distributed in physical space. Traditional formal verification techniques are well suited to analyse the temporal evolution of concurrent systems; however, properties of space are typically not explicitly taken into account. This tutorial provides an introduction to recent work on a topology-inspired approach to formal verification of spatial properties depending upon (physical) space. A logic is presented, stemming from the tradition of topological interpretations of modal logics, dating back to earlier logicians such as Tarski, where modalities describe neighbourhood. These topological definitions are lifted to the more general setting of *closure spaces*, also encompassing discrete, graph-based structures. The present tutorial illustrates the extension of the framework with a spatial *surrounded* operator, leading to the spatial logic for closure spaces SLCS, and its combination with the temporal logic CTL, leading to STLCS. The interplay of space and time permits one to define complex spatio-temporal properties. Both for the spatial and the spatio-temporal fragment efficient model-checking algorithms have been developed and their use on a number of case studies and examples is illustrated.

1 Introduction

Modal logics, model checking and static analysis enjoy an outstanding mathematical tradition, spanning over logics, abstract mathematics, artificial intelligence, theory of computation, system modelling, and optimisation. However, the *spatial* aspects of systems, that is, dealing with properties of entities that relate to their position, distance, connectivity and reachability in *space*, have never been truly emphasised in computer science. With the recent interest in the design of fully *decentralised* systems that are composed of a large number of locally interacting objects that are distributed in physical space, also called *Collective Adaptive Systems* (CAS) [2], spatial reasoning and formal spatial verification have gained renewed interest. A starting point is provided by so-called *spatial logics* [1],

Research partially funded by EU project QUANTICOL (nr. 600708).

that have been studied from the point of view of (mostly modal) logics. The field of spatial logics is well developed in terms of descriptive languages and aspects such as computability and complexity. The development dates back to work by early logicians such as Tarski, who studied possible semantics of classic modal logics, using topological spaces in place of Kripke frames. However, the frontier of current research does not yet address formal verification problems, and in particular, discrete spatial models are still a relatively unexplored field.

In this tutorial, we review some relevant current literature dealing with models where space is continuous, and start an analysis of the situation in the case of discrete structures. The interest in such an analysis comes from the conjecture that properties may be described using the same languages in the continuous, discrete, and relational (classical) case. The longer term aim is to provide a unifying view of temporal and spatial properties which is independent of the kind of models that are taken into account. The tutorial is intended to be a starting point for understanding which descriptive languages are most suitable for such an endeavour. As a next step we show how to cast the well known developments in spatio-temporal reasoning in the realm of discrete and finite structures, and develop efficient and effective verification algorithms. Their use will be illustrated on small and larger examples in the field of CAS. This development constitutes a novel, relatively unexplored research line. The lack of applications of spatial logic in the field of verification is also witnessed in the introduction to the Handbook of Spatial Logics [1], that we use as one of our main references. The tutorial is furthermore partially based on some of our previous and forthcoming joint work with various other authors (see [14, 16, 17, 19, 34, 37]).

We can distinguish broadly three main aspects in this field. First of all, the *spatial structures* with which relevant aspects of space can be modelled must be identified. Second, suitable *spatial and spatio-temporal logics* must be developed. Finally it is useful to enrich the setting to be able to take *quantitative aspects* into account such as distance or probability. Let us briefly review these three aspects:

Spatial Structures. Space can be modelled as a discrete or continuous entity. This ought to be accommodated in a general setting by choosing appropriate abstract mathematical structures. *Topological spaces* are typical examples. However, we shall see that for dealing with discrete spatial structures *closure spaces*, a generalisation of topological spaces, are a better starting point.

Spatial Logics. Spatial logics predicate on properties of entities located in the space; for example, one may be interested in entities that are *inside, outside* or on the *boundary* of regions of space where certain properties hold. Depending on the specific logical language, the entities described can be:

– Points in the space. In this case reasoning has a strongly local flavour. Global properties (e.g., a region of a space not having "holes") cannot be expressed.

- Spaces as a whole. Global properties can easily be expressed if the point of view is shifted from the behaviour of an individual in a specified setting, to the analysis of several possible global scenarios consisting of all the entities in a given space.
- Regions of space. This approach combines reasoning on multiple entities simultaneously with a focus on the interaction (e.g., overlapping or contact) between areas having different properties.

Spatio-Temporal Logics. The combination of spatial logics with other modal logics, such as temporal logics, provides an even richer language for the formulation of properties that reflect both spatial and temporal aspects at the same time. The combination of spatial and temporal logics introduces more design variables, especially for what concerns the interplay between the spatial and the temporal component. Computational properties, such as decidability and complexity, of several possible combinations are examined in detail in [31].

Quantitative Aspects. Distance-based logics extend topological logics. Formulas are indexed by intervals, which are used as constraints. Metric-topological properties are verified by a model if the topological part of the formula is verified, and the constraints are satisfied. For example, one may require that points satisfying a certain property are located at most at a specified distance from each other, or from points characterised by some other property.

This tutorial does not cover several topics that are relevant for spatial and spatio-temporal logics and model checking. A partial list of well-known approaches that are not covered in this chapter includes: *Metric Interval Temporal Logics* [33]; *Spatial Signal Temporal Logic* [36,37]; *SpaTeL* [29] a spatial logic based on Quad Trees; logics of process calculi such as the Mobile Stochastic Logic *MoSL* [22], *Ambient Logic* [12] and *Separation Logic* [38]. Finally, we will only touch upon extensions with quantitative features in the final section. Spatial representations and non-logic based spatial analysis techniques will be addressed in an other chapter of this volume [25].

The outline of the tutorial is as follows. In Sect. 2 we present and relate several classes of spatial structures that are directly relevant to the approach we follow to develop spatial logics and model-checking techniques. Section 3 reviews modal logics and their extension to space. In Sect. 4 a spatial logic for closure spaces is introduced and its operators are illustrated in a number of examples. In Sect. 6 the spatial logic is extended with temporal operators and in Sect. 7 some aspects of a spatio-temporal model-checking algorithm are presented. Section 8 presents some larger case studies in which we illustrate some more complicated spatio-temporal properties of a bike-sharing system and public bus transportation seen as a CAS. To conclude, Sect. 9 provides a discussion of open issues. The detailed proofs of theorems, lemmas and propositions stated in this tutorial can be found in [16].

2 Closure Spaces, Topology and Graphs

In this section we introduce some relevant mathematical notions and facts that are used throughout the tutorial and that form the basis for the particular spatial logics and related model checking algorithms that are the main topic of this tutorial. We briefly explain the various mathematical structures that are involved, in particular topological structures, their generalisation, also known as Closure Spaces, and two particular discrete subsets of the latter, namely *quasi-discrete closure spaces* and finite *graphs*. This section is intended as a minimal reference tailored to make the tutorial self-contained, rather than as an introduction to topology or other mathematical subjects, for which the reader is invited to consult more authoritative sources. The section on topology of [40] may be used as a gentle introduction; a comprehensive reference is [30].

2.1 Topological Spaces

Topological spaces may be presented as generalisations of Euclidean spaces by focussing on the notion of *closeness* without making reference to an explicit metric. In topology this notion is expressed in terms of relationships between sets of points instead of in terms of distance. Any topological space consists of a set of points, a set of subsets of points called *open sets* and two operators on sets, namely union and intersection. More formally, a topological space can be defined as follows.

Definition 1. *A topological space is a pair* (X, O) *of a set* X *and a collection* $O \subseteq \wp(X)$ *of subsets of* X *called* open sets, *such that* $\emptyset, X \in O$, *and subject to closure under arbitrary unions and finite intersections.*

The dual of an open set is a *closed* set, but open and close sets are not mutually exclusive, as illustrated in Example 1 below following the definition of closed set.

Definition 2. *A subset* S *of* X *is called* closed *if* $X \setminus S \in O$. *A* clopen *is a set that is both open and closed.*

Example 1. Assume the topological space X is the two dimensional Euclidean space and that it is equipped with the usual Euclidean (metric) topology. The pink shapes in Fig. 1 are subsets of this space. The shape on the left is open, that on the right is closed. An example of a set that is both open and closed is the space X itself. This can be seen as follows. First note that X is open by definition, but its complement, the empty set, is also open by definition. Therefore X is also closed. The same holds for the empty set. Furthermore, switching to the one dimensional case for simplicity, the set $[1, 2)$ is *neither* open *nor* closed because its complement is neither open nor closed. However, $[1, \infty)$ is a closed set because its complement is open.

Also the notion of *neighbourhood* of a point in space plays an important role, as well as those of *interior* and *closure* of a set of points.

open set   closed set

Fig. 1. Example of an open set (left) and a closed set (right). (Color figure online)

Definition 3. *An* open neighbourhood *of $x \in X$ is an open set o with $x \in o$. The* interior *of $S \subseteq X$, denoted by $\mathcal{I}(S)$, is the largest open set contained in S and the* closure *of S, denoted by $\mathcal{C}(S)$, is the smallest closed set containing S.*

The interior and closure are dual. Let $\overline{S}$ denote $X \setminus S$ (the complement of S in X). Then we have $\mathcal{I}(S) = \overline{\mathcal{C}(\overline{S})}$ and $\mathcal{C}(S) = \overline{\mathcal{I}(\overline{S})}$. In Fig. 1 the open set can be seen as the interior of the closed set and the closed set as the closure of the open set. In fact, the closure operator adds all *limit points* of an open set to it. A point p is called a *limit point* of a set S if every open set containing p also contains some point of S. Limit points are a fundamental notion in topological spaces and could be used to provide an alternative axiomatical definition of topological spaces known as Kuratowsky spaces. They also reflect the inherent *continuous* aspect of topological spaces. In topological spaces the closure operator is idempotent, so $\mathcal{C}(\mathcal{C}(S)) = \mathcal{C}(S)$.

2.2   Closure Spaces

Discrete spatial structures could be treated as in the continuous case, by defining a topology on top of the points of the structure. However, by doing so, one does not gain much, as the closure operator is idempotent in topological spaces. This assumption becomes too stringent for discrete structures. For example, in the case of regular grids, it is natural to interpret closure as the operation of enlarging a set of points by one step (in all possible directions) on the grid. Such interpretation is clearly not idempotent. By removing the idempotency assumption, *closure spaces* are obtained that therefore are a generalisation of topological spaces, as shown in the following definition and more explicitly in Definition 7.

Definition 4. *A* closure space *is a pair $(X, \mathcal{C})$ where X is a set, and $\mathcal{C} : 2^X \to 2^X$ assigns to each subset of X its* closure, *such that, for all $A, B \subseteq X$:*

1. *$\mathcal{C}(\emptyset) = \emptyset$;*
2. *$A \subseteq \mathcal{C}(A)$;*
3. *$\mathcal{C}(A \cup B) = \mathcal{C}(A) \cup \mathcal{C}(B)$.*

As in topological spaces, we can define the interior operator for closure spaces as well as the notions of neighbourhood, open set and closed set.

Definition 5. *In a closure space $(X, \mathcal{C})$, given $A \subseteq X$ and $x \in X$: (i) the interior $\mathcal{I}(A)$ of A is the set $\overline{\mathcal{C}(\overline{A})}$; (ii) A is a neighbourhood of $x \in X$ if and only if $x \in \mathcal{I}(A)$; and (iii) A is closed if $A = \mathcal{C}(A)$ and it is open if $A = \mathcal{I}(A)$.*

The above defined operators enjoy the following useful properties.

Proposition 1. *In a closure space $(X, \mathcal{C})$, for all $A, B \subseteq X$, the following holds:*
(i) $\mathcal{I}(A) \subseteq A$; (ii) A is open if and only if $\overline{A}$ is closed; (iii) $A \subseteq B \implies \mathcal{C}(A) \subseteq \mathcal{C}(B)$ and $\mathcal{I}(A) \subseteq \mathcal{I}(B)$; and (iv) the open sets are closed under finite intersections, and arbitrary unions.

Below, we provide a definition of the *boundary* of a set A which is entirely given in terms of closure and interior, and coincides with the definition of boundary in a topological space. Moreover, in discrete spaces (such as, grids) it sometimes makes sense to consider just the part of the boundary of a set A which lies entirely within, or outside, A itself. We also define these notions.

Definition 6. *In a closure space $(X, \mathcal{C})$, the* boundary *of $A \subseteq X$ is defined as $\mathcal{B}(A) = \mathcal{C}(A) \setminus \mathcal{I}(A)$. The* interior boundary *is $\mathcal{B}^-(A) = A \setminus \mathcal{I}(A)$, and the* closure boundary *is $\mathcal{B}^+(A) = \mathcal{C}(A) \setminus A$.*

Proposition 2. *The following equations hold in a closure space:*

$$\mathcal{B}(A) = \mathcal{B}^+(A) \cup \mathcal{B}^-(A) \tag{1}$$
$$\mathcal{B}^+(A) \cap \mathcal{B}^-(A) = \emptyset \tag{2}$$
$$\mathcal{B}(A) = \mathcal{B}(\overline{A}) \tag{3}$$
$$\mathcal{B}^+(A) = \mathcal{B}^-(\overline{A}) \tag{4}$$
$$\mathcal{B}^+(A) = \mathcal{B}(A) \cap \overline{A} \tag{5}$$
$$\mathcal{B}^-(A) = \mathcal{B}(A) \cap A \tag{6}$$
$$\mathcal{B}(A) = \mathcal{C}(A) \cap \mathcal{C}(\overline{A}) \tag{7}$$

The axioms defining a closure space are also part of the definition of a *Kuratowski closure space*, which is an alternative definition of a *topological space*. More precisely, the only missing axiom that makes a closure space Kuratowski is idempotence[1], that is $\mathcal{C}(\mathcal{C}(A)) = \mathcal{C}(A)$.

Definition 7. *A* topological space *is a closure space where the closure operator is idempotent, that is, for all $A \subseteq X$, $\mathcal{C}(\mathcal{C}(A)) = \mathcal{C}(A)$.*

The correspondence between the Kuratowski definition (Definition 7) and the open sets definition (Definition 1) can be sketched as follows. To view a topological space defined in terms of open sets as a closure space, one defines $\mathcal{C}(A)$ as the smallest closed set containing A. For the converse, one uses the definition of an open set in a closure space, as given in Definition 5 (noting that closure is already assumed to be idempotent, by the Kuratowski definition).

---

[1] When recovering the definition of a topological space via open sets from the Kuratowski definition, it is noteworthy that the preservation of binary unions is sufficient to prove that *arbitrary* unions of *open* sets are open.

2.3   Graphs as Closure Spaces

Discrete spatial structures typically come in the form of a graph. A graph is described by its set of nodes X and its *connectedness* binary relation $R \subseteq X \times X$. A closure operator $\mathcal{C}_R$ can be derived from R as follows.

Definition 8. *Given a set X and a relation $R \subseteq X \times X$, define the closure operator $\mathcal{C}_R(A)$ as follows:* $\mathcal{C}_R(A) = A \cup \{x \in X \mid \exists a \in A.(x,a) \in R\}$.

Proposition 3. *The pair $(X, \mathcal{C}_R)$ is a closure space.*

Closure operators obtained by Definition 8 are not necessarily idempotent. This is intimately related to reflexivity and transitivity of R, as shown by Lemma 11 in [26], that we rephrase below.

Lemma 1. *The operator $\mathcal{C}_R$ is idempotent if and only if the reflexive closure $R^=$ of R is transitive.*

Note that, when R is transitive, so is $R^=$, thus $\mathcal{C}_R$ is idempotent. The reverse implication is not true, as one may have $(x,y) \in R$, $(y,x) \in R$, but $(x,x) \notin R$.

Finally, we recall that in [27], a discrete variant of the topological definition of the boundary of a set A is given, for the case where a closure operator is derived by Definition 8 from a reflexive and symmetric relation. Therein, in Lemma 5, it is proved that the definition coincides with the one we provided (see Definition 6). The latter is entirely given in terms of closure and interior, and coincides with the definition of boundary in a topological space. Therefore we preferred to adopt it for the general case of a closure space in this tutorial.

2.4   Quasi-discrete Closure Spaces

We now discuss interesting structures that do not necessarily have idempotent closure. See also Lemma 9 of [26] and the subsequent statements. We shall see that there is a very strong relation between the definition of a *quasi-discrete* space, given below, and graphs.

Definition 9. *A closure space is* quasi-discrete *if and only if one of the following equivalent conditions holds:*

(i) each $x \in X$ has a minimal neighbourhood[2] N_x;
(ii) for each $A \subseteq X$, $C(A) = \bigcup_{a \in A} C(\{a\})$.

The following is proved as Theorem 1 in [26].

Theorem 1. *A closure space $(X, \mathcal{C})$ is quasi-discrete if and only if there is a relation $R \subseteq X \times X$ such that $\mathcal{C} = \mathcal{C}_R$.*

---

[2] A *minimal neighbourhood* of x is a set that is a neighbourhood of x and is included in all other neighbourhoods of x.

Example 2. Existence of minimal neighbourhoods does not depend on finiteness of the space, and they do not even depend on the existence of a "closest element" for each point. To see this, consider the rational numbers $\mathbb{Q}$, equipped with the relation $\leq$. Such a relation is reflexive and transitive, thus the closure space $(\mathbb{Q}, \mathcal{C}_\leq)$ is topological and quasi-discrete.   •

Example 3. An example of a topological closure space which is not quasi-discrete is the set of real numbers equipped with the Euclidean topology (the topology induced by arbitrary union and finite intersection of open intervals). To see that the space is not quasi-discrete, one applies Definition 9. Consider an open interval (x, y). We have $\mathcal{C}((x, y)) = [x, y]$, but for each point z, we also have $\mathcal{C}(z) = [z, z] = \{z\}$. Therefore $\bigcup_{z \in (x,y)} \mathcal{C}(z) = \bigcup_{z \in (x,y)} \{z\} = (x, y) \neq [x, y]$.   •

Example 4. The reader may think that quasi-discreteness is also related to the space having a smaller cardinality than that of the real numbers. This is not the case. To see this, just equip the real numbers with an arbitrary relation in a similar way to Example 2. The obtained closure space is quasi-discrete.   •

Summing up, whenever one starts from an arbitrary relation $R \subseteq X \times X$, the obtained closure space $(X, \mathcal{C}_R)$ enjoys minimal neighbourhoods, and the closure of a set A is the union of the closure of the singletons composing A. Furthermore, such nice properties are only verified in a closure space when there is some R such that the closure operator of the space is derived from R. In the next example, we show some aspects of quasi-discreteness.

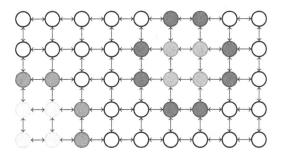

Fig. 2. A graph inducing a *quasi-discrete* closure space (Color figure online)

Example 5. Every graph induces a *quasi-discrete* closure space. For instance, we can consider the (undirected) graph depicted in Fig. 2. Let R be the (symmetric) binary relation induced by the graph edges, and let Y and G denote the set of *yellow* and *green* nodes, respectively. The closure $\mathcal{C}_R(Y)$ consists of all *yellow* and *red* nodes, while the closure $\mathcal{C}_R(G)$ contains all *green* and *blue* nodes. The interior $\mathcal{I}(Y)$ of Y contains a single node, i.e. the one located at the bottom-left in Fig. 2. On the contrary, the *interior* $\mathcal{I}(G)$ of G is empty. Finally, we have that $\mathcal{B}(G) = \mathcal{C}_R(G)$, while $\mathcal{B}^-(G) = G$ and $\mathcal{B}^+(G)$ consists of the *blue* nodes.   •

2.5 Paths in Closure Spaces

A very useful notion to reason about spatial structures is that of paths. A uniform definition of paths for all closure spaces is, however, non-trivial. It is possible, and often done, to borrow the notion of path from topology. However, the extension is not fully satisfactory. For example, the topological definition does not yield graph-theoretical paths in the case of quasi-discrete closure spaces. As a pragmatic solution, we provide a natural definition of paths for interesting classes of closure spaces[3]. In particular, in this section we introduce the definition of continuous function, which restricts to topological continuity in the setting of idempotent closure spaces, and define paths in the case of quasi-discrete closure spaces. We postpone the discussion of paths for Euclidean topological space to Sect. 9.

Definition 10. *A* continuous function $f : (X_1, \mathcal{C}_1) \rightarrow (X_2, \mathcal{C}_2)$ *is a function* $f : X_1 \rightarrow X_2$ *such that, for all* $A \subseteq X_1$, *we have* $f(\mathcal{C}_1(A)) \subseteq \mathcal{C}_2(f(A))$.

Definition 11. *A (quasi-discrete)* path *for quasi-discrete closure space* $(X, \mathcal{C})$ *is a continuous function* $p : (\mathbb{N}, \mathcal{C}_{\text{Succ}}) \rightarrow (X, \mathcal{C})$, *where* $(\mathbb{N}, \mathcal{C}_{\text{Succ}})$ *is the closure space where* $(n, m) \in \text{Succ} \iff m = n + 1$.

As a matter of notation, we call p a path *from* x, and write $p : x \rightsquigarrow \infty$, when $p(0) = x$. We write $y \in p$ whenever there is i such that $p(i) = y$.

It is worth noting that graph-theoretical and quasi-discrete paths coincide.

Proposition 4. *Given a (quasi-discrete) path* p *in a quasi-discrete space* $(X, \mathcal{C}_R)$, *for all* $i \in \mathbb{N}$ *with* $p(i) \neq p(i + 1)$, *we have* $(p(i), p(i + 1)) \in R$, *i.e., the image of* p *is a (graph theoretical, countably infinite) path in the graph of* R. *Conversely, each countable path in the graph of* R *uniquely determines a quasi-discrete path.*

2.6 Distance Spaces and Metric Spaces

For the analysis of CAS often quantitative spatial information is required as well. Closure spaces can be enriched with such information by introducing *distance spaces* and *metric spaces*. We briefly mention some of them here. The interested reader may refer to Sect. 3.1 of [31] to get further insight on distance spaces. In particular, *qualitative* notions such as "being at a short distance" can be modelled in distance spaces but not in metric spaces.

Definition 12. *A* distance space *is a pair* (X, d) *of a set* X *and a function* $d : X \times X \rightarrow \mathbb{R}$ *such that, for all* $x, y \in X$, $d(x, y) = 0 \iff x = y$, *and* $d(x, y) \geq 0$. *If, in addition also* $d(x, y) = d(y, x)$ *and* $d(x, z) \leq d(x, y) + d(y, z)$ *hold, then the space is called a* metric space.

---

[3] We leave open the possibility to change this notion, in chosen classes of closure spaces, practically making our theory dependent on such choice. The theoretical question of finding a uniform notion of path is left for future work.

Definition 13. *A metric space can be equipped with the metric topology where the open sets are induced by the basis of* open balls, *that is, B is the collection of subsets $o \subseteq X$ such that there are $k \in \mathbb{R}$, $y \in X$ with $o = \{x \in X \mid d(x,y) < k\}$.*

The above definitions naturally extend to distance/metric closure spaces, when X is equipped with a closure operator $\mathcal{C}$.

2.7 Hierarchy of Closure Spaces

In Fig. 3, the hierarchy of closure spaces with respect to quasi-discreteness is shown. All finite spaces are quasi-discrete, as closure of finite sets is determined by that of the singletons, by inductive application of the axiom $\mathcal{C}(A \cup B) = \mathcal{C}(A) \cup \mathcal{C}(B)$. Obviously, there are quasi-discrete infinite spaces (any infinite graph interpreted as a closure space is an example). A quasi-discrete space which is also topological is the space associated with any complete graph. In this case, for any set, $\mathcal{C}(A)$ is the whole space, thus closure is idempotent. One may wonder what kind of topology is associated with any complete graph, seen as a quasi-discrete closure space. This is precisely the *indiscrete* topology, which is also characterised, using the open sets definition, as the topology generated by a basis in which the only open sets are the empty set and the whole space. Indeed, there is another way to equip a set of points with a relation, in such a way that the resulting graph is an idempotent closure space. Namely, one can just consider the identity relation. Then, the closure operator of the obtained space is the identity function (which is clearly idempotent). Using the open sets definition, the obtained topology is the *discrete* topology, where the open sets are all the singletons. Finally there are closure spaces that are neither topological nor quasi discrete. The most obvious example is the *coproduct* (disjoint union) of a topological and a quasi-discrete, but not topological, closure space, which is a closure space under Definition 14.

Definition 14. *Given two closure spaces $(X, \mathcal{C}^X)$ and $(Y, \mathcal{C}^Y)$, consider the disjoint union of X and Y, represented as $X \uplus Y = X' \cup Y'$ with $X' = \{(1,x) \mid x \in X\}$ and $Y' = \{(2,y) \mid y \in Y\}$. In order to equip the set $X \uplus Y$ with a closure operator, for each $A \subseteq X \uplus Y$, let $A^X = \{x \mid (1,x) \in A\}$ and $A^Y = \{y \mid (2,y) \in A\}$. Define $\mathcal{C}(A) = \{(1,x) \mid x \in \mathcal{C}^X(A^X)\} \cup \{(2,y) \mid y \in \mathcal{C}^Y(A^Y)\}$.*

3 Modal Logics

Now that we have seen the various mathematical structures to represent space, we turn to *reasoning* about space and spatial properties. A successful way to reason about properties of structures in general is by way of a logic with which to express the properties of interest. Spatial logics have been mainly studied from the point of view of *modal* logics and there is a historical reason for that. In a seminal work of 1938, Tarski presented a spatial, and in particular topological, interpretation of modal logic, which paved the way for a new line of research

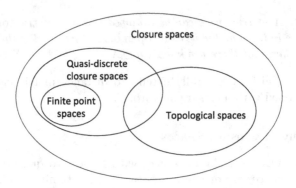

Fig. 3. The hierarchy of closure spaces.

on the relationship between topological spaces and modal logics, culminating in the proof, by Tarsky and McKinsey in 1944, that the simple (and decidable) modal logic $\mathcal{S}4$ is complete when interpreting the *possibility* modality $\Diamond$ of $\mathcal{S}4$ as *closure* on the reals or any similar metric space. The modal logics approach to spatial logic contemplates *purely spatial* logics, meaning that they deal with the spatial configuration of a system at a certain point in time, considering a particular 'snapshot', with no subsequent temporal evolution.

In this section we briefly recall the notions of modal logics that are directly relevant to the topic of this tutorial. For a more extended introduction to modal logics the reader is invited to consult more authoritative sources. A recommended reference for modal logics is [8], whereas for the study of their spatial interpretation we refer to [40].

3.1   Modal Logics

We introduce the syntax of a basic modal logic, that we denote with $\mathcal{L}$, which forms the basis for most other logics presented in this tutorial.

Definition 15. *Fix a set of proposition letters AP, also called Atomic Propositions. Let p denote an arbitrary letter. The syntax of $\mathcal{L}$ is described by the grammar:*

$$\Phi ::= p \mid \top \mid \bot \mid \neg\Phi \mid \Phi \wedge \Phi \mid \Phi \vee \Phi \mid \Box\Phi \mid \Diamond\Phi$$

The relational semantics of $\mathcal{L}$ is given using *frames* and *models*.

Definition 16. *Fixed a set AP of atomic propositions, a (Kripke) frame is a pair (X, R) of a set X and an accessibility relation $R \subseteq X \times X$. A model $\mathcal{M} = ((X, R), \mathcal{V})$ consists of a frame (X, R) and a valuation $\mathcal{V} : AP \to \wp(X)$, assigning to each atomic proposition the set of points (also called 'possible worlds') that satisfy it.*

Truth of a formula is defined at a specific point $x \in X$.

Definition 17. *Truth* $\models$ *of modal formulas in model* $\mathcal{M} = ((X, R), \mathcal{V})$ *at point* $x \in X$ *is defined by induction as follows:*

$$
\begin{aligned}
\mathcal{M}, x &\models \top && \Longleftrightarrow\ true \\
\mathcal{M}, x &\models \bot && \Longleftrightarrow\ false \\
\mathcal{M}, x &\models p && \Longleftrightarrow\ x \in \mathcal{V}(p) \\
\mathcal{M}, x &\models \neg\varphi && \Longleftrightarrow\ \mathsf{not}\ \mathcal{M}, x \models \varphi \\
\mathcal{M}, x &\models \varphi \wedge \psi && \Longleftrightarrow\ \mathcal{M}, x \models \varphi\ \mathsf{and}\ \mathcal{M}, x \models \psi \\
\mathcal{M}, x &\models \Box\varphi && \Longleftrightarrow\ \forall y \in X.(x, y) \in R \implies \mathcal{M}, y \models \varphi \\
\mathcal{M}, x &\models \Diamond\varphi && \Longleftrightarrow\ \exists y \in X.(x, y) \in R \wedge \mathcal{M}, y \models \varphi
\end{aligned}
$$

The operators $\neg$, $\wedge$ and $\vee$ are the common Boolean operators negation, conjunction and disjunction, respectively. There are two modal operators. The operator $\Box\varphi$ denotes necessity and $\Diamond\varphi$ possibility. A point $x \in X$ satisfies *necessarily* φ, denoted as $\Box\varphi$, if φ holds in all points (worlds) y that are accessible from x via the accessibility relation R. A point $x \in X$ satisfies *possibly* φ, denoted as $\Diamond\varphi$, if there exists a point (world) y, accessible from x via the accessibility relation R, such that φ holds in y.

3.2   Modal Logics of Space

The presentation in this section is dealing with modal logics of space and is mostly based on the book chapter [40]. Many variants of spatial modal logics have been proposed. For instance, in *local topological logics* modalities identify *open sets* in which some or all points ought to satisfy a given property, while in *global topological logics* it is possible, in addition, to predicate about the satisfaction of a certain property by classes of points in the space (e.g., *all* points) and in *distance logics* truth depends upon some notion of distance between entities, just to mention a few. In the following we only address the local topological variant in some more detail and illustrate the limitations of a purely topological approach when we are interested in a spatial logic for the wider class of spatial structures that are of interest for CAS, which includes discrete spatial structures.

Topo-Models and Topo-Logics. Following the approach of Tarski, a topological space (Definition 1) may be used in place of a *frame* (see Definition 16) in order to interpret the modal logic $\mathcal{L}$ (Definition 15), obtaining *topological modal logics* or simply *topo-logics*. We first need to define a topological *model*.

Definition 18. *Fixed a set AP of* Atomic Propositions, *a topological model or* topo-model $\mathcal{M} = ((X, O), \mathcal{V})$ *consists of a topological space* (X, O) *and a valuation* $\mathcal{V} : AP \rightarrow \wp(X)$, *assigning to each atomic proposition the set of points that it satisfies.*

Truth of a formula is defined at a specific point x in space X.

Definition 19. *Truth* $\models$ *of modal formulas in model* $\mathcal{M} = ((X,O),\mathcal{V})$ *at point* $x \in X$ *is defined by induction as follows:*

$$
\begin{aligned}
\mathcal{M}, x &\models \top &&\Longleftrightarrow true \\
\mathcal{M}, x &\models \bot &&\Longleftrightarrow false \\
\mathcal{M}, x &\models p &&\Longleftrightarrow x \in \mathcal{V}(p) \\
\mathcal{M}, x &\models \neg\varphi &&\Longleftrightarrow \text{not } \mathcal{M}, x \models \varphi \\
\mathcal{M}, x &\models \varphi \wedge \psi &&\Longleftrightarrow \mathcal{M}, x \models \varphi \text{ and } \mathcal{M}, x \models \psi \\
\mathcal{M}, x &\models \Box\varphi &&\Longleftrightarrow \exists o \in O.(x \in o \text{ and } \forall y \in o.\mathcal{M}, y \models \varphi) \\
\mathcal{M}, x &\models \Diamond\varphi &&\Longleftrightarrow \forall o \in O.(x \in o \text{ implies } \exists y \in o.\mathcal{M}, y \models \varphi)
\end{aligned}
$$

The usual De Morgan-style dualities hold, including $\mathcal{M}, x \models \Box\varphi \Longleftrightarrow \mathcal{M}, x \models \neg\Diamond\neg\varphi$. The interpretation of formulas identifies regions of space that depend on the valuation $\mathcal{V}$. In particular, note that the operation $\Box\varphi$ identifies the topological *interior* of the region where φ holds. Dually, $\Diamond\varphi$ denotes the topological *closure* of φ. An example formula which is widely used is the *boundary* of a property, which can be introduced as a derived operator: $\mathcal{B}\varphi \triangleq \Diamond\varphi \wedge \neg\Box\varphi$.

Example 6. We report in Fig. 4 the first example from [40]. The topological space here is the two-dimensional Euclidean plane $\mathbb{R}^2$ equipped with the metric topology. The only proposition letter is p and the valuation of p assigns to this property the shape of a "spoon" composed of a line segment and a filled ellipse. Various formulas can denote regions such as the boundary of the spoon, including or excluding the handle, the inner part of the spoon, the whole figure without the handle, etc. The boundary and the handle are drawn much thicker than they really are only to show them more clearly.

Axiomatic Aspects and Relational Semantics. From the point of view of logics, it is important to understand the axioms and the deductive power of a logic, and in particular its completeness with respect to classes of models. A logic is complete with respect to a class of models C, if all formulas that

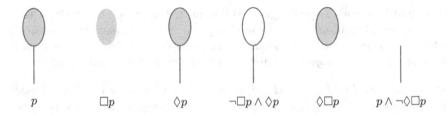

$$p \qquad \Box p \qquad \Diamond p \qquad \neg\Box p \wedge \Diamond p \qquad \Diamond\Box p \qquad p \wedge \neg\Diamond\Box p$$

Fig. 4. Topological interpretation of formulas over a topo-model. From left to right: all points satisfying atomic proposition p; the interior of the region where p holds; the closure of the region where p holds; the closure without the interior points satisfying p; the closure of the interior of the points satisfying p; those points satisfying p that are not in the closure of the interior.

are true in every model in C, i.e. they are *valid* in C, are also provable using the axioms and rules of the logic. For such a statement to make sense in the setting of topo-logics, one needs to specify that a formula φ is true in a model $\mathcal{M} = ((X, O), \mathcal{V})$ if $\mathcal{M}, x \models \varphi$ for all $x \in X$. Once this is established, various axioms are considered. As an example, we show those of the logic $\mathcal{S}4$, together with the relevant theorem. The logic $\mathcal{S}4$ is the modal logic $\mathcal{L}$ obtained when the accessibility relation of the frame is transitive and reflexive. We refer the reader to [40] for further details.

Definition 20. *The logic $\mathcal{S}4$ is $\mathcal{L}$ under the axioms K, T, 4.*

$$\Box(p \to q) \to (\Box p \to \Box q) \quad \text{(K) } distributivity$$
$$\Box p \to \Box\Box p \quad \text{(4) } transitivity$$
$$\Box p \to p \quad \text{(T) } reflexivity$$

These axioms further clarify the properties of topo-logics, as follows.

Theorem 2. *Assume the rules for* modus ponens *and* necessitation:

$$\frac{\varphi \quad \varphi \to \psi}{\psi} \qquad \frac{\varphi}{\Box\varphi}$$

The logic $\mathcal{S}4$ is complete with respect to topological models, that is, whenever φ is valid, it can be proved using the axioms $K, 4, T$, using modus ponens and necessitation.

Having seen this, and knowing that there are relational models of $\mathcal{S}4$, that is, the reflexive and transitive Kripke frames, one may wonder whether the connection is deeper. This is analysed in Sect. 2.4.1 of [40]. It is possible to derive a topological space from a frame, and the other way around, in a sound and complete way. The topological spaces that are used are the so-called *Alexandroff spaces*. These are spaces in which each point has a least open neighbourhood.

The correspondence between topological spaces and reflexive and transitive Kripke frames is not easily extended to arbitrary frames, as transitivity and reflexivity always hold in topo-logics where the basic modality is the closure.

On the other hand, requiring transitivity in all models may be a too limiting constraint, e.g., when "one-way" links are to be taken into account. This is the main reason to further investigate non-transitive concepts of spatial models in the context of closure spaces, and in particular quasi-discrete ones.

4   Spatial Logic for Quasi-discrete Closure Spaces

Whereas the local variant of modal logics of space follows a purely topological approach, in this section we lift the topological definitions to the more general setting of closure spaces and extend this framework with further spatial operators, in particular a spatial surrounded operator. The resulting logic is SLCS: a *Spatial Logic for Closure Spaces*, that we first proposed in [17]. In this section we focus our attention to quasi-discrete closure spaces; some ideas on how to generalize definitions and results to a broader class of spaces, including for instance the Euclidean topological space, will be discussed in Sect. 9.

SLCS is meant to assign to formulas a *local* meaning; for each point, formulas may predicate both on the possibility of reaching other points satisfying specific properties, or of being reached from them, along paths of the space. In [17], SLCS is equipped with two *spatial operators*: a "one step" modality, called "near" and denoted by $\mathcal{N}$, turning the closure operator $\mathcal{C}$ into a logical operator, and a binary *spatial until* operator $\mathcal{U}$, which is a spatial counterpart of the temporal *until* operator that is part of many well-known temporal logics such as CTL (for a description see e.g. [4]). In this tutorial we denote this last operator by $\mathcal{S}$ in order to avoid confusion in later sections where SLCS is further combined with temporal operators such as the temporal until operator $\mathcal{U}$ leading to a spatio-temporal logic.

$$
\begin{aligned}
\Phi ::= \ &p &&[\text{ATOMIC PROPOSITION}] \\
| \ &\top &&[\text{TRUE}] \\
| \ &\neg \Phi &&[\text{NOT}] \\
| \ &\Phi \wedge \Phi &&[\text{AND}] \\
| \ &\mathcal{N}\Phi &&[\text{NEAR}] \\
| \ &\Phi \mathcal{S} \Phi &&[\text{SURROUNDED}]
\end{aligned}
$$

Fig. 5. SLCS syntax

Assume a finite or countable set AP of *atomic propositions*. The syntax of SLCS is defined by the grammar in Fig. 5, where p ranges over AP. In Fig. 5, $\top$ denotes the truth value *true*, $\neg$ is negation, $\wedge$ is conjunction, $\mathcal{N}$ is the *closure* operator, $\mathcal{S}$ is the *surrounded* operator. From now on, with a small overload of notation, we let Φ denote the set of SLCS formulas. Next we define the interpretation of formulas.

Definition 21. *A closure model is a pair* $\mathcal{M} = ((X,\mathcal{C}),\mathcal{V})$ *consisting of a closure space* $(X,\mathcal{C})$ *and a valuation* $\mathcal{V} : AP \rightarrow 2^X$, *assigning to each atomic proposition the set of points where it holds. Whenever* $(X,\mathcal{C})$ *is quasi-discrete,* $\mathcal{M}$ *is called a* quasi-discrete closure model.

Definition 22. *Satisfaction* $\mathcal{M}, x \models \varphi$ *of formula* φ *at point* x *in quasi-discrete closure model* $\mathcal{M} = ((X,\mathcal{C}),\mathcal{V})$ *is defined, by induction on terms, as follows:*

$$
\begin{aligned}
\mathcal{M}, x \models p \quad &\Longleftrightarrow \quad x \in \mathcal{V}(p) \\
\mathcal{M}, x \models \top \quad &\Longleftrightarrow \quad true \\
\mathcal{M}, x \models \neg\varphi \quad &\Longleftrightarrow \quad \text{not } \mathcal{M}, x \models \varphi \\
\mathcal{M}, x \models \varphi \wedge \psi \quad &\Longleftrightarrow \quad \mathcal{M}, x \models \varphi \text{ and } \mathcal{M}, x \models \psi \\
\mathcal{M}, x \models \mathcal{N}\varphi \quad &\Longleftrightarrow \quad x \in \mathcal{C}(\{y \in X | \mathcal{M}, y \models \varphi\}) \\
\mathcal{M}, x \models \varphi \mathcal{S} \psi \quad &\Longleftrightarrow \quad \exists A \subseteq X. x \in A \wedge \forall y \in A.\mathcal{M}, y \models \varphi \wedge \\
&\qquad\qquad \forall z \in \mathcal{B}^+(A).\mathcal{M}, z \models \psi
\end{aligned}
$$

Atomic propositions and boolean connectives have the expected meaning. For formulas of the form $\varphi_1 \mathcal{S} \varphi_2$, the basic idea is that point x satisfies $\varphi_1 \mathcal{S} \varphi_2$ whenever there is "no way out" from a set of points, including x, and that each satisfy φ_1 unless passing by a point that satisfies φ_2. For instance, if we consider the model of Fig. 2, *yellow* nodes should satisfy *yellow* $\mathcal{S}$ *red* while *green* nodes should satisfy *green* $\mathcal{S}$ *blue*.

In Fig. 6, we present some derived operators. Besides standard logical connectives, the logic can express the *interior* ($\mathcal{I}\varphi$), the *boundary* ($\delta\varphi$), the *interior boundary* ($\delta^-\varphi$) and the *closure boundary* ($\delta^+\varphi$) of the set of points satisfying formula φ. Moreover, by appropriately using the *surrounded* operator, operators concerning *reachability* ($\varphi_1 \mathcal{R} \varphi_2$), *from-to* ($\varphi_1 \mathcal{T} \varphi_2$), *global satisfaction* ($\mathcal{E}\varphi$, *everywhere* φ) and *possible satisfaction* ($\mathcal{F}\varphi$, *somewhere* φ) can be derived.

A point x satisfies $\varphi_1 \mathcal{R} \varphi_2$ if and only if either φ_2 is satisfied by x or there exists a sequence of points after x, all satisfying φ_1, leading to a point satisfying both φ_2 and φ_1. In the second case, it is not required that x itself satisfies φ_1. For instance, both *red* and *green* nodes in Fig. 2 satisfy (*white* $\vee$ *blue*) $\mathcal{R}$ *blue*, as well as the *white* and *blue* nodes. The formula is not satisfied by the *yellow* nodes. This is so because the first node of a path leading to a *blue* node is not required to satisfy *white* or *blue*. It is easy to strengthen the notion of reachability when we want to identify all *white* nodes from which a *blue* node can be reached by requiring in addition that the first node of the path has to be *white*. This is the reason for the introduction of derived operator $\mathcal{T}$. The operator $\mathcal{T}$ is a slightly stronger version of the reachability operator $\mathcal{R}$ requiring that there is a path from a point satisfying φ_1, reaching a point satisfying φ_2 while passing only by points satisfying φ_1. Note that φ_2 is occurring also in the first argument of $\mathcal{R}$ in the definition of $\mathcal{T}$. This is because satisfaction of $\varphi_1 \mathcal{R} \varphi_2$ requires that the final node on the path satisfies both φ_1 and φ_2. We show further examples of the use of $\mathcal{T}$ shortly.

An interesting observation is that the modal spatial operators can also be characterised by definitions based on quasi-discrete paths.

Proposition 5. *We have that:*

1. $\mathcal{M}, x \models \mathcal{N}\varphi$ *if and only if there is y and $p : y \rightsquigarrow \infty$ such that $p(0) = y$ and $\mathcal{M}, y \models \varphi$ and $p(1) = x$;*
2. $\mathcal{M}, x \models \varphi_1 \mathcal{S} \varphi_2$ *if and only if $\mathcal{M}, x \models \varphi_1$ and for all $p : x \rightsquigarrow \infty. \forall l. \mathcal{M}, p(l) \models \neg\varphi_1$ implies $\exists k.0 < k \leq l. \mathcal{M}, p(k) \models \varphi_2$;*
3. $\mathcal{M}, x \models \varphi_1 \mathcal{R} \varphi_2$ *if and only if there is $p : x \rightsquigarrow \infty$ and k such that $\mathcal{M}, p(k) \models \varphi_2$ and for each j with $0 < j \leq k$, we have $\mathcal{M}, p(j) \models \varphi_1$;*
4. $\mathcal{M}, x \models \mathcal{E}\varphi_1$ *if and only if for each $p : x \rightsquigarrow \infty$ and $i \in \mathbb{N}$, $\mathcal{M}, p(i) \models \varphi_1$;*
5. $\mathcal{M}, x \models \mathcal{F}\varphi_1$ *if and only if there is $p : x \rightsquigarrow \infty$ and $i \in \mathbb{N}$ such that $\mathcal{M}, p(i) \models \varphi_1$.*

We conclude this section by pointing out that digital images are a noteworthy example of quasi-discrete closure models.

$$\bot \triangleq \neg\top \qquad\qquad \varphi_1 \vee \varphi_2 \triangleq \neg(\neg\varphi_1 \wedge \neg\varphi_2)$$
$$\mathcal{I}\varphi \triangleq \neg(\mathcal{N}\neg\varphi) \qquad\qquad \delta\varphi \triangleq (\mathcal{N}\varphi) \wedge (\neg\mathcal{I}\varphi)$$
$$\delta^-\varphi \triangleq \varphi \wedge (\neg\mathcal{I}\varphi) \qquad\qquad \delta^+\varphi \triangleq (\mathcal{N}\varphi) \wedge (\neg\varphi)$$
$$\varphi_1 \mathcal{R}\varphi_2 \triangleq \neg((\neg\varphi_2)\mathcal{S}(\neg\varphi_1)) \qquad\qquad \mathcal{E}\varphi \triangleq \varphi\mathcal{S}\bot$$
$$\varphi_1 \mathcal{T}\varphi_2 \triangleq \varphi_1 \wedge ((\varphi_1 \vee \varphi_2)\mathcal{R}\varphi_2) \qquad\qquad \mathcal{F}\varphi \triangleq \neg(\mathcal{E}\neg\varphi)$$

Fig. 6. Some SLCS derived operators

Example 7. Any digital image can be treated as a finite, thus quasi-discrete model. Consider the closure space consisting of the plane $X = \mathbb{N} \times \mathbb{N}$, equipped with the closure operator $\mathcal{C}_{4adj}$ defined by the 4-adjacency relation of digital topology, namely:

$$((x_1, y_1), (x_2, y_2)) \in 4adj \iff ((x_1 - x_2) + (y_1 - y_2))^2 = 1$$

In words, such closure space is a regular grid where each pixel, except those on the borders, has four neighbours, corresponding to the directions right, left, up and down[4]. On top of this space, atomic propositions can be interpreted as specifications of colours (such as, RGB coordinates, ranges, etc.), so that each point (x, y) satisfies precisely those specifications that include the colour of the pixel at coordinates (x, y). •

5 Spatial Model Checking

In [17] algorithms for spatial model checking of SLCS are presented. They are available as a proof-of-concept tool[5] called `topochecker`. The tool is implemented in OCaml[6], and can be invoked as a global model checker for SLCS. The time complexity of the spatial model checking algorithm is *linear* in the number of points and arcs in the space and in the size of the formula. The more interesting part of the algorithm is that for the surrounded operator, shown in Fig. 7. Function `Sat`, computed by Algorithm 1, implements the model checker for SLCS. The function takes as input a finite, quasi-discrete model $\mathcal{M} = ((X, \mathcal{C}_R), \mathcal{V})$ and a SLCS formula φ, and returns the set of all points in X satisfying φ. The function is inductively defined on the structure of φ and, following a bottom-up approach, computes the resulting set via an appropriate combination of the recursive invocations of `Sat`on the subformulas of φ. When φ is of the form $\top$, p, $\neg\varphi_1$ or $\varphi_1 \wedge \varphi_2$, the definition of `Sat`$(\mathcal{M}, \varphi)$ is straightforward. To compute the set of points satisfying $\mathcal{N}\varphi_1$, the closure operator $\mathcal{C}$ of the space is applied to the set of points satisfying φ_1. When φ is of the form $\varphi_1\mathcal{S}\varphi_2$, function `Sat` relies on the function `CheckSurr` defined in Algorithm 2.

---

[4] This notion of neighbourhood is also known as the von Neumann neighbourhood of radius 1.

[5] Web site: http://www.github.com/vincenzoml/topochecker.

[6] See http://ocaml.org.

This global flooding algorithm and its operation is illustrated in an informal way in Fig. 8 for the formula *yellow* $\mathcal{S}$ *red*. First all points that certainly do not satisfy the formula (i.e. those that are neither *yellow* nor *red*) are marked black, then yellow points that are neighbours of black ones are removed from the set of points that potentially satisfy the formula by turning them black in successive steps until a fixed point is reached. The remaining yellow points satisfy the formula. The actual algorithm incorporates several optimisations.

Function Sat($\mathcal{M}, \varphi$)
   Input: Finite, quasi-discrete
      closure model
      $\mathcal{M} = ((X, \mathcal{C}), \mathcal{V})$, formula φ
  Output: Set of points
      $\{x \in X \mid \mathcal{M}, x \models \varphi\}$
  Match φ
    case $\top$: **return** X
    case p : **return** $\mathcal{V}(p)$
    case $\neg\varphi_1$:
      let $P = $ Sat($\mathcal{M}, \varphi_1$)
      return $X \setminus P$
    case $\varphi_1 \wedge \varphi_2$:
      let $P = $ Sat($\mathcal{M}, \varphi_1$)
      let $Q = $ Sat($\mathcal{M}, \varphi_2$)
      return $P \cap Q$
    case $\mathcal{N}\varphi_1$:
      let $P = $ Sat($\mathcal{M}, \varphi_1$)
      return $\mathcal{C}(P)$
    case $\varphi_1 \mathcal{S} \varphi_2$:
      return CheckSurr
      ($\mathcal{M}, \varphi_1, \varphi_2$)

Function CheckSurr ($\mathcal{M}, \varphi_1, \varphi_2$)
   Input: Finite, quasi-discrete
      closure model
      $\mathcal{M} = ((X, \mathcal{C}), \mathcal{V})$,
      formulas φ_1, φ_2
  Output: Set of points $\{x \in X \mid$
      $\mathcal{M}, x \models \varphi_1 \mathcal{S} \varphi_2\}$
  var $V := $ Sat($\mathcal{M}, \varphi_1$)
  let $Q = $ Sat($\mathcal{M}, \varphi_2$)
  var $T := \mathcal{B}^+(V \cup Q)$
  while $T \neq \emptyset$ **do**
    var $T' := \emptyset$
    for $x \in T$ **do**
      let $N = pre(x) \cap V$
      $V := V \setminus N$
      $T' := T' \cup (N \setminus Q)$
    $T := T'$;
  return V

Algorithm 1: Decision procedure for the model checking problem of SLCS.

Algorithm 2: Checking *surrounded* formulas in a quasi-discrete closure space.

Fig. 7. Spatial model checking algorithm: surrounded operator.

As an illustration, we show the application of the spatial model checker on an image (see Fig. 9) representing a maze, in which the pixels form the points of a finite quasi-discrete closure space and in which the relation between points is given as a regular graph, much like in Fig. 2, but not shown in the image. The green area is the exit. The blue areas (rectangular and round spots) are starting points.

All model checking examples in this tutorial use **topochecker** as supporting tool. Consequently, we use the tool's frontend syntax for the SLCS formulas in such examples. The correspondence is straightforward.

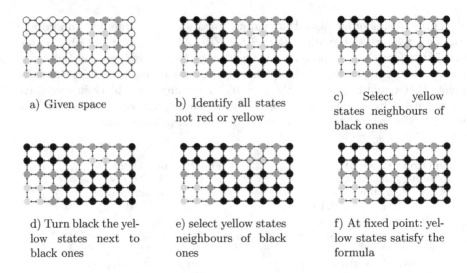

a) Given space

b) Identify all states not red or yellow

c)   Select   yellow states neighbours of black ones

d) Turn black the yellow states next to black ones

e) select yellow states neighbours of black ones

f) At fixed point: yellow states satisfy the formula

Fig. 8. Flooding algorithm evaluating *Yellow S Red* (Color figure online)

The spatial model checker can be used to identify (sets of) points, by marking them with a specific colour of choice, that satisfy particular spatial properties. In this case three formulas are used to identify interesting areas. The formulas make indirect use of the S operator, by means of the derived operators R and T, discussed earlier. In the following we are interested in three sets of points, each characterised by an *SLCS* formula.

(1) White points from which an exit can be reached.

$$\texttt{toExit} = [\texttt{white}]\, T\, [\texttt{green}]$$

The model checking result is shown in Fig. 10. Points that satisfy this formula are the yellow and orange ones[7].

(2) Regions containing a starting point (blue area) from which an exit can be reached. These are white points from which it is both possible to reach an exit and to reach a blue point (starting point).

$$\texttt{fromStartToExit} = \texttt{toExit}\,\&\,([\texttt{white}]\, T\, [\texttt{blue}])$$

Points that satisfy this formula are the orange ones in Fig. 10.

(3) Points that are starting points and from which the exit can be reached. These are the blue points depicted as a rectangular shape for easy recognition in the image.

$$\texttt{startCanExit} = [\texttt{blue}]\, T\, \texttt{fromStartToExit}$$

---

[7] Actually one colour (yellow) could have been used, but in order to show multiple verification results combined in one picture, the orange points show the points that are yellow but that also satisfy the second property.

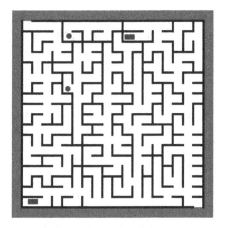

Fig. 9. A maze. (Color figure online)

Fig. 10. Model checker output. (Color figure online)

Points that satisfy this formula are shown in red in Fig. 10. These are indeed only the two small rectangular shapes (coloured red as result) and not the two round shapes (that remained blue), since from the latter it is not possible to reach an exit.

A further example is the use of spatial model checking to study the formation of patterns in bio-chemical systems. Alan Turing conjectured in [39] that pattern formation is a consequence of the coupling of reaction and diffusion phenomena involving different chemical substances distributed in the same physical space. Such behaviour can be described by a set of reaction-diffusion equations. We show here a variant in which wave-like patterns emerge. This is a variant of the models used in [3,29,37].

We use a reaction-diffusion system that is discretised, according to a Finite Difference scheme, as a system of ordinary differential equations whose variables are organised in a $K \times K$ rectangular grid. As before, such a grid is considered as a bi-directional graph, where each node $(i,j) \in L = \{1,\ldots,K\} \times \{1,\ldots,K\}$ represents a discrete location, edges connect pairs of neighbouring nodes along four directions as in the 4-adjacency relation in digital topology of Example 7.

We consider two chemical substances A and B in the $K \times K$ grid, obtaining the system:

$$\begin{cases} \frac{dx_{i,j}^A}{dt} = R_1 x_{i,j}^A x_{i,j}^B - x_{i,j}^A + R_2 + D_1(\mu_{i,j}^A - x_{i,j}^A) & i = 1..,K,\ j = 1,..,K, \\ \frac{dx_{i,j}^B}{dt} = R_3 x_{i,j}^A x_{i,j}^B + R_4 + D_2(\mu_{i,j}^B - x_{i,j}^B) & i = 1..,K,\ j = 1,..,K, \end{cases} \quad (8)$$

where: $x_{i,j}^A$ and $x_{i,j}^B$ are the concentrations of the two chemical substances in the location (i,j); R_i, $i = 1,\ldots,4$ are the parameters that define the reaction between the two species; D_1 and D_2 are the diffusion constants, i.e. constants that define the movement of the molecules between locations; $\mu_{i,j}^A$ and $\mu_{i,j}^B$ are the average concentrations of the locations adjacent to location (i,j), that is

$$\mu_{i,j}^n = \frac{1}{|\nu_{i,j}|} \sum_{\nu \in \nu_{i,j}} x_\nu^n \qquad n \in \{A, B\}, \tag{9}$$

where $\nu_{i,j}$ is the set of indices of locations adjacent to (i,j). Note that if the average concentration of a substance in the adjacent nodes of (i,j) is higher than in the node (i,j) itself, then there is a flow of the substance entering (i,j) (assuming that the diffusion parameter has a positive value). If the average concentration is lower, there is a flow going out of (i,j) to its neighbours.

The flow towards a location could happen in two ways. In the first case, all neighbours are equally "attractive", and so the location (i,j) could receive an equal proportion from each of the adjacent locations. This leads to the formation of patterns with more or less round "spots" as shown also in [3, 29, 37]. However, we can also study cases in which there is a preferred direction of such flow, for example the location receives relatively more substance from the neighbours located north-west of it than from those located south-east of it. This is just one of the possibilities and it leads to the formation of rather different patterns. We can introduce such a bias by multiplying the concentrations of the four neighbours by different weights, k_i with $i \in \{north, south, west, east\}$ in such a way that their sum is equal to 4 (as would have been the case when they would all have the same weight).

$$\mu_{i,j}^n = \frac{1}{|\nu_{i,j}|}(k_{north}x_{\nu_{i,j+1}}^n + k_{south}x_{\nu_{i,j-1}}^n + k_{west}x_{\nu_{i-1,j}}^n + k_{east}x_{\nu_{i+1,j}}^n) \tag{10}$$

An example of the evolution of the concentration of substance A is shown in Fig. 11 for snapshots of the space at various points in time. These snapshots are obtained from a numerical solution of Eq. (8) by standard tools such as Octave[8]. The following values were used for the parameters: $K = 32$, $R1 = 1$, $R2 = -12$, $R3 = -1$, $R4 = 16$, $D1 = 5.6$ and $D2 = 25.5$. Whereas $k_{north} = 1.5$, $k_{south} = 0.5$, $k_{west} = 1.5$ and $k_{east} = 0.5$. The initial condition is set randomly.

Using spatial model checking we can easily identify the points in which the concentration a of A is smaller than 2 and that are surrounded by points in which the concentration is higher than 2.

$$\texttt{pattern} = [a < 2]\texttt{S}\,[a > 2]$$

The points that satisfy property **pattern** in the snapshot at $t = 10$ are shown in Fig. 12 (left). In this figure neither the edges of the graph nor the nodes where the formula is not satisfied are shown to avoid cluttering the image.

The following formula provides an indication of the regularity of the pattern. It is satisfied by points that are at least 4 steps away from the areas with low concentration. These are shown in pink in Fig. 12 (centre and right) for $t = 10$,

---

[8] See http://octave.sourceforge.net/.

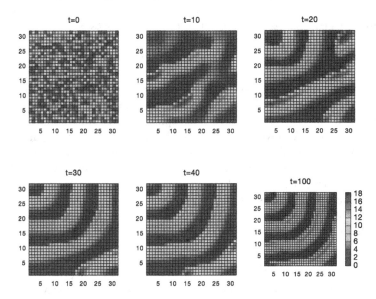

Fig. 11. Evolution of a wave-like pattern at different points in time. The colours indicate the concentration of substance A.

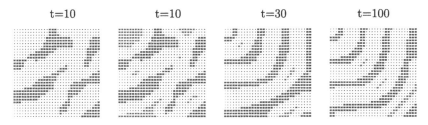

Fig. 12. Points that satisfy **pattern** in the snapshot at $t = 10$ (left); points that satisfy property **far_from_pattern** are shown in pink in the other three snapshots at $t = 10$, $t = 30$ and $t = 100$, respectively; blue points satisfy property **pattern**. (Color figure online)

$t = 30$ and $t = 100$. The points in blue are the ones satisfying property **pattern**. For more regular patterns fewer points are satisfying the property. If we would have verified the property for a 'distance' of 5 instead of 4 then in the snapshot of $t = 100$ none of the points would satisfy the property. This can been checked using the somewhere operator $\mathcal{F}$ by verifying $\neg \mathcal{F}$**far_from_pattern** (not shown in the figure).

$$\texttt{far\_from\_pattern} = !(\texttt{N}(\texttt{N}(\texttt{N}(\texttt{N}(\texttt{pattern})))))$$

6 Spatio-Temporal Logic of Closure Spaces

Starting from a spatial formalism and a temporal formalism, *spatio-temporal* logics may be defined, by introducing some mutually recursive nesting of spatial and temporal operators. Several combinations can be obtained, depending on the chosen spatial and temporal fragments, and the permitted forms of nesting of the two. For spatial logics based on *topological* spaces a large number of possibilities are explored in [31]. We investigated one such structure, in the setting of closure spaces, namely the combination of the temporal logic *Computation Tree Logic* (CTL) (see e.g. [4]) and of SLCS, resulting in the *Spatio-Temporal Logic of Closure Spaces* (STLCS). In STLCS spatial and temporal fragments may be arbitrarily and mutually nested.

First, we show the formal syntax of formulas, described by the grammar in Fig. 13, where p ranges over a finite or countable set of atomic propositions AP.

$$
\begin{array}{llll}
\Phi ::= & \top & [\text{TRUE}] & \qquad \varphi ::= \mathcal{X}\,\Phi \quad [\text{NEXT}] \\
& |\ p & [\text{ATOMIC PREDICATE}] & \qquad |\ \mathbf{F}\,\Phi \quad [\text{EVENTUALLY}] \\
& |\ \neg\Phi & [\text{NOT}] & \qquad |\ \mathbf{G}\,\Phi \quad [\text{GLOBALLY}] \\
& |\ \Phi \wedge \Phi & [\text{AND}] & \qquad |\ \Phi\,\mathcal{U}\,\Phi \quad [\text{UNTIL}] \\
& |\ \mathcal{N}\,\Phi & [\text{NEAR}] & \\
& |\ \Phi\,\mathcal{S}\,\Phi & [\text{SURROUNDED}] & \\
& |\ \mathbf{A}\,\varphi & [\text{ALL FUTURES}] & \\
& |\ \mathbf{E}\,\varphi & [\text{SOME FUTURE}] &
\end{array}
$$

Fig. 13. STLCS syntax

Besides classical Boolean connectives, and the operators of SLCS we have introduced in Sect. 4, STLCS features the CTL path quantifiers $\mathbf{A}$ ("for all paths"), and $\mathbf{E}$ ("there exists a path"). As in CTL, such quantifiers must necessarily be followed by a path-specific operator, namely $\mathcal{X}$ ("next"), $\mathbf{F}$ ("eventually"), $\mathbf{G}$ ("globally"), $\mathcal{U}$ ("until").

The definition of a model $\mathcal{M}$ of STLCS is based on the notion of Kripke frame (see Definition 16) and is given below.

Definition 23. *A model is a structure* $\mathcal{M} = ((X, \mathcal{C}), (S, \mathcal{R}), \mathcal{V}_{s \in S})$ *where* $(X, \mathcal{C})$ *is a quasi-discrete closure space,* $(S, \mathcal{R})$ *is a Kripke frame, and* $\mathcal{V}$ *is a family of valuations, indexed by* S; *for each* $s \in S$, *we have* $\mathcal{V}_s : AP \to \wp(X)$.

The truth value of a formula is defined at a point in space x at state s. Valuations of atomic propositions depend both on states and points of the space. Intuitively, there is a set of possible worlds, i.e. the states in S, and a spatial structure represented by a closure space. In each possible world there is a different valuation of atomic propositions, inducing a different "snapshot" of the spatial situation which "evolves" over time (non-deterministically); see Fig. 14 (left) where space is a two-dimensional structure, and valuations at each state are depicted by different colours. In this tutorial we assume that the spatial *structure* $(X, \mathcal{C})$ does not change over time. However, other options are possible. For instance, when space depends on S, one may consider an S-indexed family $(X_s, \mathcal{C}_s)_{s \in S}$ of closure spaces.

A *path* in the model is a sequence of *spatial models* indexed by instants of time; see Fig. 14 (right). Given that we assume that the closure space $(X, \mathcal{C})$ does not depend on S, in the sequel we will use a simplified notion of path, defined on Kripke frames, instead of on full models. In other words, we consider paths of indexes (cf. Definition 23).

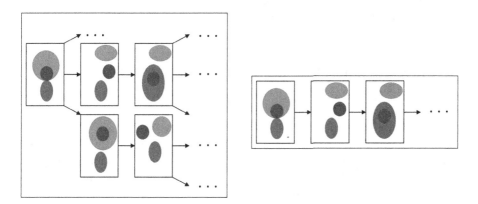

Fig. 14. In spatio-temporal logics, a temporal structure represents a computation tree of *snapshots* induced by the time-dependent valuations of the atomic propositions (left). A path in the model (right).

Definition 24. *Given Kripke frame* $\mathcal{K} = (S, \mathcal{R})$, *a* path σ *is a function from* $\mathbb{N}$ *to* S *such that for all* $n \in \mathbb{N}$ *we have* $(\sigma(i), \sigma(i+1)) \in \mathcal{R}$. *Call* $\mathcal{P}_s$ *the set of infinite paths in* $\mathcal{K}$ *rooted at* s, *that is, the set of paths* σ *with* $\sigma(0) = s$, *where, whenever for some* i *there is no* $s' \in S$ *s.t.* $(\sigma(i), s') \in \mathcal{R}$, *we let* $\sigma(j) = \sigma(i)$ *for all* $j > i$ *(self-loop completion).*

The evaluation contexts are of the form $\mathcal{M}, x, s \models \Phi$, where Φ is a STLCS formula, s is a state of a Kripke frame, and x is a point in space X. The formal semantics of the logic is given in Definition 25 and can also be found in [15].

Definition 25. *Satisfaction is defined in a model* $\mathcal{M} = ((X, \mathcal{C}), (S, \mathcal{R}), \mathcal{V}_{s \in S})$ *at point* $x \in X$ *and state* $s \in S$ *as follows:*

$$\mathcal{M}, x, s \models \top$$
$$\mathcal{M}, x, s \models p \iff x \in \mathcal{V}_s(p)$$
$$\mathcal{M}, x, s \models \neg \Phi \iff \text{not } \mathcal{M}, x, s \models \Phi$$
$$\mathcal{M}, x, s \models \Phi \wedge \Psi \iff \mathcal{M}, x, s \models \Phi \text{ and } \mathcal{M}, x, s \models \Psi$$
$$\mathcal{M}, x, s \models \mathcal{N}\Phi \iff x \in \mathcal{C}(\{y \in X | \mathcal{M}, y, s \models \Phi\})$$
$$\mathcal{M}, x, s \models \Phi \mathcal{S} \Psi \iff \exists A \subseteq X. x \in A \text{ and } \forall y \in A. \mathcal{M}, y, s \models \Phi \wedge$$
$$\text{and } \forall z \in \mathcal{B}^+(A). \mathcal{M}, z, s \models \Psi$$
$$\mathcal{M}, x, s \models \mathbf{A}\varphi \iff \forall \sigma \in \mathcal{P}_s. \mathcal{M}, x, \sigma \models \varphi$$
$$\mathcal{M}, x, s \models \mathbf{E}\varphi \iff \exists \sigma \in \mathcal{P}_s. \mathcal{M}, x, \sigma \models \varphi$$

$$\mathcal{M}, x, \sigma \models \mathcal{X}\Phi \iff \mathcal{M}, x, \sigma(1) \models \Phi$$
$$\mathcal{M}, x, \sigma \models \Phi \mathcal{U} \Psi \iff \exists n. \mathcal{M}, x, \sigma(n) \models \Psi \text{ and } \forall n' \in [0, n). \mathcal{M}, x, \sigma(n') \models \Phi$$

where $[n, n) = \emptyset$ *for all* $n \in \mathbb{N}$.

Let us proceed with a few examples. Consider the STLCS formula

$$\mathbf{E}\,\mathbf{G}\,(green\ \mathcal{S}\ blue)$$

Point x satisfies such formula in state s if there exists (**E**) a temporal path rooted at s, such that in all states (**G**), i.e. at any point in time, x satisfies atomic property *green*, and it is not possible to start from x, following edges of the spatial graph, and leave the region of points satisfying *green* in which x is located, unless passing by a point satisfying *blue*.

The mutual nesting of spatial and temporal operators permits one to express rather complex spatio-temporal properties. An example exhibiting nesting of spatio-temporal operators is the STLCS formula

$$\mathbf{E}\,\mathbf{F}\,(green\ \mathcal{S}\ (\mathbf{A}\mathcal{X}\,blue))$$

This formula is satisfied by a point x in state s if point x possibly (**E**) satisfies *green* in some future (**F**) state s', and in that state, it is not possible to leave the area of points satisfying *green* unless passing by a point that will necessarily (**A**) satisfy *blue* in the next (**X**) time step.

7   Spatio-Temporal Model Checking

Based on the formal semantics of the spatio-temporal logic STLCS, presented in the previous section, `topochecker` has been extended to deal with spatio-temporal properties. In the spatio-temporal setting, models are composed of a temporal part, which is a Kripke frame, and a spatial part, which is a finite, quasi-discrete closure space.

The model checker enriches basic STLCS by allowing users to use floating-point variables and define some atomic propositions as simple assertions on the value of such variables, e.g. comparison of such values with (floating-point) constants. It finally permits parametric macro abbreviations, that we use in the examples in the next section.

The temporal part of the model checking algorithm is a variant of the well-known Computation Tree Logic (CTL) labelling algorithm, whereas the spatial part is based on the algorithms described in Sect. 5. For more information on CTL and its model checking techniques, see e.g., [4] or [20]. Given a formula Φ and a model $\mathcal{M}$, the algorithm proceeds by induction on the structure of Φ; the output of the algorithm is the set of pairs (x, s) such that $\mathcal{M}, x, s \models \Phi$. For further details of the algorithm we refer to [15].

The complexity of the algorithm is linear in the product of three quantities: (1) the sum of the number of states and the number of transitions of the temporal model; (2) the size of the formula; (3) the sum of the number of points and the number of arcs of the space. Such efficiency is sufficient for experimenting with the logic, and it is comparable to the efficiency of classical in-memory model checking algorithms when the spatial part of the model is relatively small, as shown in the examples in the next section. Further improvements of the efficiency is part of future work.

To illustrate spatio-temporal model checking, let us continue the example of the wave pattern formation introduced in Sect. 4. We can use spatio-temporal formulas to study the evolution of the wave pattern over time. As a model we use a sequence of snapshots of the first 100 time steps of the solution of the Eq. (8) of Sect. 4. Such a sequence is a simple Kripke frame without any non-determinism. The last snapshot is repeated in an artificial way to obtain an infinite path. We can use this structure, for example, to identify which points are part of a pattern for a number of consecutive steps in the evolution. We start from the spatial property 'pattern' as in Sect. 4:

$$\texttt{pattern} = [\texttt{a<0}]\texttt{S}[\texttt{a>=0}]$$

Then we define the various periods, ranging from 3 to 10 time steps, during which a point remains part of the pattern as follows (using the frontend notation of `topochecker` also for STLCS formulas):

```
pattern2steps  = pattern & A X (A X pattern)
pattern3steps  = pattern2steps & A X (A X (A X pattern))
...

pattern10steps = pattern9steps & A X (A X (A X (··· (A X pattern)) ··· )
```

The spatio-temporal model checker can verify multiple properties simultaneously and show their results in different selected colours[9]. Note that the properties require a point to satisfy a particular spatial property to hold in the current snapshot and in several subsequent snapshots. Moreover, we can verify such

---

[9] Note that the results may involve the same points, in which case the later result overwrites the previous result.

properties starting from the initial snapshot, but also starting from any other chosen snapshot.

Figure 15 shows the evolution when the formulas are evaluated taking as initial snapshot the one at time 10, 20, 30 and 40, respectively. The results show that the pattern seems to stabilise starting from the north-western corner of the figure after which the points towards the south-eastern corner become increasingly stable, at least for 10 subsequent steps in time.

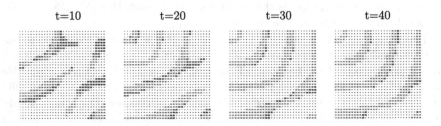

Fig. 15. Formation of a wave-like pattern; evolution in steps of 10; pink points are part of a wave for 2 subsequent steps; cyan points for 10 subsequent steps. Other colours represent the intermediate number of steps. (Color figure online)

We can also check the emergence of truly stable points, i.e. points that, once they satisfy property `pattern`, they continue to do so until the end of the simulation trace. This property can be formalised as:

$$\texttt{pattern\_permanent} = \texttt{pattern} \,\&\, \texttt{A\,G\,pattern}$$

The model checking results for this formula are shown in Fig. 16. Points that satisfy property `pattern_permanent` are orange.

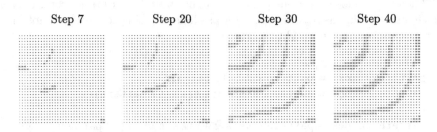

Fig. 16. Formation of a wave-like pattern; formula `pattern_permanent` evaluated starting from snapshots at times 7, 20, 30 and 40 of the simulation trace, respectively. (Color figure online)

8   Case Studies on Collective Adaptive Systems

We present several case studies to illustrate the use of spatio-temporal model checking based on closure spaces for Collective Adaptive Systems. The first case study concerns emergent behaviour in bike sharing systems. The case study builds on earlier joint work [19,34] involving some of the co-authors. The system model, based on Markov Renewal Processes, explicitly takes a part of user behaviour into account that is usually not visible in the data collected by operators of bike sharing systems, such as failure to obtain a bike when needed or difficulties to park a bike close to the desired destination. We illustrate how spatio-temporal model checking can be used to obtain a deeper insight into these problems, using a model-based approach, that complements other analysis methods.

In this first example the structures used for model checking are linear in nature. This is due to the fact that they are produced as single simulation traces of a complex behavioural model. In the second case study a richer structure is addressed in which branching time behaviour is combined with spatial aspects. Such more complex structures reflect non-deterministic choice between possible ways the behaviour may evolve in time and space and in this case this is exploited to analyse the effect of adaptation strategies in the context of public bus transportation systems. The example extends the work on spatial model checking applied to public bus transportation systems presented in [14].

Both examples use `topochecker` as supporting tool. Consequently, also in this section we use the tool's frontend syntax for the logic formulas. The correspondence to the languages defined in Sects. 4 and 6 is straightforward.

8.1   Bike Sharing

Smart bike sharing systems (BSS) have recently become a popular public transport mode in hundreds of modern cities [21,35] operating from a few (e.g. Pisa) up to hundreds or thousands of docking stations (e.g. Hangzhou, Paris, or London[10]). The principle of bike sharing is quite simple. A number of stations with docks partially filled with bicycles are placed throughout a city. Users of the service may hire any bicycle at any station at any time for private use, and must return it at some station of their choice. The initial period of, typically, thirty minutes is free of charge, after which an hourly fee is charged. The operator assumes the responsibility to maintain a high level of usage of the system.

Operating a BSS raises multiple issues such as efficiency of fuel-consuming repositioning services [23], or integration with other public transport modes. Not the least, it is important to make the service attractive to its users. An indication that user satisfaction should be addressed seriously can be found in a survey of user experience with the *Bicing* BSS in Barcelona, conducted by Froelich et al. [24]. It reports that 75 % of the users who used it for commuting between their home and study or work, stated that 'finding an available bike and parking slot' were the two

---

[10] Pisa: http://www.pisamo.it, Hangzhou: http://www.publicbike.net; Paris: http://www.velib.paris.fr, London: https://tfl.gov.uk/modes/cycling/santander-cycles.

most important problems, encountered in 76 % and 66 % cases of 212 respondents, respectively. Since a bicycle must be returned to one of the designated stations, otherwise a high fine needs to be paid, users must find a drop off station that is not full. The additional searching time, and the associated risk of undesired and unpredictable delays, and fees, are likely to affect the user's satisfaction with the system. However, user satisfaction due to such temporarily full stations is difficult to evaluate quantitatively when only data obtained from real systems is available. This is so because the cycling data alone is not sufficient to understand the *intentions* of its users, nor the predictability of the service. To investigate this issue from a different angle, a model-based approach was presented in [34], in which the point of view of 'rational agents' who participate in bike-sharing, is assumed. The results of the study reveals that cycling times in different cities, and among pairs of stations within cities, are similarly distributed[11], suggesting a possibility of a generic interpretation. The rational agent model, based on minimal assumptions about travelling and decision making, reproduces rather well the cycling time distributions in London and Pisa [34]. The analysis suggests that some features of the cycling time distribution may be related to the (un)predictability of a travel process which, as just discussed, may influence the users' satisfaction with a system. In particular, the results suggest a further analysis extending the notion of a problematic full station to the notion of a problematic *area* in which all stations are full: a full-station *cluster*. Formation of clusters and their evolution is also evident from the existing bike-sharing visualisations[12].

The main advantage of a modelling approach over data analysis is a possibility to study hypothetical cases where station configurations, traffic flows, or incentives are altered to explore the efficiency of proposed solutions to the aforementioned issues. In this example we will analyse some traces generated by a simulation model developed in [34] using `topochecker`.

The examples presented here and some further examples can be found in [19]. We are concerned with the full-station clusters as more salient to the discussion, although the extension of the same kind of formulas to clusters of empty stations is straightforward.

The bike-sharing model presented in [34] describes the dynamics of a population of rational agents, coupled to the dynamics of bicycle stations in a two-dimensional rectangle representing a city. The rational agent model is based on a Markov Renewal Process (MRP [34]). The model parameters, such as the number of stations, cycling pace and request rate, are calibrated so as to reproduce cycling times of a particular real BSS, in this particular case that of London. The result is a $7 \times 13 \, \text{km}^2$ area with a 19×38 array of stations with randomly perturbed locations, random capacities between 15 and 40 docks, and 500 agents that make, on average, 900 trips per hour. The agent behaviour is sampled randomly. However, to introduce flows, a superposition of Gaussian distributions for the origin and destination locations is used and some counter-current flows are added to improve the balance of the flow. Numerical simulation of this model

---

[11] See also [9].

[12] See, e.g. http://bikes.oobrien.com/london.

generates traces, each trace consisting of snapshots, each snapshot representing a system's state at a particular instance of time.

Agents risk, of course, that a suitable station within the area is not found. These 'bad' events affect only a small fraction of all trips if the distribution of agents' origins and destinations is spatially homogeneous, but become more relevant if some destinations are more popular than others. Presence of areas that attract more users than other areas is a reasonable assumption about real cities. An obvious consequence is that also the areas of full stations will be, as a rule, larger. However, identification and analysis of problematic areas with traditional means is not so straightforward as they dynamically evolve over time.

Spatial structure is added to the simulation model in the form of an undirected graph, whose vertices represent stations, and edges represent the nearest station connections. The graph represents 722 stations, arranged in a grid layout of 19×38 nodes as explained above. A single trace of the simulation model is used as input to the spatio-temporal model-checker. It represents the evolution of the model at specific time intervals, for a given number of steps. It provides for each station, at each step, the number of bikes parked in it, the number of free parking places and its capacity (among other information). For all the experiments described below, except the last one (related to user satisfaction), the duration of an interval is 10 min, and the number of time steps is 101. In the last experiment, we considered a time interval of 1 min and 301 time steps[13]. Starting with simple expressions of a system's state, we proceed to develop more complex formulas that nest spatial and temporal operators.

Full Stations and Clusters. First, we characterise stations that are *full*, that is, with no vacant places, and *clusters* of full stations, that is, stations that are full, and are connected only to other stations that are full in turn. These two (purely spatial) properties are formalised below.

$$\texttt{full} \quad = [\texttt{vacant}{==}0]$$
$$\texttt{cluster} = \texttt{I}\,(\texttt{full})$$

The macro abbreviation `full` uses a boolean predicate (equality), applied to the quantitative value of the atomic property [vacant]. Connectivity is expressed by the derived *interior* operator $\texttt{I}\,\varPhi = \,!\,(\texttt{N}\,(!\,\varPhi))$. Informally speaking, in an undirected graph, points satisfying $\texttt{I}\,\varPhi$ are only connected to points satisfying $\varPhi$. The smallest possible cluster is therefore composed of a full station such that its direct neighbours in the north, south, east and west directions are also full. Note that the definition of `cluster` only identifies (on purpose) these "inner" full stations and not their direct full neighbours.

Formation of Clusters. A point evolves into a cluster when it becomes full, and stays full until it becomes part of a cluster. This may be detected by the following formulas:

---

[13] The results can be reproduced using the data and scripts, provided with the source code of the tool.

$$\text{implies}(f, g) = (!\,f)\,|\,g;$$
$$\text{nextCluster} = (\text{E\,F full})\,\&$$
$$(\text{A\,G implies}(\text{full},$$
$$\text{A full\,U cluster}))$$

Here, `implies` is standard logical implication. The definition of `nextCluster` characterises points that will eventually become full and, for every future state, whenever full, they will remain full until becoming part of a cluster. This is a very strong property, that few points possess. Such points are central in cluster formation, as they represent stations that always form a cluster when they become full. In Fig. 17, these points are shown in red, in a state[14] of the simulation where there are many of them. For comparison, the boundary of the points that will become a cluster are shown in green, that is, those points satisfying $(\text{N E\,F cluster})\,\&\,(!\,\text{E\,F cluster})$.

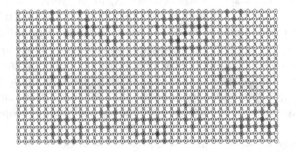

Fig. 17. Stations of time step 80 of the simulation that are on the boundary of the region of points that will eventually become a cluster (green), and stations that, whenever they are full, stay full and become part of a cluster (red). (Color figure online)

Persistence of Clusters. We can identify stations belonging to clusters that *persist* for some amount of time, that is, they last for a specific number of time steps. This situation is described for two and three time steps by the formulas:

$$\text{cluster2steps} = \text{cluster}\,\&\,(\text{A\,X cluster})$$
$$\text{cluster3steps} = \text{cluster}\,\&\,(\text{A\,X cluster2steps})$$

By combining the formulas described above with the *eventually* operator, the tool is able to detect the stations that, in any state, will eventually be part of a cluster, with specified persistence. Let us look at Fig. 18, where we show the output of a model checking session. The tool colours in red nodes that satisfy the formula `EF cluster3steps` (and thus also formulas describing shorter

---

[14] The tool is a *global* model checker, therefore it is able to produce a graph for each state of the model, related to the truth value of formulas in that particular state, even if we only show results related to one specific state.

persistence times), in blue those that satisfy `EF cluster2steps`, and in green those where formula `EF cluster` is true.

Propagation of Clusters. Another phenomenon that can be investigated using `topochecker` is the spatial propagation of clusters. Among many possible related STLCS formulas, we show how to detect points that satisfy at least one of the following: (1) they are not full, but are close to a cluster, and will necessarily become part of a cluster in the near future (`growingCluster`); or (2) they are part of a cluster, but will necessarily become not full in a short amount of time, even if still being physically close to a cluster (`shrinkingCluster`). This is achieved by the following definitions, where the macro `bdry` implements the derived 'boundary' operator δ (see Fig. 6).

$$
\begin{aligned}
\texttt{bdry(f)} &= (\texttt{N f}) \,\&\, (!\,\texttt{f}) \\
\texttt{growingCluster} &= (!\,\texttt{cluster}) \,\& \\
&\quad (\texttt{N (bdry(cluster))}) \,\&\, (\texttt{A X full}) \\
\texttt{shrinkingCluster} &= \texttt{cluster} \,\&\, (\texttt{A X (!\,full)})
\end{aligned}
$$

For instance, in time step 77 of our simulation, there are both stations that will join a cluster and stations that will leave a cluster in the next time step. We show the result in Fig. 19 using different colours for facilitating comparison of results. The stations that satisfy `cluster` are green, the stations satisfying `growingCluster` are red and stations satisfying `shrinkingCluster` are blue. The results of these formulas provide insight in the dynamics of the clusters at particular time steps, in particular the directions in which the clusters are evolving. This may be important information for the development of repositioning strategies in particular when such dynamics are repeated over time in the same areas.

User Experience. STLCS can also be used to identify specific problems related to user experience in BSSs. For example, when a user wants to leave a bike at a specific station, and that station is full, she may try to find a nearby station with

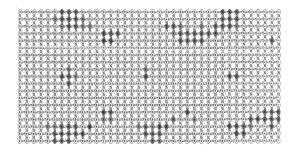

Fig. 18. Points of the initial state of the simulation that will eventually become part of a cluster (green), or of a cluster that persists for two (resp. three) time steps, coloured in blue (resp. red) (Color figure online).

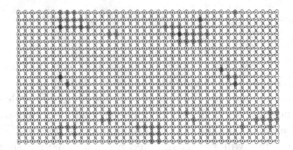

Fig. 19. Points of time step 77 of the simulation that are part of a cluster (in green), or are not full, but will become part of a cluster in one step (in red), or that are part of a cluster but will become not full in one step (in blue). (Color figure online)

available parking slots, or she may wait for some time in the same station. This behaviour may be typically sufficient to solve the problem, at the expense of a longer trip duration. One may want to check whether this procedure is effective in a few time steps. In the following formula, we check whether it is possible that, in three time steps, the user still was unable to leave the bike in the same or a nearby station, when the preferred station is full in the current state.

$$\mathtt{tripEnd} = \mathtt{full} \,\&\, (\mathtt{N}\,(\mathtt{A\,X}\,(\mathtt{full}\,\&$$
$$(\mathtt{N}\,(\mathtt{A\,X}\,(\mathtt{full}\,\&\,\mathtt{N}\,(\mathtt{A\,X}\,\mathtt{full}))))))) $$

Figure 20 shows the output from the model checker, where the red points indicate the stations where the formula is true at time step 0. The formula can be checked for various numbers of consecutive time steps, which provides an indication of how severe the problem is at a particular time of interest and a particular area. Ideally, the formula should not hold anywhere, or at most be true only in a few points and for a small number of steps.

8.2   Bus Clumping in Frequent Bus Services

The next example concerns the analysis of some specific problems in modern public transport systems such as the so-called "frequent" bus services operating in many densely populated cities. Frequent bus services are public bus services without a published timetable but with regular and frequent buses operating along pre-established routes. In such services a particular phenomenon may occur that is commonly known as *bus bunching* or *bus clumping*. Bus clumping occurs where one bus catches up with – or at least comes too close to – the bus which is in front of it. In the absence of a published timetable for frequent services the important performance metric to consider is not timetable adherence but headway, a measure of the separation between subsequent buses. This separation can be defined both in terms of distance between buses on the same route or in terms of the time between two buses on the same route passing by the same bus stop. It is possible to identify these two notions of clumping using STLCS

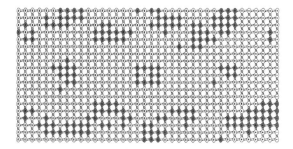

Fig. 20. Stations that are full in time step 0, where it is possible that an user applying the obvious strategy of waiting for some time, and then moving nearby, might still not find a free parking slot. (Color figure online)

on a time series of street map images on which the bus positions are projected. The problem is illustrated in Fig. 21 showing three successive "satellite view" images of the position of three different buses on the same route projected on a portion of a map of a city. The buses are shown as three small red dots on the main road. The distance between the buses is getting smaller in each successive image, resulting in the three buses being lined up in the final image as indicated by the red arrows (Fig. 21 left-bottom).

In modern public transport systems the operation depends on accurate fleet management which is supported by an automatic vehicle location (AVL) system. Since these data are used in real-time in other systems such as bus arrival prediction systems, it is very important that these data are checked for correctness. In [14] it was shown how spatial model checking can be used to check for a number of typical error conditions occurring occasionally in the data. In this approach, the data is mapped onto a digital map of a city, much like the one shown in Fig. 21, after which spatial model checking is applied on the augmented map to identify the error conditions in an automatic way. Among the typical error conditions are bus positions that are off the road, for example in a field or in water, that indicate possible errors in the transmission of the GPS data, or bus positions that are away from the planned route, which may indicate problems with the road network.

In joint follow-up work with other authors [13] we focussed on the analysis of adaptive correction strategies to mitigate the occurrence of clumping and providing better service to the public. One such strategy is based on sending a bus the instruction to wait for a short time when getting too close to the bus ahead. Such wait-requests should be only suggested by the service operators and not imposed, since there may be several circumstances in which a bus is not in the position to be able to wait, for example due to specific traffic circumstances. Also the number of wait-requests should be carefully considered. For example, if all buses on a route were sent a wait-request, and all accepted this request, then this would not solve the problem. Similarly, sending multiple wait-requests to the same bus may create unacceptable delays. Apart from these general heuristics,

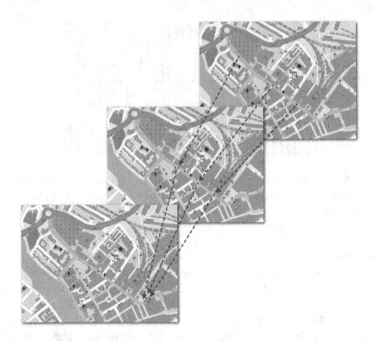

Fig. 21. Because of delays caused by boarding passengers the headway between buses is successively eroded over time until the buses are essentially 'clumped' together. (Color figure online)

there are still many options for possible strategies. In the following we illustrate how spatio-temporal model checking, based on a branching time temporal logic, can be of use to analyse the effect of such correction strategies.

First we show how spatio-temporal logic can be used to detect clumping of buses in a system trace. Such a trace consists of GPS data of a number of buses operating on the same route projected onto a digital city map. After that we show how this characterisation of clumping can be used in combination with a constructed *branching* model that captures the various ways in which wait-requests can be issued and their effect on bus positions.

Spatio-Temporal Characterisation of Clumping. Consider a single bus route, served by k buses. At each instant of time, the state of the system is completely described by a tuple of k GPS positions; therefore, a system trace is a finite sequence of such tuples. We can distinguish two different variants of clumping that differ in a subtle way. One in which two consecutive buses serving the same route are *spatially* close to each other, and one in which they pass by the same bus stop within a too short amount of time. Here we formalise the latter variant considering three buses on the same route. The input code of the model checking session is shown in Fig. 22. Formulas bus1, bus2, bus3, and busStop, are defined as colour ranges, that serve the purpose of identifying bus positions on a digital map of the city. In this example, colours are used to distinguish the different buses serving the same route, so that each bus has a specific colour. Similarly, formula busStop identifies the position of a bus stop. The formula timeConglomerate, that we explain below, is true at points of a *bus stop* whenever clumping is happening (formation of a conglomerate of buses) at that particular stop.

```
bus1 = <RED [155,155]> & <GREEN [0,0]> & <BLUE [0,0]>;
bus2 = <RED [188,188]> & <GREEN [0,0]> & <BLUE [0,0]>;
bus3 = <RED [221,221]> & <GREEN [0,0]> & <BLUE [0,0]>;
bus = bus1 | bus2 | bus3;
busStop = <RED [55,55]> & <GREEN [55,55]> & <BLUE [255,255]>;

close(x) = N^7 x;

busAtStop(x) = busStop & close(x);

busAfterBus1 = busAtStop(bus1) &
     EX busAtStop(bus2 | bus3);
busAfterBus2 = busAtStop(bus2) &
     EX busAtStop(bus1 | bus3);
busAfterBus3 = busAtStop(bus3) &
     EX busAtStop(bus1 | bus2);

timeConglomerate = (busAfterBus1 | busAfterBus2 | busAfterBus3);
```

Fig. 22. Spatio-temporal formulas for time conglomerates

A *spatio-temporal* conglomerate happens when two buses serving the same route pass by the same stop within a short amount of time. This case is subtler than the spatial one, as it does not necessary imply that the headway between two buses becomes too small. This event is described by the formula timeConglomerate, which features a combination of spatial operators (used to detect that a bus is close to a stop) and temporal operators (used to identify the spatio-temporal conglomerate). For instance, consider the formula busAfterBus1. This formula is true on points that are: (i) part of a bus stop, and close to bus1, because busAtStop must be true for bus1; (ii) such that, in the next snapshot[15], these will be part of a bus stop, and close to either bus2 or bus3. Note that the use of spatial and temporal connectives in the same formula permits one to refer to the colour of points at a specific time, and at subsequent time instants.

Figure 23 is obtained from the spatio-temporal model checker, starting from the positions of three buses serving the same route. Figure 23a–e are obtained

---

[15] More than one time step can be required. This can be achieved by repeated nesting of the EX operator. We did not do so for the sake of clarity in Fig. 23.

by mapping bus coordinates over a base map. Buses are represented by small squares of different shades of red on the roads. To make them more visible they are also highlighted by the red circles in Fig. 23b. The small dark blue square is a bus stop (see Fig. 23c). Figure 23f shows the output of the model checker when checking the formula EF timeConglomerate in the initial state shown in Fig. 23a. Indeed, Fig. 23f is the same as Fig. 23a, except for the colour of the bus stop, whose points are now turned green by the model checker, indicating that clumping happens at that stop, at some point in the future.

The Effect of Correction Strategies. Let us now address the effect of correction strategies to mitigate clumping. In particular, we use existing data (e.g. system logs) in estimating the impact of introducing new policies in a system. The spatio-temporal model checker for STLCS can be used both to detect clumping in a single system trace, as we have seen, and to analyse a *branching* model, that is, a system where at each state, non-deterministically, there may be several possible steps to different future states. Such non-deterministic models represent in a concise way a great number of possible system behaviours, depending on the choices that may be made at each execution step. We use this fact in conjunction with the idea to send buses wait-requests in order to reduce clumping. The possibility of issuing wait-requests to specific buses, or not doing so, introduces non-deterministic choice points, where some buses go ahead and others are kept on hold for a short period of time.

In more detail, consider a system trace of AVL data (e.g., provided by the bus company[16]). Let us assume that this trace reflects a typical period of the day in which clumping of buses often occurs and that the trace contains the position of the relevant buses at fixed small intervals of time of say 30 s. The trace is a sequence of tuples of k elements, where k is the number of buses serving the same route. The length of the sequence is the number of samples. Element at position i in each tuple is the position of bus i. At each step, besides the already existing transition to the next step, more transitions are added, to new states, where one or more buses wait, (therefore, their position does not change), and the other ones move as they actually did in the original system trace that was taken as starting point.

In order to illustrate the approach, a transformation algorithm was implemented. The implementation is parametric with respect to the maximum number of buses that are allowed to wait simultaneously, and the maximum number of wait instructions issued to the same bus. The input of the algorithm consists of the system trace as described above, and of a map. The state space of the branching model generated from the system presented in Fig. 23 is shown in Fig. 24; as typical in CTL model-checking, self-loops have been added to terminal states, since CTL-path formulas express properties of infinite paths, i.e. infinite sequences of states. To verify that there exist traces in which no conglomerate occurs, topochecker is applied on the formula AF timeConglomerate and

---

[16] We use artificial data for the sake of simplicity, but usage of the approach does not differ on real data.

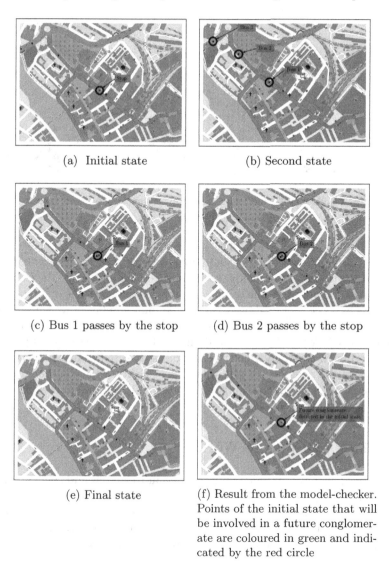

(a) Initial state

(b) Second state

(c) Bus 1 passes by the stop

(d) Bus 2 passes by the stop

(e) Final state

(f) Result from the model-checker. Points of the initial state that will be involved in a future conglomerate are coloured in green and indicated by the red circle

Fig. 23. Spatio-temporal conglomerate. (Color figure online)

the branching model. The formula is true only if a conglomerate is present on *all* (infinite) system paths. In this case no point is coloured, meaning that there exist "good" paths for each point in the model, and thus waiting-strategies that could avoid clumping in the situation represented by this trace. If the trace is indeed representative of a daily recurring situation, the conjecture is that the waiting strategies have a good chance to mitigate the clumping problem also on other days during similar periods of time.

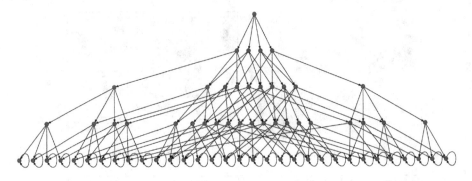

Fig. 24. Branching model obtained by augmenting a linear trace.

It is not difficult to add a facility to the model checker to generate counter examples. This can be used to obtain in an automatic way the traces that represent a waiting-strategy that avoids the emergence of clumping for the considered situation. In a second phase one could compare the quality of the different strategies and, for example, select those with the minimal number of waiting-requests.

The simple construction to consider all possible waiting strategies could be made more sophisticated by not considering a choice in *every* state of the trace, but considering choices only when a situation is detected that is known to lead to clumping in the near future with a high probability. This could be obtained, for example, using machine learning techniques. This would reduce considerably the number of states to be considered.

9   Outlook and Future Work

In this tutorial we have provided an introduction to spatial and spatio-temporal logics and model checking for closure spaces. Formal verification for continuous and discrete spatial models is still a relatively unexplored area of research. Let us conclude by briefly reviewing recent and ongoing work and open issues for the particular approach that we have addressed in this tutorial.

Further Logical Operators. Besides the basic operators and the many useful derived operators that have been presented in this tutorial, the list of operators could be further extended. Of course, this should be done in a careful way keeping in mind that the basic operators are really independent, i.e. they cannot be derived from existing operators, but also such that the properties involving these operators can be efficiently verified. Ideally, the operators have a similar semantics when interpreted on both continuous and discrete spatial models.

Along these lines, in forthcoming work [18] we investigate several new operators. The first operator $\mathcal{P}$ is capturing the notion of *propagation* and is inspired by the duality in the direction in which spatial operators can be applied. Informally, a point x satisfies $\varphi_1 \mathcal{P} \varphi_2$ if it satisfies φ_2 and it is *reachable from* a point satisfying φ_1 via a path such that all of its points, except possibly the starting point, satisfy φ_2. For instance, if we consider again the model of Fig. 2, *blue*, *green* and *white* nodes satisfy *green* $\mathcal{P}$ ¬*red* while the same formula is not satisfied by *yellow* nodes.

Furthermore, it is useful to introduce spatial operators that predicate on one or more *sets of points* rather than individual points. For example, we might want to know whether points with a particular property are actually part of a larger set of points satisfying another property, but being all connected to one another. Thus, for instance, this way one could verify whether three buses are actually on the same street segment, or whether some agents are in the same region of a maze from which they can reach the same exit.

Further extensions, involving (maybe) two sets of points, could lead to an alternative formulation of the region calculus in the setting of closure spaces.

Extensions of Spatio-Temporal Model Checking. There are various ways in which spatio-temporal model checking can be extended. First of all, further optimisations of the model checking algorithms should be investigated. In particular, in many models the spatial evolution over time could be gradual and not involve all points of the space, or the spatial evolution could be 'predictable' to some extent in specific classes of behaviours, which may allow for further optimisations.

In recent work the spatial surround operator has also been added to a different temporal logic, namely the signal temporal logic [36,37], which in turn is an extension of the Metric Interval Temporal Logic [33]. The Signal Temporal Logic is used to reason about continuous signals. Its extension with spatial modalities provides a linear time spatio-temporal logic. The two modalities are the *bounded somewhere* $\Diamond_{[w_1,w_2]}$ and the *bounded surround* operator $\mathcal{S}_{[w_1,w_2]}$, an extended adaptation of the spatial until operator of [17] to the signal models framework. The spatial somewhere operator $\Diamond_{[w_1,w_2]}\varphi$ requires φ to hold in a location reachable from the current one with a total cost (or distance) greater than or equal to w_1 and less than or equal to w_2. The surround formula $\varphi_1\,\mathcal{S}_{[w_1,w_2]}\varphi_2$ is true in a location ℓ, for the trace $\mathbf{x}$, when ℓ belongs to a set of locations A satisfying φ_1, such that its external boundary $B^+(A)$ (i.e., all the nearest neighbours external to A of locations in A) contain only locations satisfying φ_2. Furthermore, locations in $B^+(A)$ must be reached from ℓ by a shortest path of cost between w_1 and w_2. Hence, the surround operator expresses the topological notion of being surrounded by a φ_2-region, with additional metric constraints. The *everywhere* operator can be derived as the dual of the somewhere operator $\Box_{[w_1,w_2]}\varphi := \neg\Diamond_{[w_1,w_2]}\neg\varphi$ requiring φ to hold in all the locations reachable from the current one with a total cost (or distance) between w_1 and w_2.

The logic has both a boolean semantics and a quantitative semantics. The former defines when a formula is satisfied, the latter provides an indication of the robustness with which a formula is satisfied [5–7], i.e. how susceptible it is to changing its truth value, for example, as a result of a perturbation in the signals. SSTL is interpreted on spatio-temporal, real-valued signals.

Such spatio-temporal traces can be obtained by simulating a stochastic model or a deterministic model, i.e. specified by a set of differential equations. In [36] the framework of patch-based population models is discussed, which generalise population models and are a natural setting from which both stochastic and deterministic spatio-temporal traces of the considered type emerge. An alternative source of traces are measurements of real systems. Efficient monitoring algorithms have been developed for SSTL and implemented in a prototype spatio-temporal model checker.

Spatio-temporal model checking can also be conceived for models with stochastic time. For example, in the case study of the London bike sharing system we could be interested to know how likely it is that a station becomes part of a cluster of full stations. We are currently developing a statistical model checking approach for such probabilistic spatio-temporal properties. Another approach could be to extend the spatio-temporal logic itself with forms of stochasticity. There are different options. One is to add stochasticity to time, another to add it to space, a third option would be to add stochasticity to both. Furthermore, as we have seen in the Turing pattern examples, partial differential equations can be used to define large spatial collective systems featuring interaction and mobility. In some CAS it may be of interest to study the behaviour of a single individual in the context of a large spatial collective system. This can be done using for example fluid and mean field model checking [10, 11, 32]. Combining spatial model checking with such mean field model checking approaches would be very interesting but represents also a great challenge because of issues of scalability of the approach, in particular for what concerns the spatial aspects.

Path-Based Definition of SLCS Logics. As we have seen in Sect. 4, Proposition 5 allows us to reformulate the definition of the $\mathcal{S}$ using the notion of path. In other words, one could take the characterization of $\mathcal{S}$ as in Proposition 5 and use it as a definition; in the context of *quasi-discrete* closure spaces, the latter would be equivalent to Definition 22.

Interestingly, the very same, path-based, definition provides more intuitive results when interpreted on topological spaces, which instead is not the case when using Definition 22 for $\mathcal{S}$. We show this by means of an example. With this in mind, we define an *Euclidean path* in Euclidean topological space $(X, \mathcal{C}^X)$ as any continuous function $p : (\mathbb{R}_{\geq 0}, \mathcal{C}^{\mathbb{R}_{\geq 0}}) \to (X, \mathcal{C}^X)$, where $\mathbb{R}_{\geq 0}$ is the half-line $\{x \in \mathbb{R} \mid 0 \leq x\}$ and $\mathcal{C}^{\mathbb{R}_{\geq 0}}$ is the Euclidean closure operator.

Example 8. We define two models based on the Euclidean topology over $\mathbb{R}^2$, seen as a closure space $(\mathbb{R}^2, \mathcal{C})$, where $\mathcal{C}$ is the standard closure operator in $\mathbb{R}^2$. We use propositions b, w, g, depicted in Fig. 25 as black, white and grey areas, respectively. Consider the sets $H = \{(x,y) \mid x^2 + y^2 < 1\}$, $H^< = \{(x,y) \mid x^2 + y^2 = 1 \land x < 0\}$, $H^\geq = \{(x,y) \mid x^2 + y^2 = 1 \land x \geq 0\}$. Let $\mathcal{M}_i = ((\mathbb{R}^2, \mathcal{C}), \mathcal{V}_i)$, for $i \in \{1, 2\}$. Fix valuations as follows: $\mathcal{V}_1(b) = H \cup H^<, \mathcal{V}_1(w) = \mathbb{R}^2 \setminus \mathcal{V}_1(b), \mathcal{V}_1(g) = \emptyset$, $\mathcal{V}_2(b) = \mathcal{V}_1(b), \mathcal{V}_2(w) = H^\geq \setminus H, \mathcal{V}_2(g) = \mathbb{R}_2 \setminus (H \cup H^< \cup H^\geq)$. Let $x \in H$. Using the path-based definition of $\mathcal{S}$ in Proposition 5, we have $\mathcal{M}_1, x \models b\,\mathcal{S}\,w$, and $\mathcal{M}_2, x \not\models b\,\mathcal{S}\,w$, as there are paths starting at a black point in $\mathcal{M}_2$ and reaching a grey point, which does not satisfy b, without passing by white points. If we consider the closure space based semantics in Definition 22, the expectation is that $b\,\mathcal{S}\,w$ holds at x in $\mathcal{M}_1$, which is true by the choice $A = H \cup H^<$, but note that $\mathcal{B}^+(A) = H^{\geq}$. For this reason, we also have $\mathcal{M}_2, x \models b\,\mathcal{S}\,w$ by the choice $A = H \cup H^<$, which is not what one would expect when thinking of the area H being "surrounded" by white points.

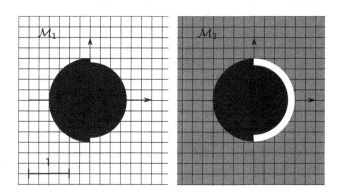

Fig. 25. Two continuous closure models (boundaries are deliberately represented as very tick, but the reader should think of them as infinitely thin).

The study of appropriate compatibility conditions, determining a universal notion of path for certain classes of closure spaces is an open issue. One of the major difficulties in finding a unifying notion is that Euclidean paths are not directed, whereas quasi-discrete paths are directed. Directed paths in topology are a highly non-trivial topic by themselves, and give rise to the subject of *directed algebraic topology* [28]. Generalising directed algebraic topology to work in the setting of closure spaces could be a relevant strategy to face these issues. We refer the interested reader to [18] for a more detailed discussion on these issues.

Topo-Bisimilarity and Completeness. A natural question is what structures are logically equivalent, that is, how fine-grained is the logic. It turns out that in topological spaces logical equivalence for $\mathcal{L}$ (see Definition 15) coincides with the notion of *topological bisimilarity*. More precisely, fix two models $\mathcal{M}_1 = ((X_1, O_1), \mathcal{V}_1)$ and $\mathcal{M}_2 = ((X_2, O_2), \mathcal{V}_2)$.

Definition 26. *A topological bisimulation, or simply topo-bisimulation, is a relation $\mathcal{R} \subseteq X_1 \times X_2$ such that, for all $(x_1, x_2) \in \mathcal{R}$,*

- *For all $p \in AP$, $x_1 \in \mathcal{V}_1(p)$ if and only if $x_2 \in \mathcal{V}_2(p)$, and*
- *for all $o_1 \in O_1$, whenever $x_1 \in o_1$, there is $o_2 \in O_2$ such that $x_2 \in o_2$ and for all $y_2 \in o_2$ there is $y_1 \in o_1$ with $(y_1, y_2) \in \mathcal{R}$, and*

– *for all $o_2 \in O_2$, whenever $x_2 \in o_2$, there is $o_1 \in O_1$ such that $x_1 \in o_1$ and for all $y_1 \in o_1$ there is $y_2 \in o_2$ with $(y_1, y_2) \in \mathcal{R}$.*

Two points x_1, x_2 are topo-bisimilar *if there is a topo-bisimulation relating them.*

Bisimilarity equates points based on their local properties. Two points are bisimilar when, first of all, they satisfy the same properties in their respective models. Then, it is required that, for every open set o_1 on one side, there is a choice of an open set o_2 on the other side, all points of which have a corresponding bisimilar point in o_1. This distinguishes points that are on the boundary of some property from points that are in its interior. Furthermore, the precise shape and size of properties in a model does not affect bisimilarity, which is only driven by the existence of open sets covering each property.

Topo-bisimilarity establishes the same relation between points in a topological space as topo-logical equivalence, i.e. $\mathcal{L}$ is fully abstract with respect to topo-bisimilarity. This is proved in [40], Theorems 5.4 and 5.5. We summarise the result as follows.

Theorem 3. *Two points $x_1 \in X_1$ and $x_2 \in X_2$ are topo-bisimilar if and only if they are logically equivalent, that is, for all formulas φ, it holds $\mathcal{M}_1, x_1 \models \varphi$ if and only if $\mathcal{M}_2, x_2 \models \varphi$.*

A situation where Theorem 3 is useful is when one wants to prove that two models are logically equivalent or to minimise the model to check. Then instead of verifying equivalence over all formulas (e.g., by induction), one can exhibit a topo-bisimulation. The development of a suitable notion of bisimilarity for SLCS is part of future research. The same holds for the understanding of the axiomatic aspects of SLCS and the development of a relational semantics.

Acknowledgments. The authors like to thank Luca Bortolussi, Stephen Gilmore, Gianluca Grilletti, Laura Nenzi and Rytis Paškauskas who are involved in the Quanticol project and who are co-authors of the various articles on which this tutorial has been based. We like to thank Ezio Bartocci for sharing with us an earlier Matlab version of the Turing model.

References

1. Aiello, M., Pratt-Hartmann, I., van Benthem, J. (eds.): Handbook of Spatial Logics. Springer, Heidelberg (2007)
2. Anderson, S., Bredche, N., Eiben, A.E., Kampis, G., van Steen, M.: Adaptive Collective Systems: Herding Black Sheep. BookSprints (2013)
3. Aydin Gol, E., Bartocci, E., Belta, C.: A formal methods approach to pattern synthesis in reaction diffusion systems. In: Proceedings of the CDC (2014)
4. Baier, C., Katoen, J.P.: Principles of Model Checking. MIT Press, Cambridge (2008)
5. Bartocci, E., Bortolussi, L., Milios, D., Nenzi, L., Sanguinetti, G.: studying emergent behaviours in morphogenesis using signal spatio-temporal logic. In: Abate, A., Safranek, D., et al. (eds.) HSB 2015. LNCS, vol. 9271, pp. 156–172. Springer, Switzerland (2015). doi:10.1007/978-3-319-26916-0_9

6. Bartocci, E., Bortolussi, L., Nenzi, L., Sanguinetti, G.: On the robustness of temporal properties for stochastic models. In: HSB. EPTCS, vol. 125, pp. 3–19 (2013)
7. Bartocci, E., Bortolussi, L., Nenzi, L., Sanguinetti, G.: System design of stochastic models using robustness of temporal properties. Theor. Comput. Sci. **587**, 3–25 (2015)
8. Blackburn, P., de Rijke, M., Venema, Y.: Modal Logic. Cambridge University Press, New York (2001)
9. Borgnat, P., Abry, P., Flandrin, P., Robardet, C., Rouquier, J.B., Fleury, E.: Shared bicycles in a city: a signal processing and data analysis perspective. Adv. Complex Syst. **14**(3), 415–438 (2011)
10. Bortolussi, L., Hillston, J.: Model checking single agent behaviours by fluid approximation. Inf. Comput. **242**, 183–226 (2015)
11. Bortolussi, L., Hillston, J., Latella, D., Massink, M.: Continuous approximation of collective system behaviour: a tutorial. Perform. Eval. **70**(5), 317–349 (2013). http://www.sciencedirect.com/science/article/pii/S0166531613000023
12. Cardelli, L., Gordon, A.D.: Anytime, anywhere: modal logics for mobile ambients. In: Proceedings of the 30th SIGPLAN-SIGACT Symposium on Principles of Programming Languages (POPL 2000), pp. 365–377 (2000)
13. Ciancia, V., Gilmore, S., Grilletti, G., Latella, D., Loreti, M., Massink, M.: On spatio-temporal model-checking of vehicular movement in transport systems - preliminary version. Technical report TR-QC-02-2016, QUANTICOL (2016)
14. Ciancia, V., Gilmore, S., Latella, D., Loreti, M., Massink, M.: Data verification for collective adaptive systems: spatial model-checking of vehicle location data. In: Eighth IEEE International Conference on Self-Adaptive and Self-Organizing Systems Workshops, SASOW 2014, London, United Kingdom, 8–12 September, 2014, pp. 32–37. IEEE Computer Society (2014). http://dx.doi.org/10.1109/SASOW.2014.16
15. Ciancia, V., Grilletti, G., Latella, D., Loreti, M., Massink, M.: An experimental spatio-temporal model checker. In: Bianculli, D., et al. (eds.) SEFM 2015 Workshops. LNCS, vol. 9509, pp. 297–311. Springer, Heidelberg (2015). doi:10.1007/978-3-662-49224-6_24. Extended version of QC-TR- 10-2014, http://milner.inf.ed.ac.uk/wiki/pages/J8N4c8/QUANTICOLTechnicalReports.html
16. Ciancia, V., Latella, D., Loreti, M., Massink, M.: Specifying and verifying properties of space. Technical report TR-QC-06-2014, QUANTICOL (2014). http://blog.inf.ed.ac.uk/quanticol/technical-reports/
17. Ciancia, V., Latella, D., Loreti, M., Massink, M.: Specifying and verifying properties of space. In: Diaz, J., Lanese, I., Sangiorgi, D. (eds.) TCS 2014. LNCS, vol. 8705, pp. 222–235. Springer, Heidelberg (2014)
18. Ciancia, V., Latella, D., Loreti, M., Massink, M.: Model checking spatial logics for closure spaces. (submitted, 2016)
19. Ciancia, V., Latella, D., Massink, M., Paškauskas, R.: Exploring spatio-temporal properties of bike-sharing systems. In: Beal, J., Hillston, J., Viroli, M. (eds.) Spatial and COllective PErvasive Computing Systems. Workshop at IEEE SASO 2015, MIT, Cambridge, MA, USA, 21 September, 2015, pp. 74–79. IEEE Computer Society Press, Cambridge (2015). doi:10.1109/SASOW.2015.17
20. Clarke, E.M., Grumberg, O., Peled, D.: Model checking. MIT Press, Cambridge (2001). http://books.google.de/books?id=Nmc4wEaLXFEC
21. De Maio, P.: Bike-sharing: its history, impacts, models of provision, and future. J. Public Transp. **12**(4), 41–56 (2009)

22. De Nicola, R., Katoen, J.P., Latella, D., Loreti, M., Massink, M.: Model checking mobile stochastic logic. Theor. Comput. Sci. **382**(1), 42–70 (2007)
23. Fishman, E., Washington, S., Haworth, N.L.: Bike share's impact on car use: evidence from the United States, Great Britain, and Australia. In: Proceedings of the 93rd Annual Meeting of the Transportation Research Board (2014)
24. Froehlich, J., Neumann, J., Oliver, N.: Sensing and predicting the pulse of the city through shared bicycling. In: IJCAI, pp. 1420–1426 (2009)
25. Galpin, V.: Spatial representations and analysis techniques. In: Bernardo, M., De Nicola, R., Hillston, J. (eds.) SFM 2016. LNCS, vol. 9700, pp. 120–155. Springer, Switzerland (2016)
26. Galton, A.: A generalized topological view of motion in discrete space. Theor. Comput. Sci. **305**(1–3), 111–134 (2003). http://www.sciencedirect.com/science/article/pii/S0304397502007016
27. Galton, A.: The mereotopology of discrete space. In: Freksa, C., Mark, D.M. (eds.) COSIT 1999. LNCS, vol. 1661, pp. 251–266. Springer, Heidelberg (1999). http://dx.doi.org/10.1007/3-540-48384-5_17
28. Grandis, M.: Directed Algebraic Topology: Models of Non-Reversible Worlds. Cambridge University Press, Cambridge (2009)
29. Haghighi, I., Jones, A., Kong, J.Z., Bartocci, E., Gros, R., Belta, C.: SpaTeL: a novel spatial-temporal logic and its applications to networked systems. In: Proceedings of the HSCC (2015)
30. Johnstone, P.T.: Sketches of an Elephant: A Topos Theory Compendium. Oxford Logic Guides, vol. 1. Clarendon Press, Oxford (2002). http://opac.inria.fr/record=b1107183. autre tirage: 2008
31. Kontchakov, R., Kurucz, A., Wolter, F., Zakharyaschev, M.: Spatial logic + temporal logic = ? In: Aiello et al. [1], pp. 497–564
32. Latella, D., Loreti, M., Massink, M.: On-the-fly PCTL fast mean-field approximated model-checking for self-organising coordination. Sci. Comput. Program. **110**, 23–50 (2015)
33. Maler, O., Nickovic, D.: Monitoring temporal properties of continuous signals. In: Lakhnech, Y., Yovine, S. (eds.) FORMATS/FTRTFT 2004. LNCS, vol. 3253, pp. 152–166. Springer, Heidelberg (2004)
34. Massink, M., Paškauskas, R.: Model-based assessment of aspects of user-satisfaction in bicycle sharing systems. In: Sotelo Vazquez, M., Olaverri Monreal, C., Miller, J., Broggi, A. (eds.) 18th IEEE International Conference on Intelligent Transportation Systems, pp. 1363–1370. IEEE Computer Society Press (2015). doi:10.1109/ITSC.2015.224
35. Midgley, P.: Bicycle-sharing schemes: enhancing sustainable mobility in urban areas. In: 19th Session of the Commission on Sustainable Development. CSD19/2011/BP8, United Nations (2011)
36. Nenzi, L., Bortolussi, L.: Specifying and monitoring properties of stochastic spatio-temporal systems in signal temporal logic. In: Haviv, M., Knottenbelt, W.J., Maggi, L., Miorandi, D. (eds.) 8th International Conference on Performance Evaluation Methodologies and Tools, VALUETOOLS 2014, ICST, Bratislava, Slovakia, 9–11 December, 2014. http://dx.doi.org/10.4108/icst.valuetools.2014.258183
37. Nenzi, L., Bortolussi, L., Ciancia, V., Loreti, M., Massink, M.: Qualitative and quantitative monitoring of spatio-temporal properties. In: Bartocci, E., Majumdar, R. (eds.) RV 2015. LNCS, vol. 9333, pp. 21–37. Springer, Switzerland (2015). http://dx.doi.org/10.1007/978-3-319-23820-3_2

38. Reynolds, J.: Separation logic: a logic for shared mutable data structures. In: Proceedings of the 17th IEEE Symposium on Logic in Computer Science (LICS 2002), Copenhagen, Denmark, 22–25 July 2002, pp. 55–74. IEEE Computer Society (2002). http://dx.doi.org/10.1109/LICS.2002.1029817

39. Turing, A.M.: The chemical basis of morphogenesis. Philos. Trans. R. Soc. Lond. B Biol. Sci. **237**(641), 37–72 (1952). doi:10.1098/rstb.1952.0012

40. van Benthem, J., Bezhanishvili, G.: Modal Logics of Space. In: Aiello, M., Pratt-Hartmann, I., Van Benthem, J. (eds.) Handbook of Spatial Logics, pp. 217–298. Springer, Heidelberg (2007)

Quantitative Abstractions for Collective Adaptive Systems

Andrea Vandin and Mirco Tribastone[✉]

IMT School for Advanced Studies Lucca, Lucca, Italy
{andrea.vandin,mirco.tribastone}@imtlucca.it

Abstract. Collective adaptive systems (CAS) consist of a large number of possibly heterogeneous entities evolving according to local interactions that may operate across multiple scales in time and space. The adaptation to changes in the environment, as well as the highly dispersed decision-making process, often leads to emergent behaviour that cannot be understood by simply analysing the objectives, properties, and dynamics of the individual entities in isolation.

As with most complex systems, modelling is a phase of crucial importance for the design of new CAS or the understanding of existing ones. Elsewhere in this volume the typical workflow of formal modelling, analysis, and evaluation of a CAS has been illustrated in detail. In this chapter we treat the problem of efficiently analysing *large-scale* CAS for quantitative properties. We review algorithms to automatically reduce the dimensionality of a CAS model preserving modeller-defined state variables, with focus on descriptions based on systems of ordinary differential equations. We illustrate the theory in a tutorial fashion, with running examples and a number of more substantial case studies ranging from crowd dynamics, epidemiology and biological systems.

1 Introduction

Distinctive features of collective adaptive systems (CAS) are the presence of a large number of entities with their own properties, objectives, and behaviour, that interact with each other and with the environment in such a way that the resulting global dynamics arises as an emergent property that cannot be directly inferred from the study of individuals in isolation. To ensure that a CAS design meets the desired properties, or to accurately understand the behaviour of existing CAS, it is of crucial importance to be able to reason about a (possibly huge) system as a whole. In this context, the modelling phase clearly plays an important role, as it does with any system characterised by high complexity.

Quantitative Abstractions. The focus of this chapter is on quantitative modelling of CAS. Due to their heterogeneity and scale, CAS introduce a number of difficult challenges, the most notable of which is the problem of state space explosion that is typically incurred when analysing large collectives of entities. Elsewhere in this volume are contributions to a prototypical design and modelling workflow for CAS which take scalability and accuracy of the analysis into account.

© Springer International Publishing Switzerland 2016
M. Bernardo et al. (Eds.): SFM 2016, LNCS 9700, pp. 202–232, 2016.
DOI: 10.1007/978-3-319-34096-8_7

The process algebra CARMA (cf. [9]) is explicitly designed to study collectives of agents evolving stochastically according to a continuous-time Markov chain (CTMC) model [57]; approximate analysis techniques based on hybrid or differential equation approximations are presented in [10]; effective approaches for dealing with the spatial dimension of CAS are reviewed in [38]; finally, model checking for spatial and temporal properties are discussed in [40]. Here, instead, we focus on techniques that crosscut the above phases of the modelling workflow. Namely, we consider the problem of obtaining suitable *abstractions* of dynamical models for CAS. We are motivated by the fact that, in real-world scenarios, the inherent system's complexity is so high that it may even defeat typically compact and effective model descriptions, such as those based on ordinary differential equations (ODEs).

Let us consider, for instance, the case of a bike-sharing system (BSS). This is a prototypical CAS [30], an instance of which has been also used as running example of [57]. Its quantitative analysis may be based on a CTMC model, which will however grow unfeasibly large in realistic settings since the state space has to cover (at least) all of the possible combinations of bike availabilities at each station. Deterministic approximations based on ODEs may come to the rescue, by more compactly associating one equation for each station and each possible link between two stations. In this case they would capture an estimate of the average number of bikes available at each station as well as of those in transit [37]. Clearly, if instantiated to a real-world large BSS such as London's, with over 700 stations, it would yield an ODE system of many thousands of equations, which is likely to drastically impact on the practical feasibility of the analysis. Furthermore, the analysis would become prohibitive if the modeller wished to track higher order moments than the averages, since the ODE system size grows polynomially with the number of variables of the original system (e.g., [32]).

Abstraction techniques may help tackle the dimensionality problem further. The basic idea is to obtain a representation of the original model projected onto a lower dimensional state space so as to allow a more efficient analysis. Due to the large scale involved in CAS models, there are four main desirable properties for an effective method:

P1. The abstraction should come with formal guarantees on the relationship between the abstract dynamics and the original one. This enables the modeller to use the abstract model with full confidence in the results of the analysis.

P2. The construction of the abstract model should be fully automatic, since the original model is likely to be unintelligible due to size.

P3. The method should be generic in order to be applicable to as a wide range of CAS models as possible.

P4. The abstract model should preserve user-defined observables of the original system. For instance, it should be possible to fully recover the dynamics of selected variables of the original model.

In this chapter we consider abstraction techniques that satisfy the above requirements for quantitative CAS models based on ODEs. However they can be applicable also to CTMC models by studying the CTMC's equations of motion, which is a linear ODE system (e.g., [61]). In fact, we will discuss that these techniques can be somewhat seen as a generalisation of aggregation algorithms specific to CTMCs, based on the well-known notion of lumpability [14]. Languages and equivalences have been extensively studied for models based on CTMC semantics (e.g., [27]).

Since ODEs are a universal dynamical model, featuring in many diverse scientific branches including organic and inorganic chemistry, ecology, economics, epidemiology, systems biology, and control theory, reducing large scale ODEs has also a long-standing tradition (e.g., [2, 48, 62]). Here we offer a specific computer-science viewpoint on this subject, looking at ODE reduction as the problem of finding an appropriate equivalence relation over the ODE's state variables, borrowing ideas from the programming languages community and concurrency theory. Most of the results discussed here, summarised from [19–21], concern exact notions of aggregation. These may be lossy in that the dynamics of some original variables cannot be recovered in the abstract model, yet all the information in the abstract model is exactly related to the original variables; approximate notions of aggregation are an exciting future research direction.

Differential Equivalences. The problem of minimising ODEs is interpreted as a quotienting up to some equivalence, akin to more classical models of computation based on labelled transition systems (LTS). We put forward the analogy between states of an LTS and ODE variables. The starting point is that of *differential equivalences*, relations between ODE variables that preserve their corresponding solutions in some appropriate sense. Here we consider two variants of differential equivalence, as first presented in [21].

In *forward differential equivalence* (FDE), an ODE system can be written for the variables that represent the equivalence classes, giving the sum of the solutions of its members at all time points t. Let us consider the example:

$$\dot{x}_1 = -x_1, \qquad \dot{x}_2 = k_1 \cdot x_1 - x_2, \qquad \dot{x}_3 = k_2 \cdot x_1 - x_3, \qquad (1)$$

where k_1 and k_2 are constants and the 'dot' operator denotes the derivative.[1] It can be shown that $\{\{x_1\}, \{x_2, x_3\}\}$ is an FDE quotienting. Indeed, exploiting basic properties one gets

$$\dot{x}_1 = -x_1, \qquad (x_2 \overset{\cdot}{+} x_3) = \dot{x}_2 + \dot{x}_3 = (k_1 + k_2) \cdot x_1 - (x_2 + x_3). \qquad (2)$$

By the change of variable $y = x_2 + x_3$, this is equivalent to writing

$$\dot{x}_1 = -x_1 \qquad\qquad \dot{y} = (k_1 + k_2) \cdot x_1 - y.$$

This quotient ODE model recovers the sum of the solutions of the variables in each equivalence class. Thus, setting the *initial condition* $y(0) = x_2(0) + x_3(0)$ yields that the solution satisfies $y(t) = x_2(t) + x_3(t)$ at all time points t.

---

[1] Throughout the paper we will work with *autonomous* ODE systems, which are not explicitly dependent on time.

Backward differential equivalence (BDE) equates variables that have the same solutions at all time points. In (1), $\{\{x_1\}, \{x_2, x_3\}\}$ is also a BDE provided that $k_1 = k_2$. In this case, we obtain a quotient ODE by removing either equation between x_2 and x_3, say x_3, and rewriting every occurrence of x_3 as x_2:

$$\dot{x}_1 = -x_1 \qquad\qquad \dot{x}_2 = k_1 x_1 - x_2.$$

Both FDE and BDE satisfy P1, since a differential equivalence will yield an abstract model that can be exactly related to the original one. However, we observe that BDE is lossless, because every variable in the same equivalence class has the same solution. Therefore, the original model solution can be fully recovered at the expense of the side condition that equivalent variables have to be initialised equally. Instead, with FDE one cannot recover the original solutions in general; on the other hand FDE has no restrictions on the initial conditions.

When the ODE comes from a CTMC model, in [21] it is shown that FDE and BDE correspond to ordinary and exact lumpability of CTMCs [14], respectively. Incidentally, this also implies that FDE and BDE are not comparable in general. The terms "forward" and "backward" are motivated by a rather established tradition in the literature to call these two notions of CTMC lumpability (e.g., [19,36,70]), due to the fact that they involve conditions on the outgoing and incoming arcs of the CTMC state transition diagram, respectively.

Differential equivalence can be in principle defined for any ODE system. However, in order to satisfy P2 and obtain a minimisation algorithm, it is necessary to impose some restrictions on the kind of admissible ODE systems. In this chapter we review two alternatives that trade off expressiveness for scalability.

Symbolic Minimisation Algorithms. The first approach, presented in [21], interprets each ODE variable directly as a real function. Establishing an equivalence between two variables thus amounts to relating two functions for all their possible assignments, which involves reasoning over uncountable state spaces. The first step in [21] is to encode the equivalence conditions into logical formulae containing ODE variables, and check them *symbolically* through a satisfiability modulo theories (SMT) solver [4]. Actually, it turns out that differential equivalences can be encoded into the quantifier-free fragment of first-order logic. By appropriately restricting the admissible ODE systems to those for which an SMT solver — in our implementation, the well-known Z3 [26] — is a decision procedure for such formulae, we obtain a rigorous way of checking the existence of a differential equivalence. The language IDOL (Intermediate Drift-oriented Language) of [21] does so by essentially excluding trigonometric functions. On the other hand, it can encode polynomials of any degree, rational expressions, minima and maxima, enough to cover affine systems, chemical reaction networks with frequently used kinetics such as the law of mass action and Hill's, and the deterministic semantics of process algebra. Thus, it can satisfy P3 to some extent.

The SMT checks can be embedded into an algorithm that finds the coarsest refinement of a given input partition up to a differential equivalence. This exploits the ability of the SMT solver to produce a *witness*, i.e., a variable assignment that falsifies the hypothesis that the current partition is a differential equivalence. The partition is then refined iteratively until a fixed point is found.

We note that the algorithm meets the requirement of property P4. Indeed, suppose that the modeller wishes to keep track of an ODE variable x in the abstract model. Then, starting the FDE algorithm with the trivial partition where all variables are in the same block might clearly lead to an equivalence where x is related to other variables. As a result, x's individual solution cannot be recovered. However, since the input partition may be chosen arbitrarily, it is possible to isolate the desired observable variables into singleton initial blocks. Similarly, to be able to fully reconstruct the original model from the abstract one when using BDE, it is necessary to construct an initial partition *consistent* with the initial conditions of the original model (that is, two variables are in the same initial block if their initial conditions are the same).

Syntax-Driven Minimisation. The second approach takes a different perspective that offers a trade off between expressiveness of the language and efficiency of the minimisation algorithm. It is based on a finitary representation of an ODE system by means of a so-called *reaction network* (RN) [20]. This is a slight extension of a formal chemical reaction network (CRN) which allows rate parameters to be also negative. Assuming *elementary* reactions only, i.e., reactions with at most two reagents, a reaction network gives rise to an ODE system with derivatives that are multivariate polynomials of degree at most two. The advantage in using this construction is that it is possible to use bisimulation-style equivalences for model reduction, originally developed in [19] for CRNs, over a state space that is discrete because it only concerns finitely many "species" (corresponding to the ODE variables) and reactions (each representing a monomial in the ODE's right-hand side, as discussed in Sect. 2.3).

The notions of bisimulation for RNs are closely related to the differential equivalences in [21]. In particular, *forward bisimulation* (FB) is a partition of an RN's set of species which represents a sufficient condition for an FDE of the corresponding ODE variables. Instead, *backward bisimulation* (BB) fully characterises BDE (for multivariate polynomials of degree at most two). The main contribution of [20] is to exploit the fact that FB and BB can be written in the Larsen-Skou style of probabilistic bisimulation [55]. This enables us to cast the computation of the largest FB/BB into Paige and Tarjan's famous coarsest refinement problem [63]. In particular, in [20] a partition refinement algorithm is developed along the lines of efficient analogues for Markov chain lumping such as [31,77], and for probabilistic transition systems [3].

Tool Support. Both families of symbolic and syntactic minimisation techniques are tool supported. The former has been implemented in ERODE, a tool offering SMT-based automatic Exact Reduction of Ordinary Differential Equations. The tool is available at http://sysma.imtlucca.it/tools/erode/, together with installation and usage instructions. ERODE is a Java tool which interacts with Z3 to perform automatic minimisation of IDOL programs up to FDE and BDE. More details on the implemented procedures are provided in Sects. 2.1 and 2.2. ERODE currently supports the continuous-state semantics based on the law of mass action of CRNs given in the .net format generated with the well-established tool BioNetGen [8], version 2.2.5-stable. This allowed us to validate our differential equivalences against a wide set of existing models in the literature. Support for the entire IDOL language in under development.

The syntactic minimisation techniques have been implemented in CRNReducer, a Java tool offering automatic exact reduction of (chemical) reaction networks. It is available at http://sysma.imtlucca.it/tools/crnreducer/. CRNReducer performs the syntactic checks necessary to minimize an input RN up to forward and backward bisimulations. More details on the implemented algorithms are given in Sects. 2.3 and 2.4. CRNReducer currently supports CRNs given in the BioNetGen's .net format, and CTMCs in the .tra/.lab format of the state-of-the-art model checker MRMC [50]. In addition, it accepts a compact CSV-like representation of linear systems of equations in the form $A \cdot x = b$, where x is the vector of unknowns. Stationary iterative methods such as Jacobi's can be seen as discrete-time dynamical system that converges to the solution. To such a system, CRNReducer can apply FB/BB (see [20] for details and benchmarks). Support for other languages which can be encoded as reaction networks is currently under development.

As part of a larger effort, a new tool collecting both symbolic and syntactic minimization techniques is currently under development. The tool will be provided with a modern integrated development environment, will offer full support for the IDOL language, and will be equipped with importing capabilities from a number formats.

Paper Structure. The paper is organized as follows. Section 2 presents our symbolic (Sects. 2.1 and 2.2) and syntactic (Sects. 2.3 and 2.4) reduction techniques. Then, Sect. 3 shows how they can be applied to crowd dynamics models (Sect. 3.1), to multi-community epidemiology models (Sect. 3.2), as well as to models from the realm of evolutionary biology (Sect. 3.3) and biochemistry (Sect. 3.4). Finally, Sect. 4 discusses related works, while Sect. 5 concludes the paper.

2 Background

2.1 Differential Equivalences

Although differential equivalences can be in principle defined for a larger class of ODE models, here we consider a fragment, identified by a formal kernel language called Intermediate Drift-oriented Language (IDOL), which guarantees decidability for the problem of computing a differential equivalence.

Definition 1 (IDOL Syntax). *The syntax of programs of the* intermediate drift oriented language *(IDOL) is given by*

$$p ::= \varepsilon \mid \dot{x}_i = f, \, p$$
$$f ::= n \mid x_i \mid f + f \mid f \cdot f \mid f^{\frac{1}{m}}$$

where $x_i \in \mathbb{V}$ and $n, m \in \mathbb{Z}$ and $m \neq 0$.

The set $\mathbb{V}$ represents ODE variables. A program is a list of elements $\dot{x}_i = f$ where each element gives the drift f for ODE of the variable x_i. The "dot" operator indicates the derivative with respect to time. Given an IDOL program p, we define $V_p = \{x_1, \ldots, x_n\}$ as the set of variables in p. We say that p is *well-formed* if for every $x_i \in V_p$ there exists a unique term $\dot{x}_i = f$ in p. We denote its drift by f_i. Throughout this paper we will consider well-formed programs only.

We remark that IDOL can cover frequently used dynamics such as:

- the law of mass action for CRNs, using drifts such as $x_1 \cdot x_2$;
- the Hill kinetics for CRNs, with drifts such as $x_1^2/(1 + x_1^2)$;
- and the *minimum* function for threshold based drifts, where

$$\min(x_1, x_2) := \frac{1}{2}(x_1 + x_2 - |x_1 - x_2|), \quad \text{with } |x| := (x \cdot x)^{\frac{1}{2}}.$$

The semantics of IDOL is given denotationally through the ODE solution of an initial value problem, starting from an initial condition $\hat{\sigma}$. For an IDOL program p, we denote by $\Theta(p)$ the logical formula that encodes the appropriate domain where the solution lives (which must be regular enough, cf. [21] for details). Furthermore, we slightly ease notation with respect to [21] by representing the solution for variable x_i simply by $x_i(t)$.

FDE is a partition over IDOL variables satisfying the property that sums of variables can be factored out from the cumulative derivatives that sum across the drifts of all variables belonging to each block, e.g. (2). This property can be captured by replacing each variable as a scaled sum of the corresponding variables of its block, such that all scaling factors are non-negative and sum to one; in the example (1), we would keep x_1 as is (it is a singleton block), and replace x_2 with $s_1 \cdot (x_2 + x_3)$ and x_3 with $s_2 \cdot (x_2 + x_3)$, where s_1 and s_2 are the scaling factors. Then, FDE amounts to proving that the aggregated drifts do not depend on the assignments of the scaling factors. For instance, in (1) we would rewrite the aggregated drift $f_2 + f_3$ as follows

$$\begin{aligned}
f_2 + f_3 &= k_1 \cdot x_1 - x_2 + k_2 \cdot x_1 - x_3 \\
&= k_1 \cdot x_1 - s_1 \cdot (x_2 + x_3) + k_2 \cdot x_1 - s_2 \cdot (x_2 + x_3) \\
&= (k_1 + k_2) \cdot x_1 - (s_1 + s_2) \cdot (x_2 + x_3) \\
&= (k_1 + k_2) \cdot x_1 - (x_2 + x_3)
\end{aligned}$$

Indeed it does not depend on the choice of s_1 and s_2, since $s_1 + s_2 = 1$.

We now appeal to a fundamental result from [72], which shows that it is enough to check this for a particular choice. For technical reasons discussed in [21], FDE checks this through a *uniform* scaling (for instance $s_1 = s_2 = 1/2$ in the example).

Definition 2 (FDE). *Let p be an IDOL program and $\mathcal{Z}$ a partition of V_p. Then, $\mathcal{Z}$ is a* forward differential equivalence *if the following formula is valid:*

$$\Theta(p) \to \bigwedge_{H \in \mathcal{Z}} \left(\sum_{x_i \in H} f_i = \sum_{x_i \in H} f_i \left[x_j \bigg/ \frac{\sum_{x_k \in H'} x_k}{|H'|} : H' \in \mathcal{Z}, \; x_j \in H' \right] \right) \qquad (\Phi^{\mathcal{Z}})$$

As usual, we have denoted by $\psi[t/s]$ the term where each occurrence of t in ψ is replaced by s.

Definition 3 (FDE Quotient). *Let p be an IDOL program and $\mathcal{Z}$ an FDE partition. Then, the* forward quotient *of p with respect to $\mathcal{Z}$, denoted by $\overrightarrow{p_{\mathcal{Z}}}$, is:*

$$\dot{y}_H = \sum_{x_i \in H} f_i\left[x_j \middle/ \frac{y_{H'}}{|H'|} : H' \in \mathcal{Z},\ x_j \in H'\right], \quad for\ all\ H \in \mathcal{Z}.$$

We now state a crucial *dynamical characterization* theorem: A partition of IDOL variables is FDE if and only if the ODEs of the quotient program preserve the sums of the original trajectories in each equivalence class. Hence the largest FDE represents the best possible aggregation that can be obtained in this sense.

Theorem 1. *Let p be an IDOL program with initial condition $\hat{\sigma}$, $\mathcal{Z}$ a partition of $\mathcal{V}_p$. Then, $\mathcal{Z}$ is an FDE partition with forward quotient $\overrightarrow{p_{\mathcal{Z}}}$ if and only if*

$$y_H(t) = \sum_{x_i \in H} x_i(t)$$

for all t for which the solutions exist and for an initial condition of the quotient program $\hat{\sigma}_{\mathcal{Z}}$ that satisfies $\hat{\sigma}_{\mathcal{Z}}(y_H) = \sum_{x_i \in H} \hat{\sigma}(x_i)$ for all $H \in \mathcal{Z}$.

One extra step is needed to make FDE usable in a minimisation algorithm. We need to be able to refer FDE to properties enjoyed by the single variables, as opposed to blocks of variables in the original definition. If a candidate partition is not FDE, the algorithm needs to "split" the partition blocks in such a way that it isolates such variables that prevent the partition from being an FDE. For this we consider an alternative characterisation of FDE in terms of *binary* conditions.

Theorem 2 (Binary FDE Characterization). *Let p be an IDOL program, $\mathcal{R}$ be an equivalence relation on $\mathcal{V}_p$, and $\mathcal{Z} = \mathcal{V}_p/\mathcal{R}$. Then $\mathcal{Z}$ is an FDE if and only if for all distinct $x_i, x_j \in \mathcal{V}_p$ we have that $(x_i, x_j) \in \mathcal{R}$ implies that the following formula is valid:*

$$\Theta(p) \rightarrow \bigwedge_{H \in \mathcal{Z}} \left(\sum_{x_k \in H} f_k = \sum_{x_k \in H} f_k \left[x_i/s \cdot (x_i + x_j), x_j/(1-s) \cdot (x_i + x_j) \right] \right) \quad (\Phi^{\mathcal{Z}}_{\mathbf{x_i},\mathbf{x_j}})$$

We now turn to BDE. The fact that IDOL variables have the same solutions at all time points is characterized by the property that variables with the same assignment are mapped to equal drifts.

Definition 4 (BDE). *Let p be an IDOL program and $\mathcal{Z}$ a partition of $\mathcal{V}_p$. Then $\mathcal{Z}$ is a* backward differential equivalence *if the following formula is valid:*

$$\Theta(p) \rightarrow \left(\bigwedge_{H \in \mathcal{Z}} (x_{H,1} = \ldots = x_{H,|H|}) \rightarrow \bigwedge_{H \in \mathcal{Z}} (f_{H,1} = \ldots = f_{H,|H|}) \right) \quad (\Psi^{\mathcal{Z}})$$

Now, similarly to FDE it is possible to define a notion of quotient IDOL program, and state the dynamical characterization theorem.

Definition 5 (BDE Quotient). *Let p be an IDOL program and $\mathcal{Z}$ a BDE partition of V_p. The backward quotient of p with respect to $\mathcal{Z}$, denoted by $\overleftarrow{p_{\mathcal{Z}}}$, is given by*

$$\dot{y}_H = f_{H,1}\left[x_{H',1}/y_{H'}, \ldots, x_{H',|H'|}/y_{H'} : H' \in \mathcal{Z}\right], \quad for\ H \in \mathcal{Z}.$$

Theorem 3 (Dynamical BDE Characterization). *Let p be an IDOL program and $\mathcal{Z}$ a partition of V_p. Then, $\mathcal{Z}$ is a BDE partition with backward quotient $\overleftarrow{p_{\mathcal{Z}}}$ if and only if $\hat{\sigma}_{\mathcal{Z}}(y_H) = \hat{\sigma}(x_{H,1}) = \ldots = \hat{\sigma}(x_{H,|H|})$ for all $H \in \mathcal{Z}$ implies*

$$y_H(t) = x_{H,1}(t) = \ldots = x_{H,|H|}(t)$$

for all $H \in \mathcal{Z}$ and all t for which the solutions exist.

2.2 Symbolic Minimisation

The first step toward a symbolic minimisation algorithm is to be able to check whether a candidate partition is a differential equivalence. The problem amounts to establishing the validity of the (quantifier-free) formulae $\Phi^{\mathcal{Z}}$, $\Phi^{\mathcal{Z}}_{x_i,x_j}$ and $\Psi^{\mathcal{Z}}$, which are decidable by Tarski's famous result. To check them, we encode the problem into the unsatisfiability of their negations, i.e., by computing $\mathsf{sat}(\neg\Phi^{\mathcal{Z}})$, $\mathsf{sat}(\neg\Phi^{\mathcal{Z}}_{x_i,x_j})$, and $\mathsf{sat}(\neg\Psi^{\mathcal{Z}})$. These can be decided using the decision procedure nlsat [49], which is implemented in Z3 v4.0 [26]. Thus, a partition $\mathcal{Z}$ is FDE (resp., BDE) if and only if $\mathsf{sat}(\neg\Phi^{\mathcal{Z}})$ (resp., $\mathsf{sat}(\neg\Psi^{\mathcal{Z}})$) returns "unsatisfiable".

Example 1. Consider the ODE system given in Eq. (1), the partition of its species $\mathcal{Z}_1 = \{\{x_1\}, \{x_2, x_3\}\}$, and $\neg\Psi^{\mathcal{Z}_1}$, i.e., the formula to check if $\mathcal{Z}_1$ is a BDE. Listing 1 provides the encoding of $\neg\Psi^{\mathcal{Z}_1}$ in the standard SMT-LIB v2.0 [5]. Given that Eq. (1) is parametric with respect to two real variables k_1 and k_2, we declare them in Lines 2–3. We consider two cases: either $k_1 = k_2 = 1$ (Lines 6), or $k_1 = 1$ and $k_2 = 2$ (commented out in Lines 7). We have three ODE variables: x_1, x_2 and x_3, declared as real variables in Lines 10–12, paired with the three corresponding drifts f_1, f_2 and f_3, defined as functions in Lines 20–28. The three functions implicitly take k_1, k_2 and the three ODE variables as arguments, and evaluate in a real number. In this example we assume that the domain Θ of interest is $\mathbb{R}^3_{\geq 0}$, as encoded in Lines 15–17. After having specified the ODE system of interest, in Lines 31–32 we can provide the actual encoding of $\neg\Psi^{\mathcal{Z}_1}$. By applying simple transformations we can rewrite $\neg\Psi^{\mathcal{Z}_1} \equiv \neg((x_2 = x_3) \implies (f_2 = f_3))$ as $(x_2 = x_3) \wedge (f_2 \neq f_3)$. The first conjunct $(x_2 = x_3)$ imposes that the ODE variables are *constant on* $\mathcal{Z}_1$ (i.e., the ODE variables in the same block have same value). Instead, the second conjunct $(f_2 \neq f_3)$ imposes that the drifts are not constant on $\mathcal{Z}_1$. Note that the given SMT-encoding has 3 free variables: x_1, x_2 and x_3. If there exists an assignment for them that satisfies Listing 1, then $\mathcal{Z}_1$ is not a BDE. The command to check the satisfiability is given in Line 35, while Line 36 asks the solver to return one of the satisfying assignments (if any).

```
1   ;Declare a real constant per parameter k1 and k2
2   (declare-const k1 Real)
3   (declare-const k2 Real)
4
5   ;We consider k₁ = 1, and either k₂ = 1 or k₂ = 2
6   (assert (= k1 1)) (assert (= k2 1))
7   ;(assert (= k1 1)) (assert (= k2 2))
8
9   ;Declare a real constant per ODE variable
10  (declare-const x1 Real)
11  (declare-const x2 Real)
12  (declare-const x3 Real)
13
14  ;We assume to have ℝ³≥0 as domain Θ.
15  (assert (>= x1 0))
16  (assert (>= x2 0))
17  (assert (>= x3 0))
18
19  ;Define the drift fᵢ of each ODE variable xᵢ.
20  (define-const f1 Real
21    (* -1 x1)
22  )
23  (define-const f2 Real
24    (+ (* k1 x1) (* -1 x2))
25  )
26  (define-const f3 Real
27    (+ (* k2 x1) (* -1 x3))
28  )
29
30  ;We encode ¬Ψᶻ¹ in the equivalent form (x₂ = x₃) ∧ (f₂ ≠ f₃)
31  (assert (= x2 x3))
32  (assert (not (= f2 f3)))
33
34  ;Check if the formula is satisfiable, and return a witness if so
35  (check-sat)
36  (get-model)
```

Listing 1. SMT-LIB v2.0 encoding of $\neg\Psi^{\mathcal{Z}_1}$ to check that $\mathcal{Z}_1$ is a BDE for (1)

Listing 1 can be solved using any of the SMT solvers supporting the SMT-LIB v2.0 standard. The executable Z3 encoding of Listing 1 for both the cases $k_1 = k_2$ and $k_1 \neq k_2$ is available via the *rise4fun* web interface at http://rise4fun.com/Z3/lW7d1. For the case $k_1 = k_2$ we obtain "unsatisfiable", because $\mathcal{Z}_1$ is a BDE partition if $k_1 = k_2$, as discussed. Instead, for the case $k_1 \neq k_2$ we obtain "satisfiable", and the assignment $\sigma_w = \{x_1 = 1, x_2 = 0, x_3 = 0\}$. In fact, we have $[\![f_2]\!](\sigma_w) = 1$ and $[\![f_3]\!](\sigma_w) = 2$, where by $[\![f]\!](\sigma)$ we have denoted the interpretation of f as a real function, evaluated with the assignment σ.

The steps $\mathsf{sat}(\Phi^{\mathcal{Z}}_{x_i, x_j})$ and $\mathsf{sat}(\Psi^{\mathcal{Z}})$ can be embedded into an algorithm that computes the coarsest FDE/BDE refinement of a given input partition, shown in Algorithm 1, and parametrised by the differential equivalence of interest (by setting $\chi = F$ and $\chi = B$ for FDE and BDE, respectively).

The refinement step for FDE (Algorithm 2) exploits its binary characterization, relating two variables whenever they do not prevent the current partition from being an FDE.

Algorithm 1. Construction of the largest FDE and BDE.

Require: Program p, partition $\mathcal{G}$ of $\mathcal{V}_p$ and $\chi \in \{F, B\}$.

  $\mathcal{Z} \leftarrow \mathcal{G}$
  while true do
    $\mathcal{Z}' \leftarrow \mathit{refine}_\chi(\mathcal{Z})$
    if $\mathcal{Z}' = \mathcal{Z}$ **then**
      return $\mathcal{Z}$
    else
      $\mathcal{Z} \leftarrow \mathcal{Z}'$
    end if
  end while

Algorithm 2. Routine refine_F

Require: Program p and a partition $\mathcal{Z}$ of $\mathcal{V}_p$.

  $\mathcal{Z}' \leftarrow \emptyset$
  for all $H \in \mathcal{Z}$ **do**
    $\mathcal{R} \leftarrow \{(x_i, x_j) : x_i, x_j \in H \text{ and } (x_i = x_j \text{ or } \Phi^{\mathcal{Z}}_{x_i, x_j} \text{is valid})\}$
    $\mathcal{Z}' \leftarrow \mathcal{Z}' \cup (H/\mathcal{R})$
  end for
  return $\mathcal{Z}'$

Algorithm 3. Routine refine_B

Require: Program p and a partition $\mathcal{Z}$ of $\mathcal{V}_p$.

  if $\Psi^{\mathcal{Z}}$ is valid **then**
    $\mathcal{Z}' \leftarrow \mathcal{Z}$
  else
    $\sigma_w \leftarrow \mathit{getWitness}(\mathsf{sat}(\neg \Psi^{\mathcal{Z}}))$
    $\mathcal{Z}' \leftarrow \emptyset$
    for all $H \in \mathcal{Z}$ **do**
      $\mathcal{R} \leftarrow \{(x_i, x_j) : x_i, x_j \in H \text{ and } [\![f_i]\!](\sigma_w) = [\![f_j]\!](\sigma_w)\}$
      $\mathcal{Z}' \leftarrow \mathcal{Z}' \cup (H/\mathcal{R})$
    end for
  end if
  return $\mathcal{Z}'$

The refinement step for BDE (Algorithm 3) exploits the fact that, when the current partition is not a BDE, i.e., $\neg \Psi^{\mathcal{Z}}$ is satisfiable, then the SMT solver can produce a witness assignment, σ_w, as shown in Example 1. This can be interpreted as a counterexample with respect to BDE, since it provides evidence that an equal assignment of variables within the same block of the candidate partition gives different values of the corresponding drifts, denoted by $[\![f_i]\!](\sigma_w)$ and $[\![f_j]\!](\sigma_w)$ in the algorithm. The idea of the refinement is to preserve variables in the same block whenever the corresponding drifts are not distinguished by the witness assignment.

Example 2. Let us consider the following IDOL program:

$$\dot{x}_1 = -\min(x_1, x_3) + x_2$$
$$\dot{x}_2 = -\min(x_2, x_3) + x_1$$
$$\dot{x}_3 = -\min(x_1, x_3) - \min(x_2, x_3)$$

We show that $\{\{x_1, x_2\}, \{x_3\}\}$ is the coarsest BDE that refines the initial partition $\mathcal{Z} = \{\{x_1, x_2, x_3\}\}$. Indeed, by applying the partition refinement algorithm, at the first iteration the formula $\Psi^{\mathcal{Z}}$ reads

$$x_1 = x_2 = x_3 \rightarrow$$
$$-\min(x_1, x_3) + x_2 = -\min(x_2, x_3) + x_1 = -\min(x_1, x_3) - \min(x_2, x_3)$$

where we have omitted the encoding of the domain $\Theta(p)$ since we assume the whole of $\mathbb{R}^3$. Its negation $\neg\Psi^{\mathcal{Z}}$ is satisfiable. Indeed, a witness assignment is $\sigma_w = \{x_1 = 1, x_2 = 1, x_3 = 1\}$, which yields a drift evaluation $[\![f_1]\!](\sigma_w) = [\![f_2]\!](\sigma_w) = 0$, and $[\![f_3]\!](\sigma_w) = -2$. This triggers a new iteration with a refined partition that preserves variables whenever their corresponding drifts evaluated for the witness are equal. In this case, we obtain the partition $\mathcal{Z}' = \{\{x_1, x_2\}, \{x_3\}\}$. Then, at the next iteration $\Psi^{\mathcal{Z}'}$ reads

$$x_1 = x_2 \rightarrow -\min(x_1, x_3) + x_2 = -\min(x_2, x_3) + x_1$$

Now, its negation is unsatisfiable, thus terminating the algorithm.

2.3   Reaction Networks

An RN (S, R) is a pair of a finite set of *species* S and a finite set of *reactions* R. A reaction is a triple written in the form $\rho \xrightarrow{k} \pi$, where ρ and π are multisets of species, called *reactants* and *products*, respectively, and $k \neq 0$ is the *reaction rate*. We restrict to *elementary* reactions where $|\rho| \leq 2$ (while no restriction is posed on the products). We denote by $\rho(X)$ the multiplicity of species X in the multiset ρ, and by $\mathcal{MS}(S)$ the set of finite multisets of species in S. The operator $+$ denotes multiset union, e.g., $X + Y + Y$ (or just $X + 2Y$) is the multiset $\{\!| X, Y, Y |\!\}$. We also use X to denote either the species X or the singleton $\{\!| X |\!\}$.

The semantics of an RN (S, R) is given by the ODE system $\dot{V} = f(V)$, with $f : \mathbb{R}^S \rightarrow \mathbb{R}^S$, where each component f_X, with $X \in S$ is defined as:

$$f_X(V) := \sum_{\rho \xrightarrow{\alpha} \pi \in R} (\pi(X) - \rho(X)) \cdot \alpha \cdot \prod_{Y \in S} V_Y^{\rho(Y)} .$$

This ODE satisfies a unique solution $V(t) = (V_X(t))_{X \in S}$ for any initial condition $V(0)$. The restriction to elementary reactions ensures that the monomials are of degree at most 2. A standard CRN with mass-action semantics (where reaction speeds are proportional to the product of the concentrations of the reactants) is recovered by restricting to positive reaction rates and non-negative initial conditions. Instead, an arbitrary ODE system with multivariate polynomials can be encoded according to the following.

Lemma 1.   *Consider the ODE system $\dot{y} = G(y)$ with components*

$$\dot{y}_k = G_k(y) := \sum_{1 \le i,j \le n} \alpha_{i,j}^{(k)} \cdot y_i \cdot y_j + \sum_{1 \le i \le n} \alpha_i^{(k)} \cdot y_i + \beta^{(k)}, \quad 1 \le k \le n, \quad (3)$$

and with $\alpha_{i,j}^{(k)}, \alpha_i^{(k)}, \beta^{(k)} \in \mathbb{R}$. Then, then RN (S_G, R_G), with $S_G := \{1, \dots, n\}$ and

$$R_G := \left\{ i+j \xrightarrow{\alpha_{i,j}^{(k)}} i+j+k \mid \alpha_{i,j}^{(k)} \ne 0 \right\}$$

$$\cup \left\{ i \xrightarrow{\alpha_i^{(k)}} i+k \mid \alpha_i^{(k)} \ne 0 \right\} \cup \left\{ \emptyset \xrightarrow{\beta^{(k)}} k \mid \beta^{(k)} \ne 0 \right\},$$

has ODEs $\dot{V}_k = G_k(V)$, for $1 \le k \le n$.

This encoding gives one reaction for each monomial in the ODE.

FB and BB are relations over the species of an RN defined only through properties that concern the reactions in which they are involved. Thus we say that they are *syntax-based* in that the ODE system is never analysed directly, in contrast to the symbolic checks performed with IDOL. FB is a sufficient condition for FDE, defined in terms of *reaction* and *production* rates.

Definition 6 (Reaction and Production Rates). *Let (S, R) be an RN, $X, Y \in S$, and $\rho \in S \cup \{\emptyset\}$. The ρ-reaction rate of X, and the ρ-production rate of Y-elements by X are defined respectively as*

$$\mathbf{crr}[X, \rho] := (\rho(X) + 1) \sum_{X + \rho \xrightarrow{k} \pi \in R} k, \quad \mathbf{pr}(X, Y, \rho) := (\rho(X) + 1) \sum_{X + \rho \xrightarrow{k} \pi \in R} k \cdot \pi(Y)$$

Finally, for $H \subseteq S$ we define $\mathbf{pr}[X, H, \rho] := \sum_{Y \in H} \mathbf{pr}(X, Y, \rho)$.

Definition 7. *Let (S, R) be an RN, $\mathcal{R}$ an equivalence relation over S and $\mathcal{Z} = S/\mathcal{R}$. Then, $\mathcal{R}$ is a forward RN bisimulation (FB) if for all $(X, Y) \in \mathcal{R}$, all $\rho \in S \cup \{\emptyset\}$, and all $H \in \mathcal{Z}$ it holds that*

$$\mathbf{crr}[X, \rho] = \mathbf{crr}[Y, \rho] \quad and \quad \mathbf{pr}[X, H, \rho] = \mathbf{pr}[Y, H, \rho] \quad (4)$$

This definition, originally proposed in [19] for chemical reaction networks, carries over to RNs. An important observation that is instrumental for the development of an efficient partition refinement algorithm is that, as discussed, FB is in the Larsen-Skou style of probabilistic bisimulation, whereby species are related with respect to their aggregate behaviour toward the equivalence classes, parametrised by a further object ρ which plays a role akin to "action labels" in probabilistic transition systems.

Example 3. Consider the RN with species $\{X_1, X_2, X_3, X_4, X_5\}$ and reactions

$$X_1 \xrightarrow{1} X_2 \qquad\qquad X_1 + X_3 \xrightarrow{3} X_4$$
$$X_2 \xrightarrow{1} X_1 \qquad\qquad X_2 + X_3 \xrightarrow{3} X_5$$
$$X_4 \xrightarrow{1} X_1 + X_3 \qquad\qquad X_5 \xrightarrow{1} X_2 + X_3$$
$$X_4 + X_3 \xrightarrow{3} X_5 \qquad\qquad X_5 + X_3 \xrightarrow{3} X_4$$

Then, it holds that $\{\{X_1, X_2\}, \{X_3\}, \{X_4, X_5\}\}$ is an FB. For instance, we have

$$\mathbf{crr}[X_1, \emptyset] = \mathbf{crr}[X_2, \emptyset] = 1$$
$$\mathbf{crr}[X_4, X_3] = \mathbf{crr}[X_2, X_3] = 3$$

As regards **pr**, we have

$$\mathbf{pr}[X_1, \emptyset, \{X_1, X_2\}] = \mathbf{pr}[X_1, \emptyset, \{X_1, X_2\}] = 1$$
$$\mathbf{pr}[X_4, \emptyset, \{X_3\}] = \mathbf{pr}[X_5, \emptyset, \{X_3\}] = 1$$

We now provide a version of BB developed in [20] in the same style.

Definition 8 (Cumulative Splitter Flux Rate). *Let (S, R) be an RN, $X, Y \in S$, $\mathcal{Z}$ a partition of S, $H \in \mathcal{Z}$ and $H' \in \mathcal{Z} \cup \{\{\emptyset\}\}$. We define*

$$\mathbf{sr}(X, Y, H') := \sum_{\rho' \in H'} \sum_{\substack{\rho \xrightarrow{\alpha} \pi \in R \\ \rho = Y + \rho'}} (\pi(X) - \rho(X)) \cdot \alpha', \quad \mathbf{sr}[X, H, H'] := \sum_{Y \in H} \mathbf{sr}(X, Y, H').$$

with $\alpha' = \frac{\alpha}{2}$ if $Y \neq \rho'$ and $Y \in H'$, or $\alpha' = \alpha$ otherwise. We call the quantity $\mathbf{sr}[X, H, H']$ the cumulative (H, H')-splitter flux rate of X.

Note that we account for summands that are counted twice due to the two summations over H' in $\mathbf{sr}[X, H', H']$ by choosing $\alpha' \in \{\alpha, \frac{\alpha}{2}\}$ in the above definition.

Theorem 4. *Let (S, R) be an RN, $\mathcal{R}$ an equivalence relation over S and $\mathcal{Z} = S/\mathcal{R}$. Then $\mathcal{R}$ is a BB if and only if for all $(X, Y) \in \mathcal{R}$, all $H \in \mathcal{Z}$ and all $H' \in \mathcal{Z} \cup \{\{\emptyset\}\}$ it holds that $\mathbf{sr}[X, H, H'] = \mathbf{sr}[Y, H, H']$.*

Example 4. The partition $\{\{X_1, X_2\}, \{X_3\}, \{X_4, X_5\}\}$ of Example 3 is also a BB. For instance, due to the reactions $X_1 \xrightarrow{1} X_2$ and $X_2 \xrightarrow{1} X_1$ we have

$$\mathbf{sr}[X_1, \{X_1, X_2\}, \emptyset] = \mathbf{sr}(X_1, X_1, \emptyset) + \mathbf{sr}(X_1, X_2, \emptyset) = -1 + 1 = 0$$

Similarly, we have

$$\mathbf{sr}[X_2, \{X_1, X_2\}, \emptyset] = \mathbf{sr}(X_2, X_1, \emptyset) + \mathbf{sr}(X_2, X_2, \emptyset) = 1 - 1 = 0$$

2.4   Partition-Refinement Algorithms for RNs

The minimisation algorithms for FB and BB are partition-refinement algorithms based on Paige and Tarjan's approach, iteratively refining an input partition based on a *splitter* block that tells apart the behaviour of two species toward that block. We omit the technical details of the minimisation algorithm, which can be found in [20]. Here we remark that the coarsest FB and BB partitions of an arbitrary polynomial ODE system can be computed in $\mathcal{O}(r \cdot s \cdot \log s)$ time and $\mathcal{O}(r \cdot s)$ space, where r is the number of monomials and s is the number of species. Instead, here we provide a step-by-step illustration of the algorithms on a simple example.

For FB, we first observe that the **crr**-condition of FB can be implemented as an initialization step that pre-partitions the species according to the values of **crr**. This is because **crr** is a "global" property of the RN, i.e., it does not depend on the current partition. Then, as discussed, the conditions on **pr** require the iterative partition-refinement treatment, where ρ plays the role of the label as discussed. An important property is that, at each iteration, the blocks of the current partition are used as potential splitters. This ensures that the list of splitters can be updated at no additional cost while splitting the blocks.

Example 5. Let us consider again the RN in Example 3 and compute the coarsest FB refinement of the trivial partition $\{\{X_1, X_2, X_3, X_4, X_5\}\}$. The initialization step that computes the pre-partitioning with respect to the values of **crr** leads to the refinement $\{\{X_1, X_2, X_4, X_5\}, \{X_3\}\}$. Now, both blocks $\{X_1, X_2, X_4, X_5\}$ and $\{X_3\}$ will be considered as potential splitters. The former does not cause any splitting because, for any species X_i and any label ρ, the values of $\mathbf{pr}[X_i, \{X_1, X_2, X_4, X_5\}, \rho]$ are the same. Instead, $\{X_3\}$ will split the block $\{X_1, X_2, X_4, X_5\}$ into two blocks $\{X_1, X_2\}$ and $\{X_4, X_5\}$, because, e.g., it holds

$$\mathbf{pr}[X_4, \{X_3\}, \emptyset] = \mathbf{pr}[X_5, \{X_3\}, \emptyset] = 1$$
$$\mathbf{pr}[X_1, \{X_3\}, \emptyset] = \mathbf{pr}[X_2, \{X_3\}, \emptyset] = 0$$

Since $\{X_1, X_2, X_4, X_5\}$ has already been used as a splitter, following the principle of *ignoring the largest part* [63], the sub-block with maximal size is not added to the list of potential splitters. In this case, the algorithm will add $\{X_4, X_5\}$, which remains the only splitter to be considered. Since it does not refine any of the existing blocks, the algorithm terminates with the partition $\{\{X_1, X_2\}, \{X_3\}, \{X_4, X_5\}\}$ being the coarsest FB refinement.

For BB, instead, the third argument of **sr** can be seen as a label. However, while in FB this ranges over the set of species (together with the distinguished species $\emptyset$ to indicate unary reactions), in BB it ranges over blocks of the candidate BB partition to be checked (again, together with the distinguished set $\{\emptyset\}$ for unary reactions). When used within the partition refinement algorithm, splitting a partition block leads to a refinement of the BB labels. In other words, unlike for FB the set of labels must be updated at every iteration. However, it can be shown that this incurs no additional computational cost [20].

Example 6. Let us consider once more the RN in Example 3 and compute the coarsest BB refinement of the trivial partition $\{\{X_1, X_2, X_3, X_4, X_5\}\}$. In the first iteration the block $H = \{X_1, X_2, X_3, X_4, X_5\}$ is used to split itself, computing $\mathbf{sr}[X_i, H, \emptyset]$ and $\mathbf{sr}[X_i, H, H]$ for all $i \in \{1, 2, 3, 4, 5\}$. This leads to the partition $\{\{X_1, X_2\}, \{X_3\}, \{X_4, X_5\}\}$. Similarly to the FB case, since H has already been used as a splitter, only $\{X_4, X_5\}$ and $\{X_3\}$ are added as potential splitters, while $\{X_1, X_2\}$ is ignored. The two candidate splitters do not lead to any refinement, and thus the previously computed partition is returned.

3   Case Studies

This section presents four case studies of CAS models. We begin in Sect. 3.1 with a crowd dynamics scenario, where the emergent behaviour of a population arises from decisions made locally by individuals. Then, in Sect. 3.2 we consider an epidemiological model, where the emergent phenomenon of an infection spreading is the result of individual opportunistic contacts between agents. Incidentally, these two case studies feature space and locality as first-class citizen, with increasing complexity. In the crowd dynamics model, individuals do not have an internal status, and dynamics are restricted only to movements among locations. The epidemiological model, instead, does account for individuals' internal states, affected by *local interactions* with other individuals in the same location. In both cases, we start from specifications given by co-authors of this volume in two formal languages, namely BioPEPA [22] and PALOMA [33], from which (together with PEPA [44] and SCEL [28]), originates the CARMA language described elsewhere in this volume.

Sections 3.3 and 3.4 present case studies of biological relevance. Specifically, Sect. 3.3 deals with adaptation in biological systems through evolution of simple structures into more complex ones that retain some of the original behaviour. This is formally captured by means of suitable differential equivalences between CRNs. Section 3.4 presents reductions of a number of CRN models of protein interaction networks presented in the literature, which are well known to the problem of ODEs with combinatorial complexity (e.g., [23]; see also Sect. 4 for further related work).

3.1   Crowd Dynamics

Our first case study regards a crowd scenario in which individuals move among the squares of a city according to certain policies. Our starting point is the famous "El Botellón" model [66], used to describe the spontaneous self-organization of drinking parties in the squares of Spanish cities. The model considers four squares connected in a ring by streets. The movements of a single individual are dictated by a simple rule: if no *friend* (or partner to talk to) can be found in its current square, the individual randomly moves to one of the two connected squares. The model assumes that an individual in square i moves with probability $(1-c)^{s_i-1}$, where s_i is the number of people currently in square i, and c is the *chat probability*, i.e. the probability that an individual finds a friend. The model has been also studied in [58] by co-authors of this volume using related analysis techniques.

More recently, a variant of El Botellón has been proposed in [12] and further analysed in [11], where the chat probability is not a constant, but it depends on two parameters: (i) The *socialisation factor* of the population (*soc*), i.e. the average number of friends of each individual; (ii) The total number of considered individuals (N). The socialisation-driven chat probability is then given by $c = soc/N$. The intuition is that people tend to have a limited number of friends, *soc*, hence the larger is the considered population, the lower is the probability of meeting a friend.

Inspired by the El Botellón model and its socialisation-based variant, we hereby propose a sort of dual scenario where individuals do not move across the squares on their own, e.g. because streets are not safe, but move only if they are able to meet a friend to share the path with. Also, we assume that movements follow a biologically-inspired dynamics: movements from a square i to a square j happen with a rate proportional to the power of the number of people in square i (s_i^2), modelling the probability of two individuals to meet, multiplied by the socialisation-driven chat probability. This is reminiscent of the already discussed law of mass action, which states that the firing rate of a chemical reaction $X_i + X_j \xrightarrow{k} Y_r + Y_l$ is proportional to the concentration of the reacting species of the reaction (X_i, X_j), times the kinetic constant k. Considering n squares, we assume to have an $n \times n$ routing matrix Q, where each $Q_{i,j}$ entry stores the probability that an individual moves from square i to square j. The evolution of the population of each square is governed by an ODE system defined as, for all $i \in \{1 \ldots n\}$:

$$\dot{s}_i = - \sum_{1 \leq j \leq n} 2 \cdot c \cdot Q_{i,j} \cdot s_i^2 + \sum_{1 < j \leq n} 2 \cdot c \cdot Q_{j,i} \cdot s_j^2 \tag{5}$$

For example, the ODE system for the case of four cities ($n = 4$) is

$$\dot{s}_1 = 2 \cdot c \cdot \left(- \sum_{1 \leq j \leq 4} Q_{1,j} \cdot s_1^2 + \sum_{1 < j \leq 4} Q_{j,1} \cdot s_j^2 \right)$$

$$\dot{s}_2 = 2 \cdot c \cdot \left(- \sum_{1 \leq j \leq 4} Q_{2,j} \cdot s_2^2 + \sum_{1 < j \leq 4} Q_{j,2} \cdot s_j^2 \right)$$

$$\dot{s}_3 = 2 \cdot c \cdot \left(- \sum_{1 \leq j \leq 4} Q_{3,j} \cdot s_3^2 + \sum_{1 < j \leq 4} Q_{j,3} \cdot s_j^2 \right)$$

$$\dot{s}_4 = 2 \cdot c \cdot \left(- \sum_{1 < j \leq 4} Q_{4,j} \cdot s_4^2 + \sum_{1 < j \leq 4} Q_{j,4} \cdot s_j^2 \right)$$

The same dynamics can be expressed also in terms of a reaction network defined as, for all $i, j \in \{1 \ldots n\}$ such that $i \neq j$:

$$s_i + s_i \xrightarrow{Q_{i,j} \cdot c} s_j + s_j \tag{6}$$

Fig. 1. Probabilities of movements among squares in the crowd dynamics model.

This reaction models the fact that two individuals in square i meet and move together to a target square j. The deterministic firing rate of the reaction is $[s_i] \cdot [s_i] \cdot Q_{i,j} \cdot c$, where $[s_i]$ is the number of individuals in square i. The term $[s_i] \cdot [s_i]$ accounts for the number of meetings[2], while c restricts to the successful ones (i.e. those among friends), and $Q_{i,j}$ for those leading to movements towards square j. For example, the reaction network for the case of four cities ($n = 4$) is

$$s_1 + s_1 \xrightarrow{Q_{1,2} \cdot c} s_2 + s_2 \qquad\qquad s_3 + s_3 \xrightarrow{Q_{3,4} \cdot c} s_4 + s_4$$

$$s_1 + s_1 \xrightarrow{Q_{1,4} \cdot c} s_4 + s_4 \qquad\qquad s_3 + s_3 \xrightarrow{Q_{3,2} \cdot c} s_2 + s_2$$

$$s_2 + s_2 \xrightarrow{Q_{2,3} \cdot c} s_3 + s_3 \qquad\qquad s_4 + s_4 \xrightarrow{Q_{4,1} \cdot c} s_1 + s_1$$

$$s_2 + s_2 \xrightarrow{Q_{2,1} \cdot c} s_1 + s_1 \qquad\qquad s_4 + s_4 \xrightarrow{Q_{4,3} \cdot c} s_3 + s_3$$

This model shows an interesting property in case Q is symmetric, i.e. $Q_{i,j} = Q_{j,i}$: independently from how the individuals are initially distributed among the squares, on the long run they will be evenly distributed. The same property is found also in the models of [12,66]. To show this, Fig. 2 depicts the evolution of 200000 individuals among the squares (s_1, s_2, s_3, and s_4) with symmetric routing matrix Q defined such that $Q_{1,2} = Q_{2,1} = \frac{1}{4}$, $Q_{1,4} = Q_{4,1} = \frac{3}{4}$, $Q_{2,3} = Q_{3,2} = \frac{3}{4}$ and $Q_{3,4} = Q_{4,3} = \frac{1}{4}$, as depicted in Fig. 1. The socialisation factor soc is set to 2. In the left plot all individuals are initially located in square s_1, while in the right plot they are evenly divided among s_1 and s_2. After some time, in both plots individuals equi-distribute in the four squares, as expected. We notice that more time is required in the case in which all individuals are initially located in the first square.

Figure 2 shows a further interesting property of the crowd scenario. From Fig. 2 (right) we note that the populations in squares s_1 and s_2, as well as those in s_3 and s_4, evolve in the same way if individuals are initially evenly distributed in s_1 and s_2. Instead, such symmetries do not appear in Fig. 2 (left). This can be proven using our backward reductions. The model has the following property:

---

[2] Note that $[s_i] \cdot [s_i]$ should actually be $[s_i] \cdot ([s_i] - 1)$, since an individual cannot meet itself. However, this is irrelevant for large populations, and hence, as for existing ODE-based semantics in the biological context [78], we approximate it to $[s_i] \cdot [s_i]$.

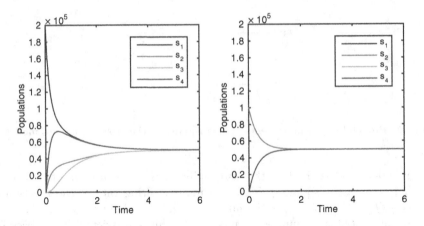

Fig. 2. Crowd scenario with 4 squares and 200000 individuals. All individuals initially in square 1 (left) or evenly divided among squares 1 and 2 (right) (Color figure online).

Whenever the initial number of individuals in s_1 and s_3 is equal that of s_2 and s_4, respectively, then the number of individuals in s_1 and s_3 will be equal to that of s_2 and s_4, respectively, at any point in time.

This property can be verified by reducing the model up to BDE (or equivalently up to BB) using a pre-partition coherent with the required initial conditions, i.e., $\mathcal{Z}_{BDE} = \{\{s_1, s_2\}, \{s_3, s_4\}\}$. The algorithm returns $\mathcal{Z}_{BDE}$ itself, confirming that it is a BDE. Instead, by using an initial partition coherent with Fig. 2 (left), i.e. $\{\{s_1\}, \{s_2, s_3, s_4\}\}$ we obtain no reduction, as expected.

The model also allows for forward reductions, even though they are less interesting. It can be shown that the only forward differential equivalence of the model is $\mathcal{Z}_{FDE} = \{\{s_1, s_2, s_3, s_4\}\}$. This one-block partition is typical of *mass-preserving systems*, i.e. where the total number of entities does not change. In fact, the corresponding FDE-reduced model is $\dot{s}_{FDE} = 0$, meaning that the cumulative population $s = s_1 + s_2 + s_3 + s_4$ is an invariant of the system. No reduction can be instead computed using FB. This is because, as discussed in [19], FB distinguishes among incoming and outgoing flow.

3.2 Multi-community Epidemiology

Our second case study is inspired from the well-known epidemiology model SIR [51], describing the spreading of an infection in a population from *infected* individuals (I), to *susceptible* individuals (S), considering the possibility of *recovering* (R) from infection after some time. We hereby consider a multi-community SIR model extended with spatial features as considered in [33]. Intuitively, individuals move among a number of *communities*, similarly to our crowd model. In addition, individuals in the same location might interact spreading the infection.

More in particular, the authors of [33] use PALOMA (the Process Algebra of Located Markovian Agents), a predecessor of the CARMA language described

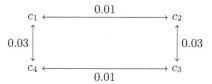

Fig. 3. Rates of movements between communities in the multi-community SIR model.

in this volume, to formalize a simplified model of the 1918–1919 flu epidemic in central Canada originally described in [68]. The model consists of m communities, with a routing matrix Q used to store the rates (rather than probabilities as in the previous crowd model) at which individuals travel between communities. While moving between communities, individuals might interact with locals, spreading the flu. Interactions might happen at different rates in each community (e.g., to distinguish among residential and business areas). Each community c is thus associated with three populations of susceptible (S_c), infected (I_c) and recovered (R_c) individuals. Upon a *contact* between an S_c and I_c individual, the former gets infected with a given probability p. In $1/\gamma$ days on average, an I_c will recover, becoming immune from the flu. The two parameters p and γ are system-dependent, while the rate of contact might change in each community.

In the rest of this section we will consider two variants of the model. In the first one we assume that contacts happen with rate 0.03 in all communities, while in the second variant we have contact rate equal to 0.03 in communities c_1 and c_2, and to 0.04 in the others. In both models we have $p = 0.5$, $\gamma = 0.2$, and, as for the crowd protocol we consider four locations (i.e., communities) connected in a ring by streets according to the symmetric routing matrix Q defined as in Fig. 3. Also, we assume that each community initially has 150000 susceptible individuals, 11000 infected ones and 12000 recovered ones.

The actual PALOMA specification (up to slight changes in the parameters) can be found at http://groups.inf.ed.ac.uk/paloma/SIR.paloma. We refer the interested reader to [33] for more details about the considered PALOMA specification, as well as PALOMA's syntax and tool support. Thanks to the tool support of PALOMA, it is possible to generate an ODE system whose solution gives an approximation of the expected values and the variances of the three populations (S, I and R) in each of the four locations, for a total of 24 measures of interest [32]. In total, 90 ODEs are generated.

The obtained ODE system belongs to the IDOL language, allowing us to apply our symbolic reduction techniques. The coarsest BDE of the model variant with homogeneous contact rates consists of 27 blocks, 6 of which contain all and only the 24 measures of interest, while the other 21 blocks contain the additional variables. In particular, we have three blocks containing the expected values of the three populations in each community: $\{y_{E[S_{c_1}]}, y_{E[S_{c_2}]}, y_{E[S_{c_3}]}, y_{E[S_{c_4}]}\}$, $\{y_{E[I_{c_1}]}, y_{E[I_{c_2}]}, y_{E[I_{c_3}]}, y_{E[I_{c_4}]}\}$, $\{y_{E[R_{c_1}]}, y_{E[R_{c_2}]}, y_{E[R_{c_3}]}, y_{E[R_{c_4}]}\}$. This tells us that (the approximation of) each population evolves in the same way in all communities if initialized equally. This might be expected in a sense, due to the fact that interaction rates do not depend on the community of residence. However, it is interesting to note that populations remain evenly distributed among communities despite having different inter-community transition rates. This can be explained using similar arguments to those of the crowd scenario. The other three blocks are similar, but refer to the second-order moments. Hence, not just the expected values, but also the variances of the populations evolve equally. The obtained BDE partition does not change even if starting with an initial partition coherent with the discussed initial populations. Hence, when the populations of each of S, I and R are initially evenly divided among the four communities, we have that the same information contained in the original ODE system can be recovered from one with 30 % of its original size. Similarly to the crowd scenario, FDE does not produce notable reductions.

We now focus on the model variant having 0.03 as contact rate in communities c_1 and c_2, and 0.04 in c_3 and c_4. By applying BDE starting from the trivial partition with one block only, or from the one coherent with the initial populations, we obtain a partition of 48 blocks. This is actually a refinement of the BDE partition obtained from the homogeneous model variant. In particular, the 6 blocks of interest are split to separate the populations of the communities c_1 and c_2 from those of c_3 and c_4; e.g., the 3 blocks about the average populations are split in $\{y_{E[S_{c_1}]}, y_{E[S_{c_2}]}\}, \{y_{E[S_{c_3}]}, y_{E[S_{c_4}]}\}, \{y_{E[I_{c_1}]}, y_{E[I_{c_2}]}\}, \{y_{E[I_{c_3}]}, y_{E[I_{c_4}]}\}, \{y_{E[R_{c_1}]}, y_{E[R_{c_2}]}\}, \{y_{E[R_{c_3}]}, y_{E[R_{c_4}]}\}$. As a result, an ODE system of size of about 50 % the original one can be used to study the measures of interest of the model.

3.3   Evolutionary Biology

A major subject of investigation in evolutionary biology is to understand how simple structures may evolve into more complex ones as a result of their adaptation to the environment. It has been argued, for instance, that basic cellular switches have evolved in order to increase robustness in their capacity to perform certain functionality by reducing sensitivity to noise [18].

Recently, Cardelli has proposed the notion of *emulation* as a formal way of comparing two CRN models of biological systems in order to postulate an evolutionary path between them [17]. A simpler CRN, i.e. a CRN with fewer species, is said to emulate a larger CRN if in the latter it is possible to find appropriate initial conditions such that the trajectories exactly correspond to those of the simpler CRN. Since the CRN semantics associates an ODE variable with each species, the presence of an emulation will imply that in the larger CRN two or more species' ODE trajectories will overlap, and match one of the simpler CRN as well whenever the initial conditions are equal. The intuitive interpretation given to this dynamical property is that the more complex CRN might possess richer behaviour than the simpler CRN from which it descends, but that the evolution is *conservative* in the sense that under special initial conditions it may collapse onto the original one.

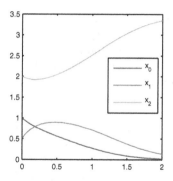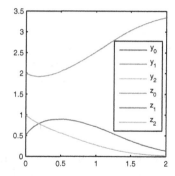

Fig. 4. ODE solutions of AM (left) and MI (right), showing equivalent trajectories with equal initial conditions (Color figure online).

Example 7. The following two mass-action CRNs describe the behaviour of AM, a basic biological switch (left) and MI, a mutual inhibition mechanism (right) [17]:

$$Y_0 + Z_0 \xrightarrow{\alpha_1} Z_0 + Y_1$$

$$Y_1 + Z_0 \xrightarrow{\alpha_2} Z_0 + Y_2$$

$$X_0 + X_2 \xrightarrow{\alpha_1} X_2 + X_1 \qquad Y_2 + Y_0 \xrightarrow{\alpha_3} Y_0 + Y_1$$

$$X_1 + X_2 \xrightarrow{\alpha_2} X_2 + X_2 \qquad Y_1 + Y_0 \xrightarrow{\alpha_4} Y_0 + Y_0$$

$$X_2 + X_0 \xrightarrow{\alpha_3} X_0 + X_1 \qquad Z_2 + Z_0 \xrightarrow{\alpha_1} Z_0 + Z_1$$

$$X_1 + X_0 \xrightarrow{\alpha_4} X_0 + X_0 \qquad Z_1 + Z_0 \xrightarrow{\alpha_2} Z_0 + Z_0$$

$$Z_0 + Y_0 \xrightarrow{\alpha_3} Y_0 + Z_1$$

$$Z_1 + Y_0 \xrightarrow{\alpha_4} Y_0 + Z_2$$

Consider the following mappings:

- Trajectories of Y_0 and Z_2 correspond to that of X_0;
- Trajectories of Y_1 and Z_1 correspond to that of X_1;
- Trajectories of Y_2 and Z_0 correspond to that of X_2.

Indeed, it can be shown that if one sets equal initial conditions for related species (e.g., by setting equal initial conditions for Y_0, Z_2, and X_0) then the trajectories will coincide at all time points (see Fig. 4). It is clear that emulation is closely related to BDE — and to BB since it has been considered for mass-action CRNs. In fact, it can be shown that an emulation is an appropriate BDE on the "union CRN" [21]. For instance, in the example above the BDE is given by $\mathcal{Z}_{EMU} = \{\{X_0, Y_0, Z_2\}, \{X_1, Y_1, Z_1\}, \{X_2, Y_2, Z_0\}\}$. This can be checked using the executable Z3 encoding available at http://rise4fun.com/Z3/bgVv, which is similar to Listing 1, but regards the nine ODEs of (the union CRN of) AM and MI. Note that $\{\{X_0\}, \{X_1\}, \{X_2\}\}$ is a BDE form AM, while $\{\{Y_0, Z_2\}, \{Y_1, Z_1\}, \{Y_2, Z_0\}\}$ is a BDE for MI.

Here we show how to exploit the expressiveness of IDOL to strengthen the idea of an evolutionary relationship between networks, by studying whether it carries over to non-mass-action kinetics as well. The possibility of reasoning using different hypotheses for the reaction kinetics is of biological relevance because in different situations one may find mass-action mechanisms (e.g., phosphotransfers) or Hill-type mechanisms (e.g., enzymes) [78]. For instance, much of the utility of Hill kinetics is owed to supporting non-integer exponents. Famously, this ranges in 2.3–3.0 for haemoglobin. Furthermore, biologists often consider exponents less than 1 in order to describe "anticooperative" behaviour [60]. Since any rational exponent can be expressed in IDOL we consider the question whether the mappings are preserved by a BDE for CRNs with Hill semantics.

We discuss an IDOL encoding of CRNs according to the Hill kinetics (e.g., [78]) in the case of *catalytic* reactions, i.e., reactions which are in the form $B + C \xrightarrow{l} D + C$ with $B \neq D$. We remark that both AM and MI are in this form. Here, C plays the role of a catalyst, a species promoting the reaction but which is not affected by it. Species B is the *substrate* that is modified, becoming D, when the reaction occurs. Each reaction is labelled with a triple $(\beta_1, \beta_2, \nu) \in \mathbb{Q}^3_{>0}$.

Definition 9 (see [21]). *A Hill CRN is a pair (S, R_S) where R_S is a finite set of catalytic reactions with $R_S \subseteq \mathbb{N}_0^S \times \mathbb{N}_0^S \times \mathbb{Q}^3_{>0}$.*

Definition 10. *The IDOL program p_S of a Hill CRN is*

$$\dot{x}_A = h_A := \sum_{\substack{\rho \xrightarrow{(\beta_1,\beta_2,\nu)} \pi \in R_S \\ \rho = B+C, \pi = D+C}} (\pi_A - \rho_A) \frac{\beta_1 x_B^\nu}{\beta_2 + x_B^\nu}, \quad \text{for all} A \in S.$$

By replacing equal mass-action rates with equivalent Hill triplets, it can be shown that the BDE carries over. The following are the Hill CRNs obtained from AM and MI by replacing each original mass action reaction $\rho \xrightarrow{\alpha} \pi$ with the corresponding Hill reaction $\rho \xrightarrow{\alpha,\alpha,\nu} \pi$:

$$X_0 + X_2 \xrightarrow{\alpha_1,\alpha_1,\nu} X_2 + X_1$$
$$X_1 + X_2 \xrightarrow{\alpha_2,\alpha_2,\nu} X_2 + X_2$$
$$X_2 + X_0 \xrightarrow{\alpha_3,\alpha_3,\nu} X_0 + X_1$$
$$X_1 + X_0 \xrightarrow{\alpha_4,\alpha_4,\nu} X_0 + X_0$$

$$Y_0 + Z_0 \xrightarrow{\alpha_1,\alpha_1,\nu} Z_0 + Y_1$$
$$Y_1 + Z_0 \xrightarrow{\alpha_2,\alpha_2,\nu} Z_0 + Y_2$$
$$Y_2 + Y_0 \xrightarrow{\alpha_3,\alpha_3,\nu} Y_0 + Y_1$$
$$Y_1 + Y_0 \xrightarrow{\alpha_4,\alpha_4,\nu} Y_0 + Y_0$$
$$Z_2 + Z_0 \xrightarrow{\alpha_1,\alpha_1,\nu} Z_0 + Z_1$$
$$Z_1 + Z_0 \xrightarrow{\alpha_2,\alpha_2,\nu} Z_0 + Z_0$$
$$Z_0 + Y_0 \xrightarrow{\alpha_3,\alpha_3,\nu} Y_0 + Z_1$$
$$Z_1 + Y_0 \xrightarrow{\alpha_4,\alpha_4,\nu} Y_0 + Z_2$$

For example, the first reaction of the Hill variant of MI introduces the terms $\frac{\alpha_1 \cdot Y_0}{\alpha_1 + Y_0}$ and $-\frac{\alpha_1 \cdot Y_0}{\alpha_1 + Y_0}$ in the drifts of Y_1 and Y_0, respectively. It can be shown that the coarsest BDE partition of the above Hill variant of MI remains $\{\{Y_0, Z_2\}, \{Y_1, Z_1\}, \{Y_2, Z_0\}\}$. Also, $\psi_{EMU} = \{\{X_0, Y_0, Z_2\}, \{X_1, Y_1, Z_1\}, \{X_2, Y_2, Z_0\}\}$ is an emulation among the two Hill CRNs. Similarly to the mass action case, we provide an executable Z3 encoding available at http://rise4fun.com/Z3/f90U to confirm this.

3.4   Protein Interaction Networks

We hereby consider three of the biochemical networks considered in our previous work [19] (and also in [20]): a model of pheromone signalling (M1, [71]); a model of a tumour suppressor protein (M2, [6]); and a MAPK model (M3, [52]). These are three biologically meaningful chemical reaction networks taken from the literature given in .net format of the widely used BioNetGen tool [8], version 2.2.5-stable. In [19,20] we have proved that FB and BB can be successfully applied to these and other BioNetGen models, providing the reduction times, the size of the obtained reduced models, and the speed-up obtained by analysing them. For each RN (S, R), in the case of FB reductions we considered the trivial partition $\{S\}$ (thus yielding the largest bisimulation). Instead, for BB an initial partition coherent with the initial conditions was chosen, due to the side condition of BB: two species were put in the same initial block in case of equal initial conditions, read from the original model specification.

The BioNetGen tool allows the modeller to specify *observables* of interest, given in the form of sums of species. When solving the ODEs underlying the considered model, a plot containing a line per observable is generated, showing the evolution of the specified cumulative concentrations. Differently from [19,20], we now study the FB partitions obtained when using initial partitions coherent with the user-specified observables. These are partitions which guarantee that the information of interest to the modeller is preserved, hence de facto obtaining "lossless" FB reductions. We remark that BioNetGen observables might not specify a partition of the species because: (i) Some species might not appear in the observables; (ii) Others might appear in more than one observable. However, it is easy to obtain an observables-preserving initial partition as follows:

1. All species not appearing in any observable are put in a single (sink) partition block;
2. For each observable, its subset of species not appearing in any other observable is turned into a partition block;
3. The set of species appearing in more than one observable is partitioned in blocks of species appearing in (all and only) the same blocks.

As shown in column $|Prep.|$ of Table 1, the 14531 species of M1 are pre-partitioned in 1345 blocks, the 796 species of M2 are pre-partitioned in 18 blocks, while the 85 species of M3 are pre-partitioned in 4 blocks. From the table we also note that, both in terms of reduction time and size of the reduced model, the pre-partitioning does not affect the FB reduction of M2, while it affects that of the other two models only slightly.

Table 1. FB reductions with and without observations-coherent pre-partitioning.

Original model				FB reduction			FB reduction with prep.			
Id	Ref.	$|R|$	$|S|$	Red.(s)	$|R|$	$|S|$	$|Prep.|$	Red.(s)	$|R|$	$|S|$
M1	[71]	194054	14531	3.88E–1	142165	10855	1345	3.28E–1	147797	12037
M2	[6]	5797	796	1.90E–2	4210	503	18	4.10E–2	4210	503
M3	[52]	487	85	2.00E–3	264	56	4	2.00E–3	362	69

4   Related Work

FDE/FB are special cases of exact ODE lumpability [62], which concerns ODE aggregations through a linear projection of the state space. While the general theory is well-established, in particular for ODEs arising from mass-action CRNs [72], there are no algorithms for computing these projections, unlike with the partition refinement algorithms of FDE/FB. As discussed, when the ODE represents the forward equations of motion of a CTMC, both FDE and FB correspond to ordinary CTMC lumpability [14]. In addition, in that case the partition refinement algorithm of FB yields the same time and space complexity of state-of-the-art algorithms for CTMCs [31,77]. FDE/FB are also related to a recently proposed notion of equivalence called *differential bisimulation* [47]. This is developed for a fragment of Hillston's PEPA process algebra that is equipped with an ODE semantics with non-linear minimum-based drifts that approximate the average evolution of underlying CTMCs with massively parallel computations [43,45,73]. Differential bisimulation is a relation over the set of constants of a PEPA model, defined in terms of conditions on the sequential behaviour and on the compositional structure of processes. It can be shown that differential bisimulation is a special case of FDE for the ODEs induced by a PEPA process.

BDE/BB are generalisations of the notion of *label equivalence* for process algebra with fluid semantics [74]. It relates processes that are equivalent whenever their ODE solutions are equal at all time points. Label equivalence is only a sufficient condition for ODE reduction since it works at a coarser level of granularity. Indeed, it relates sets of ODE variables, each corresponding to the behaviour of a sequential process. Instead, BDE/BB relate individual ODE variables. In addition, no algorithm for computing label equivalence is available. Analogously to FDE/FB, for ODEs that represent a CTMC we have that BDE and BB correspond to exact CTMC lumpability [14].

Model reductions have been extensively studied for CRNs in systems biology. In particular, for protein interaction networks, the combinatorial explosion of the state space has motived considerable research, e.g., [15,16,19,23–25,34,35]. The fragmentation approach for the rule-based language κ identifies a coarse-grained ODE system for models with mass-action semantics through sums of variables; this is weaker than an equivalence relation over species, because one variable may appear in more than one block (a *fragment*) [25,35]. Using the terminology of [62], fragmentation is a form of *improper lumping*, as opposed to the notions of equivalence presented here where a species belongs to a single block.

SMT has become a cornerstone in the programming languages and in the verification community, with contributions to program synthesis [41], constraint programming [53], and symbolic optimization [56]. The combination of SMT and equivalence relations has been the subject of recent investigations. In [7] partition-refinement algorithms are proposed to compute equivalences between terms over arbitrary theories inferred from a set of axioms. Applied to equivalences presented here, these partition-refinement algorithms could be used to check if a candidate partition is a differential equivalence, but not to compute the largest equivalence for an IDOL program. In [29] the authors present an SMT-based approach for the computation of the coarsest ordinary lumpable partition of a Markov chain, but for a fragment of the PRISM language [54].

Finally, links between ODEs and SMT are established in the formal verification community, especially for hybrid systems (e.g., [39,59,67]); however none of these works considers ODE comparisons and minimizations through equivalence relations. Bisimulation for dynamical systems have been studied by Pappas [64] and van der Schaft [69]. These works are similar in spirit to ours, but the setting is different because the focus is on *control systems*, i.e., dynamical systems with internal states, external inputs, and output maps. In that context, bisimulation relates internal states mapped to the same output, i.e., they cannot be told apart by an external observer. The largest bisimulation is therefore related to the maximal unobservability subspace of a control system (e.g., [69, Corollary 6.4]) while our largest differential equivalences provide the coarsest partition of ODE variables that preserves the dynamics.

5 Conclusion

This paper has presented a number of techniques for the automatic reduction of systems of ordinary differential equations (ODEs), motivated by their popularity in the modelling and analysis of large-scale dynamical systems such as collective adaptive systems. The symbolic approach of differential equivalences and the syntax-driven minimisation through reaction-network (RN) bisimulations offer a trade-off between expressiveness and efficiency.

Differential equivalences support a rather rich class of non-linear ODE, which can be analysed by using satisfiability solvers as the underlying engine. In general, it is well known that such solvers are more efficient in providing a positive "sat" result than a negative "unsat", which is however required to check that a candidate partition is a differential equivalence. Nevertheless, the current technology allows us to analyse models of realistic size (see also [21] for further examples). In the current prototype implementation the SMT solver is used as a black-box; it would be interesting in the future to consider the development of domain-specific heuristics that improve the search.

RN bisimulations are particularly efficient since the partition-refinement algorithms run in polynomial time and space; however, currently they support ODE with derivatives given by multivariate polynomials of order at most two. Nevertheless, they cover an interesting class of systems, including CRNs and affine systems (see also [20] for experiments in large-scale benchmarks). To further improve efficiency it would be interesting to consider parallelisation techniques; on a more theoretical viewpoint, an obvious direction for future research is to extend the bisimulations of higher-order multivariate polynomials.

The forward and backward variants of the presented equivalences are not comparable in general. This suggests a possible combined use, which has however not been investigated so far. A better understanding of the relationship between these two variants may help achieve further reductions.

Much of the efficiency in computing ODE reductions is owed to the fact that the largest differential equivalences and bisimulations exist and can be computed via partition-refinement. We argue, however, that there are situations of practical interest that cannot be cast into this framework. For example, the notion of emulation that is instrumental to investigate evolutionary aspects of CRNs, amounts to finding a particular backward bisimulation where each equivalence class contains exactly one species of the small CRN and at most one species of the larger CRN. This condition cannot be expressed as a suitable initial partition to be refined; hence, one is left with having to enumerate all possible partitions that satisfy these conditions in order to find emulations automatically. However, this is feasible only for very simple models. Further research is needed to develop algorithms that aggregate according to more liberal constraints on the desired equivalence classes.

Finally, we remark that all the techniques presented in this paper are concerned with exact aggregations. In some cases, these may be too strong because even small perturbations may discriminate ODE variables that have nearby trajectories in practice. This has motivated a large body of work into approximate notions of equivalence [1,13,42,65]. Preliminary work for models based on ODE semantics has been carried out in [76] in the case of process algebra; more general ODE systems are treated in [46,75]. However all these approaches still lack an algorithm for automatic reduction. Furthermore, they provide a priori bounds on the approximate aggregation that tend to grow fast with time. Future research work will be aimed at tackling these two issues.

Acknowledgement. This work was partially supported by the EU project QUANTI-COL, 600708. The authors thank Luca Cardelli and Max Tschaikowski who co-authored the papers [19–21] used as background material in this chapter.

References

1. Aldini, A., Bravetti, M., Gorrieri, R.: A process-algebraic approach for the analysis of probabilistic noninterference. J. Comput. Secur. **12**(2), 191–245 (2004)
2. Aoki, M.: Control of large-scale dynamic systems by aggregation. IEEE Trans. Autom. Control **13**(3), 246–253 (1968)
3. Baier, C., Engelen, B., Majster-Cederbaum, M.E.: Deciding bisimilarity and similarity for probabilistic processes. J. Comput. Syst. Sci. **60**(1), 187–231 (2000)
4. Barrett, C., Sebastiani, R., Seshia, S.A., Tinelli, C.: Satisfiability Modulo Theories. In: Handbook of Satisfiability. Frontiers in Artificial Intelligence and Applications, vol. 185, IOS Press, Amsterdam (2009)

5. Barrett, C., Stump, A., Tinelli, C.: The SMT-LIB standard: version 2.0. Technical report, Department of Computer Science, The University of Iowa (2010). www. SMT-LIB.org

6. Barua, D., Hlavacek, W.S.: Modeling the effect of APC truncation on destruction complex function in colorectal cancer cells. PLoS Comput. Biol. **9**(9), e1003217 (2013)

7. Berdine, J., Bjørner, N.: Computing all implied equalities via SMT-based partition refinement. In: Demri, S., Kapur, D., Weidenbach, C. (eds.) IJCAR 2014. LNCS, vol. 8562, pp. 168–183. Springer, Heidelberg (2014)

8. Blinov, M.L., Faeder, J.R., Goldstein, B., Hlavacek, W.S.: BioNetGen: software for rule-based modeling of signal transduction based on the interactions of molecular domains. Bioinformatics **20**(17), 3289–3291 (2004)

9. Bortolussi, L., De Nicola, R., Galpin, V., Gilmore, S., Hillston, J., Latella, D., Loreti, M., Massink, M.: CARMA: collective adaptive resource-sharing Markovian agents. In: QAPL, pp. 16–31 (2015)

10. Bortolussi, L., Gast, N.: Scalable quantitative analysis: fluid and hybrid approximations. In: SFM (2016)

11. Bortolussi, L., Latella, D., Massink, M.: Stochastic process algebra and stability analysis of collective systems. In: De Nicola, R., Julien, C. (eds.) COORDINATION 2013. LNCS, vol. 7890, pp. 1–15. Springer, Heidelberg (2013)

12. Bortolussi, L., Le Boudec, J.Y., Latella, D., Massink, M.: Revisiting the limit behaviour of "El Botellon". Technical report (2012). http://infoscience.epfl.ch/record/179935/

13. van Breugel, F., Worrell, J.: Approximating and computing behavioural distances in probabilistic transition systems. Theor. Comput. Sci. **360**(1–3), 373–385 (2006)

14. Buchholz, P.: Exact and ordinary lumpability in finite Markov chains. J. Appl. Probab. **31**(1), 59–75 (1994)

15. Camporesi, F., Feret, J.: Formal reduction for rule-based models. Electron. Notes Theoret. Comp. Sci. **276**, 29–59 (2011)

16. Camporesi, F., Feret, J., Koeppl, H., Petrov, T.: Combining model reductions. Electron. Notes Theoret. Comp. Sci. **265**, 73–96 (2010)

17. Cardelli, L.: Morphisms of reaction networks that couple structure to function. BMC Syst. Biol. **8**(1), 84 (2014)

18. Cardelli, L., Csikász-Nagy, A., Dalchau, N., Tribastone, M., Tschaikowski, M.: Noise reduction in complex biological switches. Scientific reports 6, 20214 EP (February 2016). http://dx.doi.org/10.1038/srep20214

19. Cardelli, L., Tribastone, M., Tschaikowski, M., Vandin, A.: Forward and backward bisimulations for chemical reaction networks. In: CONCUR (2015)

20. Cardelli, L., Tribastone, M., Tschaikowski, M., Vandin, A.: Efficient syntax-driven lumping of differential equations. In: TACAS (2016, to appear)

21. Cardelli, L., Tribastone, M., Tschaikowski, M., Vandin, A.: Symbolic computation of differential equivalences. In: POPL (2016)

22. Ciocchetta, F., Hillston, J.: Bio-PEPA: a framework for the modelling and analysis of biological systems. TCS **410**(33–34), 3065–3084 (2009)

23. Conzelmann, H., Fey, D., Gilles, E.: Exact model reduction of combinatorial reaction networks. BMC Syst. Biol. **2**(1), 78 (2008)

24. Conzelmann, H., Saez-Rodriguez, J., Sauter, T., Kholodenko, B., Gilles, E.: A domain-oriented approach to the reduction of combinatorial complexity in signal transduction networks. BMC Bioinform. **7**(1), 34 (2006)

25. Danos, V., Feret, J., Fontana, W., Harmer, R., Krivine, J.: Abstracting the differential semantics of rule-based models: exact and automated model reduction. In: LICS, pp. 362–381 (2010)

26. de Moura, L., Bjørner, N.S.: Z3: an efficient SMT solver. In: Ramakrishnan, C.R., Rehof, J. (eds.) TACAS 2008. LNCS, vol. 4963, pp. 337–340. Springer, Heidelberg (2008)

27. De Nicola, R., Latella, D., Loreti, M., Massink, M.: Rate-based transition systems for stochastic process calculi. In: Albers, S., Marchetti-Spaccamela, A., Matias, Y., Nikoletseas, S., Thomas, W. (eds.) ICALP 2009, Part II. LNCS, vol. 5556, pp. 435–446. Springer, Heidelberg (2009)

28. De Nicola, R., Loreti, M., Pugliese, R., Tiezzi, F.: A formal approach to autonomic systems programming: the SCEL language. TAAS 9(2), 7:1–7:29 (2014)

29. Dehnert, C., Katoen, J.-P., Parker, D.: SMT-based bisimulation minimisation of Markov models. In: Giacobazzi, R., Berdine, J., Mastroeni, I. (eds.) VMCAI 2013. LNCS, vol. 7737, pp. 28–47. Springer, Heidelberg (2013)

30. DeMaio, P.: Bike-sharing: history, impacts, models of provision, and future. J. Publ. Transp. 14(4), 41–56 (2009)

31. Derisavi, S., Hermanns, H., Sanders, W.: Optimal state-space lumping in Markov chains. Inf. Process. Lett. 87(6), 309–315 (2003)

32. Feng, C., Hillston, J.: Automatic moment-closure approximation of spatially distributed collective adaptive systems. ACM Trans. Model. Comput. Simul. (to appear)

33. Feng, C., Hillston, J.: PALOMA: a process algebra for located Markovian agents. In: Norman, G., Sanders, W. (eds.) QEST 2014. LNCS, vol. 8657, pp. 265–280. Springer, Heidelberg (2014)

34. Feret, J.: Fragments-based model reduction: some case studies. Electron. Notes Theoret. Comput. Sci. 268, 77–96 (2010)

35. Feret, J., Danos, V., Krivine, J., Harmer, R., Fontana, W.: Internal coarse-graining of molecular systems. Proc. Nat. Acad. Sci. 106(16), 6453–6458 (2009)

36. Feret, J., Henzinger, T., Koeppl, H., Petrov, T.: Lumpability abstractions of rule-based systems. Theoret. Comput. Sci. 431, 137–164 (2012)

37. Fricker, C., Gast, N.: Incentives and redistribution in homogeneous bike-sharing systems with stations of finite capacity. EURO J. Transp. Logist. 1–31 (2014). doi:10.1007/s13676-014-0053-5

38. Galpin, V.: Spatial representations and analysis techniques. In: SFM (2016)

39. Gao, S., Kong, S., Clarke, E.: Satisfiability modulo ODEs. In: FMCAD, pp. 105–112 (2013)

40. Grosu, R., Bartocci, E.: Spatio-temporal model checking. In: SFM (2016)

41. Gulwani, S., Jha, S., Tiwari, A., Venkatesan, R.: Synthesis of loop-free programs. In: PLDI, pp. 62–73 (2011)

42. Gupta, V., Jagadeesan, R., Panangaden, P.: Approximate reasoning for real-time probabilistic processes. Log. Methods Comput. Sci. 2(1) (2006)

43. Hayden, R.A., Bradley, J.T.: A fluid analysis framework for a Markovian process algebra. Theoret. Comput. Sci. 411(22–24), 2260–2297 (2010)

44. Hillston, J.: A compositional approach to performance modelling. In: CUP (1996)

45. Hillston, J.: Fluid flow approximation of PEPA models. In: QEST, pp. 33–43, September 2005

46. Iacobelli, G., Tribastone, M.: Lumpability of fluid models with heterogeneous agent types. In: DSN, pp. 1–11 (2013)

47. Iacobelli, G., Tribastone, M., Vandin, A.: Differential bisimulation for a Markovian process algebra. In: Italiano, G.F., Pighizzini, G., Sannella, D.T. (eds.) MFCS 2015. LNCS, vol. 9234, pp. 293–306. Springer, Heidelberg (2015)
48. Iwasa, Y., Andreasen, V., Levin, S.: Aggregation in model ecosystems. I. Perfect aggregation. Ecol. Model. **37**(3–4), 287–302 (1987)
49. Jovanović, D., de Moura, L.: Solving non-linear arithmetic. In: Gramlich, B., Miller, D., Sattler, U. (eds.) IJCAR 2012. LNCS, vol. 7364, pp. 339–354. Springer, Heidelberg (2012)
50. Katoen, J., Khattri, M., Zapreev, I.: A Markov reward model checker. In: QEST, pp. 243–244 (2005)
51. Kermack, W.O., McKendrick, A.: Contribution to the mathematical theory of epidemics. Proc. Roy. Soc. Lond. Ser. A Containing Papers Math. Phys. Charact. **115**(772), 700–721 (1927)
52. Kocieniewski, P., Faeder, J.R., Lipniacki, T.: The interplay of double phosphorylation and scaffolding in MAPK pathways. J. Theor. Biol. **295**, 116–124 (2012)
53. Köksal, A.S., Kuncak, V., Suter, P.: Constraints as control. In: POPL, pp. 151–164 (2012)
54. Kwiatkowska, M., Norman, G., Parker, D.: PRISM 4.0: verification of probabilistic real-time systems. In: Gopalakrishnan, G., Qadeer, S. (eds.) CAV 2011. LNCS, vol. 6806, pp. 585–591. Springer, Heidelberg (2011)
55. Larsen, K.G., Skou, A.: Bisimulation through probabilistic testing. Inf. Comput. **94**(1), 1–28 (1991)
56. Li, Y., Albarghouthi, A., Kincaid, Z., Gurfinkel, A., Chechik, M.: Symbolic optimization with SMT solvers. In: POPL, pp. 607–618 (2014)
57. Loreti, M., Hillston, J.: Modeling and analysis of collective adaptive systems with CARMA and its tools. In: SFM (2016)
58. Massink, M., Latella, D., Bracciali, A., Hillston, J.: Modelling non-linear crowd dynamics in Bio-PEPA. In: Giannakopoulou, D., Orejas, F. (eds.) FASE 2011. LNCS, vol. 6603, pp. 96–110. Springer, Heidelberg (2011)
59. Mover, S., Cimatti, A., Tiwari, A., Tonetta, S.: Time-aware relational abstractions for hybrid systems. In: EMSOFT, pp. 1–10 (2013)
60. Nelson, D.L., Cox, M.M.: Lehninger Principles of Biochemistry, 6th edn. Palgrave Macmillan, Basingstoke (2013)
61. Norris, J.: Markov Chains. Cambridge Series in Statistical and Probabilistic Mathematics. Cambridge University Press, Cambridge (1998)
62. Okino, M.S., Mavrovouniotis, M.L.: Simplification of mathematical models of chemical reaction systems. Chem. Rev. **2**(98), 391–408 (1998)
63. Paige, R., Tarjan, R.: Three partition refinement algorithms. SIAM J. Comput. **16**(6), 973–989 (1987)
64. Pappas, G.J.: Bisimilar linear systems. Automatica **39**(12), 2035–2047 (2003)
65. Di Pierro, A., Hankin, C., Wiklicky, H.: Quantitative relations and approximate process equivalences. In: Amadio, R.M., Lugiez, D. (eds.) CONCUR 2003. LNCS, vol. 2761, pp. 508–522. Springer, Heidelberg (2003)
66. Rowe, J.E., Gomez, R.: El Botellon: modeling the movement of crowds in a city. Complex Syst. **14**, 363–370 (2003)
67. Sankaranarayanan, S., Tiwari, A.: Relational abstractions for continuous and hybrid systems. In: Gopalakrishnan, G., Qadeer, S. (eds.) CAV 2011. LNCS, vol. 6806, pp. 686–702. Springer, Heidelberg (2011)
68. Sattenspiel, L., Herring, D.A.: Simulating the effect of quarantine on the spread of the 1918–19 flu in Central Canada. Bull. Math. Biol. **65**(1), 1–26. http://dx.doi.org/10.1006/bulm.2002.0317

69. van der Schaft, A.J.: Equivalence of dynamical systems by bisimulation. IEEE Trans. Autom. Control **49**(12), 2160–2172 (2004)

70. Sproston, J., Donatelli, S.: Backward bisimulation in Markov chain model checking. IEEE Trans. Softw. Eng. **32**(8), 531–546 (2006)

71. Suderman, R., Deeds, E.J.: Machines vs. ensembles: effective MAPK signaling through heterogeneous sets of protein complexes. PLoS Comput. Biol. **9**(10), e1003278 (2013)

72. Toth, J., Li, G., Rabitz, H., Tomlin, A.S.: The effect of lumping and expanding on kinetic differential equations. SIAM J. Appl. Math. **57**(6), 1531–1556 (1997)

73. Tribastone, M., Gilmore, S., Hillston, J.: Scalable differential analysis of process algebra models. IEEE Trans. Softw. Eng. **38**(1), 205–219 (2012)

74. Tschaikowski, M., Tribastone, M.: Exact fluid lumpability for Markovian process algebra. In: Koutny, M., Ulidowski, I. (eds.) CONCUR 2012. LNCS, vol. 7454, pp. 380–394. Springer, Heidelberg (2012)

75. Tschaikowski, M., Tribastone, M.: Approximate reduction of heterogenous nonlinear models with differential hulls. IEEE Trans. Autom. Control. (2015, to appear)

76. Tschaikowski, M., Tribastone, M.: A unified framework for differential aggregations in Markovian process algebra. J. Log. Algebraic Methods Program. **84**(2), 238–258 (2015)

77. Valmari, A., Franceschinis, G.: Simple $O(m \log n)$ time Markov chain lumping. In: Esparza, J., Majumdar, R. (eds.) TACAS 2010. LNCS, vol. 6015, pp. 38–52. Springer, Heidelberg (2010)

78. Voit, E.O.: Biochemical systems theory: a review. ISRN Biomath. 1–53 (2013). http://dx.doi.org/10.1155/2013/897658897658

Aggregate Programming: From Foundations to Applications

Jacob Beal[1]([✉]) and Mirko Viroli[2]

[1] Raytheon BBN Technologies, Cambridge, MA 02138, USA
jakebeal@bbn.com
[2] Alma Mater Studiorum–Università di Bologna, Cesena, Italy

Abstract. We live in a world with an ever-increasing density of computing devices, pervading every aspect of our environment. Programming these devices is challenging, due to their large numbers, potential for frequent and complex network interactions with other nearby devices, and the open and evolving nature of their capabilities and applications. Aggregate programming addresses these challenges by raising the level of abstraction, so that a programmer can operate in terms of collections of interacting devices. In particular, field calculus provides a safe and extensible model for encapsulation, modulation, and composition of services. On this foundation, a set of resilient "building block" operators support development of APIs that can provide resilience and scalability guarantees for any service developed using them. We illustrate the power of this approach by discussion of several recent applications, including crowd safety at mass public events, disaster relief operations, construction of resilient enterprise systems, and network security.

Keywords: Aggregate programming · Pervasive computing · Field calculus · Distributed systems · Domain-specific languages

1 Introduction

For some time now, our world has been undergoing a dramatic transition in how we relate to computing, as the number of computing devices rises and more and more of these devices become embedded into our environment (Fig. 1). In the past, it was reasonable to use a programming model that focused on the individual computing device, and its relationship with one or more users. Now, however, it is typically the case that many computing devices are involved in the provision of any given service, and that each machine may participate in many overlapping instances of such collective services. Moreover, the increasing mobility and wireless capabilities of some computing devices (e.g., wearable devices, smart phones, car systems, drones, electronic tags, etc.), means that many devices have the opportunity to accomplish part or all of their assigned tasks through peer-to-peer local interactions, rather than by going through fixed infrastructure such as cellular wireless or the Internet, thereby lowering latency

© Springer International Publishing Switzerland 2016
M. Bernardo et al. (Eds.): SFM 2016, LNCS 9700, pp. 233–260, 2016.
DOI: 10.1007/978-3-319-34096-8_8

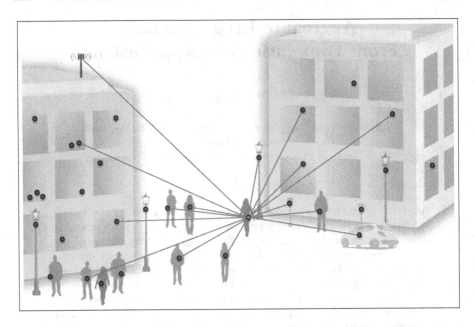

Fig. 1. Our world is increasingly filled with large numbers of computing devices, embedded into the environment and with many opportunities for local interaction as well as for more traditional location-agnostic interactions over fixed network infrastructure. Figure adapted from [8].

and increasing resilience to issues with inadequate or unavailable infrastructure, e.g., during civic emergencies or mass public events.

To effectively program such systems, we need to be able to reliably engineer collective aggregate behaviors. Ordinary programming approaches typically focus on individual devices, entangling application design with various aspects of distributed system design (e.g., efficient and reliable communication, robust coordination, composition of capabilities, etc.), as well as confronting the programmer with the notoriously difficult and generally intractable "local-to-global" problem of generating a specified emergent collective behavior from the interactions of individual devices. These problems tend to limit our ability to make use of the potential of the modern computing environment, as complex distributed services developed using device-centric programming paradigms tend to suffer from design problems, lack of modularity and reusability, deployment difficulties, and serious test and maintenance issues.

Aggregate programming provides an alternate approach, which simplifies the design, creation, and maintenance of complex distributed systems by raising the abstraction level from individual devices to potentially large aggregations of devices. This survey presents an introduction to aggregate programming and a survey of key points on the current state of the art, updating and synthesizing several prior surveys [7,8,10,11]. Aggregate programming has roots in many different communities, all of which have encountered their own versions of the aggregate programming problem and which have between them developed a vast profusion of domain-specific programming models to address it, which are briefly surveyed in Sect. 2. Recently, however, there have been a number of unifying results regarding field-based computational models, which are reviewed in Sect. 3. These results lay the foundation for a more principled approach, in which general mechanisms for robust and adaptive coordination are composed and refined to build domain-specific APIs, following the layered engineering approach reviewed in Sect. 4. Ultimately, this can provide distributed systems engineers with a simple interface for development of safe, resilient, and scalable distributed applications, some examples of which are presented in Sect. 5 before turning to discussion of future directions in Sect. 6.

2 Background and General Approach

In many ways, aggregate programming is not a new idea: the importance of raising the abstraction level for distributed programming has been recognized previously in a number of different fields, motivating work toward aggregate programming across a variety of domains, including biology, reconfigurable computing, high-performance computing, sensor networks, agent-based systems, and robotics, as surveyed in [7].

Despite the wide degree of heterogeneity in applications and context across these antecedents, the common problems in organizing aggregates have led such approaches to cluster around a few main strategies: making device interaction implicit (e.g., TOTA [31], MPI [32], NetLogo [41], Hood [47]), providing means to compose geometric and topological constructions (e.g., Origami Shape Language [33], Growing Point Language [17], ASCAPE [28]), providing means for summarizing from space-time regions of the environment and streaming these summaries to other regions (e.g., TinyDB [30], Regiment [34], KQML [23]), automatically splitting computational behaviour for cloud-style execution (e.g., MapReduce [21], BOINC [2], Sun Grid Engine [25]), and providing generalizable constructs for space-time computing (e.g., Protelis [37], Proto [5], MGS [26]).

These many prior efforts have also evidenced some commonalities in their strengths and weaknesses, which suggest that, when programming large-scale distributed systems, it is useful to conform to the following three principles: *(i)* mechanisms for robust coordination should be hidden "under-the-hood" where programmers are not *required* to interact with them, *(ii)* composition of modules and subsystems must be simple, transparent, and with consequences that can be readily predicted, and *(iii)* large-scale distributed systems typically comprise a number of different subsystems, which need to use different coordination mechanisms for different regions and times.

From these observations and the commonalities amongst the various prior approaches has come the generalized approach that we discuss in this paper, based on field calculus [19,20,46] and its practical instantiation in Protelis [37], which takes the following view of distributed systems engineering:

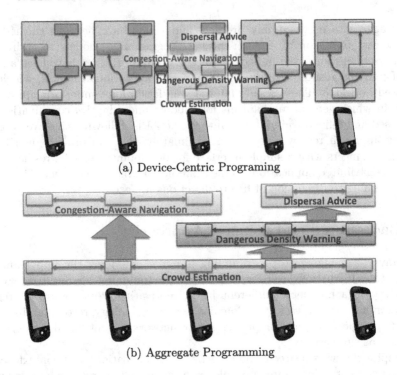

(a) Device-Centric Programing

(b) Aggregate Programming

Fig. 2. Comparison of device-centric programming of distributed algorithms (a) versus aggregate programming (b): device-centric programming designs a distributed system in terms of the (often complex) behaviors and interactions of individual devices; with aggregate programming, simpler algorithmic building blocks can be scoped and composed directly for the aggregate. Figure adapted from [8,11].

1. the "machine" being programmed is a region of the computational environment whose specific details are abstracted away (perhaps even to a pure spatial continuum);
2. the program is specified as manipulation of data constructs with extent across that region (where regions may be defined either regarding network structure or regarding continuous space and time); and
3. these manipulations are actually executed by the individual devices in the region, through local operations and interactions with (spatial or network) neighbors.

2.1   Example: Distributed Crowd Management

Consider, for example, the architecture of a crowd-safety service, such as might be distributed on the cell phones of people attending a very large public event, such as a marathon or a major city festival, as in the scenarios described in [3,36]. Figure 2 compares a traditional device-centric architecture versus an aggregate programming approach to building a crowd-safety service with four functionalities: estimation of crowd density and distribution based on interaction between phones and observation of the local wireless environment, alerting of people in or near dangerously large and dense regions of the crowd (where there is risk of trampling or panic), providing advice for people in the interior of such regions on how to move to help disperse the dangers, and crowd-aware navigation that can help other people move around the event while simultaneously avoiding dangerous areas.

With traditional device-centric approaches (Fig. 2(a)), the programmer needs to simultaneously reason about composition of services within a device, protocols for local device interactions, and also about how the desired complex global behavior will be produced from such local interactions. With aggregate programming, on the other hand, the system can be readily approached in terms of a set of distributed modules. A programmer can then compose these modules incrementally to form complete applications simply by specifying where they should execute and how information should flow between them (Fig. 2(b)). Here, for example, crowd estimation produces as output a distributed data structure—a "computational field" [19, 20]—mapping from location to crowd density, which is then an input for both crowd-aware navigation and for the alerting service. These then produce their own distributed data structures, respectively vectors for recommended travel and a map of warnings (which is in turn an input for producing dispersal advice). The details of protocol and implementation can then be automatically generated from such compositions of data structures and services. Aggregate programming thus promotes the construction of more complex, reusable, resilient, and composable distributed systems by separating the question of which services should be executed and where, from the implementation details of those services and their coordination.

2.2   Aggregate Programming Layers

Figure 3 shows how aggregate programming can hide the complexity of the underlying distributed network environment and the problems of distributed coordination with a sequence of abstraction layers. At the foundation of this approach is *field calculus* [19, 20], a core set of constructs modeling device behavior and interaction, which is terse enough to enable mathematical proof of equivalence between aggregate specifications and local implementations, yet expressive enough to be universal. The notion of "computational field," adapted from physics, makes this particularly well suited for environments with devices embedded in space and communicating with others in close physical proximity, though it is more generally suitable for any sparsely connected network. Upon this foundation, we can identify key coordination "building blocks" with desirable resilience properties, each being a simple and generalized basis element generating a broad set of programs with desirable resilience properties. Finally, common patterns for using and composing these building blocks can be captured to produce both general and domain-specific APIs for common application needs like sensing, decision, and action, together forming a collective behavior API

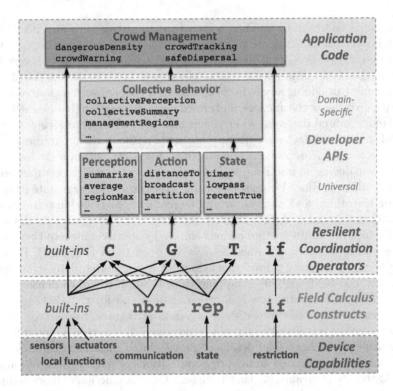

Fig. 3. Aggregate programming takes a layered approach to distributed systems development: the software and hardware capabilities of particular devices are abstracted by using them to implement a small universal calculus of aggregate-level field calculus constructs. This calculus is then used to implement a limited set of "building block" coordination operations with provable resilience properties, which are in turn wrapped and combined together to produce user-friendly APIs, both general and domain-specific, for developing distributed systems. Figure adapted from [8].

suitable for transparent implementation of complex networked services and applications [8,9,45].

Engineering distributed systems with this approach can thus allow construction of complicated resilient distributed systems with rather simple specifications, as we will see in the application examples in Sect. 5. From such a terse specification, the full complexity of the system is then automatically elaborated: from the set of resilient coordination operators that were chosen to be used, to the various ways in which resilience is actually achieved via particular building blocks or their functional equivalents, then how the aggregate specification implements each of those building blocks and maps to actions by individual devices, and finally how particular devices in the potentially heterogeneous network environment actually implement capabilities like sensing, communication, and localization.

Here, we discuss the incarnation of this approach using Protelis, a field calculus implementation with Java-like syntax and support for first-class aggregate functions. For full details on Protelis, see its presentation in [37].

3   Field Calculus

The *field calculus* [19,20,46] is an attempt to capture a set of essential features that appear across many different aggregate programing approaches. In particular, this "core calculus" approach captures these features in the syntax and semantics of a tiny programming language, expressive enough to be universal [12] yet small enough to be tractable for mathematical analysis. With regards to the overall view presented in Fig. 3, field calculus forms the second lowest layer, which is also the point where aggregate programming interfaces with the highly heterogeneous world of device infrastructure and non-aggregate software services (together comprising the lowest layer).

At its core is the notion of *computational field*, a widely-used space-time programming concept [7] inspired by the notion of fields in physics. In physics, a field is a function that maps each point in some space-time domain to a scalar value, such as the temperatures in a room, or a vector value, such as the currents in the ocean. Computational fields generalize this notion to allow the values to be arbitrary computational objects on either continuous or discrete domains, such as a set of messages to be delivered at each device in a network, or XML descriptors for a set of inventory items to be stocked on the shelves of a store.

Such spatially-extended fields, with values potentially dynamically changing over time, are then the basic "aggregate" units of values that may be distributed across many devices in the network. More precisely, a *field value* ϕ is a function $\phi : D \to \mathcal{L}$ that maps each device δ in domain D to a local value ℓ in range $\mathcal{L}$. Similarly, a *field evolution* is a dynamically changing field value, i.e., a function mapping each point in time to a field value (evolution is used here in the physics sense of "time evolution"). A field computation, then, is any function that takes field evolutions as input (e.g., from sensors or device information) and produces another field evolution as its output, from which field values are snapshots. For example, given an input of a Boolean field mapping a set of "source" devices to *true*, an output field containing the estimated distance from each device to the nearest source device can be constructed by iterative spreading and aggregation of information, such that the output always rapidly adjusts to the correct values for the current input and network structure. The field calculus [19,20] succinctly captures the essence of field computations, much as λ-calculus [14] does for functional computation and FJ [27] does for object-oriented programming.

3.1   Syntax of Field Calculus

Figure 4 presents field calculus syntax. Following the convention of [27], over-bar notation denotes metavariables over sequences, e.g., $\bar{e}$ is shorthand for the sequence of expressions $e_1, e_2, \ldots e_n$ $(n \geq 0)$. A local value ℓ represents the

$$\begin{array}{lr}
\ell ::= \mathtt{c}\langle\overline{\ell}\rangle \mid \lambda & \text{local value} \\
\lambda ::= \mathtt{o} \mid \mathtt{f} \mid (\mathtt{fun}\ (\overline{\mathtt{x}})\ \mathtt{e}) & \text{function value} \\
\mathtt{e} ::= \ell \mid \mathtt{x} \mid (\mathtt{e}\ \overline{\mathtt{e}}) & \text{expression} \\
\quad\mid (\mathtt{rep}\ \mathtt{x}\ \mathtt{w}\ \mathtt{e}) & \\
\quad\mid (\mathtt{nbr}\ \mathtt{e}) & \\
\quad\mid (\mathtt{if}\ \mathtt{e}\ \mathtt{e}\ \mathtt{e}) & \\
\mathtt{w} ::= \mathtt{x} \mid \ell & \text{variable or local value} \\
\mathtt{F} ::= (\mathtt{def}\ \mathtt{f}(\overline{\mathtt{x}})\ \mathtt{e}) & \text{function declaration} \\
\mathtt{P} ::= \overline{\mathtt{F}}\ \mathtt{e} & \text{program}
\end{array}$$

Fig. 4. Syntax of (higher-order) field calculus, as presented in [20].

value of a field at a given device, which can be any *data value* $\mathtt{c}\langle\ell_1,\cdots,\ell_m\rangle$ (written $\mathtt{c}$ when $m = 0$), such as Booleans $\mathtt{true}$ and $\mathtt{false}$, numbers, strings, or structured values like $\mathtt{Pair}\langle 3, \mathtt{Pair}\langle\mathtt{false}, 5\rangle\rangle$ or $\mathtt{Cons}\langle 2, \mathtt{Cons}\langle 4, \mathtt{Null}\rangle\rangle$. Such a value may also be a *function value* λ, i.e. a built-in operator $\mathtt{o}$, a user-defined function $\mathtt{f}$, or an anonymous function $(\mathtt{fun}\ (\overline{\mathtt{x}})\ \mathtt{e})$, where $\overline{\mathtt{x}}$ are arguments and $\mathtt{e}$ is the body, in which we assume no free variables exist. Finally, a device δ can also hold a *neighboring field value* ϕ, which maps each neighbor of δ to a local value ℓ (neighboring field values cannot be expressed directly, only appearing dynamically during computations such as with operator $\mathtt{nbr}$, so they do not appear in the syntax).

The main entities of the calculus are expressions, each of which defines a field. An expression can be a local value ℓ, representing a field mapping its entire domain to value ℓ, a variable $\mathtt{x}$ used as function parameter or state variable, or a composite created using the following constructs:

- *Built-in operator call:* $(\mathtt{e}\ \mathtt{e}_1 \cdots \mathtt{e}_n)$, where $\mathtt{e}$ evaluates to a "point-wise" built-in operator $\mathtt{o}$, involving neither state nor communication, e.g. mathematical functions like addition, comparison, and sine, or an environment-dependent function such as reading a temperature sensor or the 0-ary $\mathtt{nbr\text{-}range}$ operator returning a neighboring field mapping each neighbor to an estimate of its current distance from δ. Expression $(\mathtt{o}\ \mathtt{e}_1 \cdots \mathtt{e}_n)$ produces a field mapping each δ to the result of applying $\mathtt{o}$ to the values at δ of its $n \geq 0$ arguments $\mathtt{e}_1, \ldots, \mathtt{e}_n$.
- *User-defined function call:* $(\mathtt{e}\ \mathtt{e}_1 \cdots \mathtt{e}_n)$, where $\mathtt{e}$ evaluates to a user-defined function $\mathtt{f}$, with corresponding declaration $(\mathtt{def}\ \mathtt{f}(\mathtt{x}_1 \ldots \mathtt{x}_n)\ \mathtt{e})$. Evaluating $(\mathtt{f}\ \mathtt{e}_1 \cdots \mathtt{e}_n)$ provides a standard (possibly recursive) call-by-value abstraction.
- *Anonymous function call:* $(\mathtt{e}\ \mathtt{e}_1 \cdots \mathtt{e}_n)$, has the same semantics as user-defined function calls, except that $\mathtt{e}$ evaluates to an anonymous function $(\mathtt{fun}\ (\mathtt{x}_1 \cdots \mathtt{x}_n)\ \mathtt{e})$.
- *Time evolution:* $(\mathtt{rep}\ \mathtt{x}\ \mathtt{w}\ \mathtt{e})$ is a "repeat" construct for dynamically changing fields, using a model in which each device evaluates expressions repeatedly in asynchronous rounds. State variable $\mathtt{x}$ initialises to the value of $\mathtt{w}$, then

updates at each step by computing e against the prior value of x. For instance, (rep x 0 (+ x 1)) counts how many rounds have been computed at each device.

- *Neighboring field construction:* (nbr e) captures device-to-device interaction, returning a field ϕ that maps each device δ to a neighboring field value, which in turn maps each neighbor to its most recent available value of e (realizable e.g., via periodic broadcast). Such neighboring field values can then be manipulated and summarized with built-in operators, e.g., (min-hood (nbr e)) maps each device to the minimum value of e amongst its neighbors.
- *Domain restriction:* (if e_0 e_1 e_2) is a branching construct, computing e_1 in the restricted domain where e_0 is true, and e_2 in its complement.

To better illustrate the various constructs of field calculus, consider the following code, which estimates distance to devices where source is true, while avoiding devices where obstacle is true:

```
(def distance-avoiding-obstacle (source obstacle)
  (if obstacle infinity
    (rep d infinity (mux source 0
      (min-hood+ (+ (nbr-range) (nbr d))))))
```

coloring field calculus keywords red, built-in functions green, and user-defined functions blue. In the region outside the obstacle (with the partition conducted by if), a distance estimate d (established by rep) is computed using built-in "multiplexing" selector mux to set sources to 0 and other devices to an updated distance estimate computed using the triangle inequality, taking the minimum value obtained by adding the distance to each neighbor to its estimate of d (obtained by nbr). In particular, min-hood+ takes the minimum of all neighbors' values (excluding the device itself), and mux multiplexes between its second and third inputs, returning the second if the first is true and the third otherwise.

3.2 Local Semantics and Properties

Any field calculus program can be equivalently interpreted either as an aggregate-level computation on fields or as an equivalent "compiled" version implemented as a set of single-device operations and message passing. The full semantics may be found in [19,20], but the key ideas are simple enough to sketch briefly here.

Each field calculus program P consists of a set of user-defined function definitions and a single main expression e_0. Given a network of interconnected devices D that runs a program P, "device δ fires" means that device $\delta \in D$ evaluates e_0. The output of a device computation is a *value-tree*: an ordered tree of values tracking the result of computing each sub-expression encountered during evaluation of e_0. Each expression evaluation on device δ is performed against the most recently received value-trees of its neighbors, and the produced value-tree is conversely made available to δ's neighbors (e.g., via broadcast in compressed form) for their next firing: (nbr e) uses the most recent value of e at the same position in its neighbors' value-trees, (rep x w e) uses the value of x from the previous round, and (if e_0 e_1 e_2) completely erases the non-taken branch in the value-tree (allowing interactions through construct nbr with only neighbors that took the same branch, called "aligned neighbors"). A complete formal description of this semantics is presented in [19, 20].

A type system based on the Hindley-Milner type system [18] can then be built for this calculus [19], which has two kinds of types: local types (for local values) and field types (for field values), associating to each local value a type L, and type field(L) to a neighboring field of elements of type L, and correspondingly a type T to any expression. This type system can then be used to detect semantic errors in a program (e.g., first expression of a call not evaluating to a function, incorrect argument types for a call, first argument of if not a Boolean), ensuring that these localized versions of field calculus programs are guaranteed to observe correct domain alignment and to terminate locally if the aggregate form terminates, i.e., to faithfully implement the desired equivalence relation.

The syntax and semantics of field calculus thus form a provably sound foundation for aggregate programming, ensuring that distributed services expressed in field calculus can be safely and predictably composed and modulated. At the same time, the small set of constructs also aids in portability, infrastructure independence, and interaction with non-aggregate services: the field calculus abstraction can be supported on any device or infrastructure where these simple constructs can be implemented, including heterogeneous mixtures of devices with different sensor, actuator, computation, and communication capabilities, so long as the devices have some means of interacting. Likewise, non-aggregate software services, such as other local applications or cloud services, are often complementary to aggregate services and can be connected with aggregate services simply by importing their APIs into the aggregate programming environment [37].

4   From Theory to Pragmatic System Engineering

Field calculus may be universal, but it is also too low level to be readily used for building complex distributed services. First, like other core calculi, in order to be compact enough to be readily manipulated for mathematical results, field calculus is extremely terse and generalized, as well as lacking any of the "syntactic sugar" features that make a language more usable for programming. Second, because it is universal, field calculus can express any program, including many that have none of the safety or resilience properties that we desire in distributed systems.

To make aggregate programming practically usable as an approach, we must further raise the level of abstraction. This is done first by implementing field calculus into a full programming language, Protelis [37], which makes it more usable via syntactic sugar and methods for interfacing with other existing libraries and frameworks. Protelis contains a complete implementation of the field calculus, hosted in Java via the Xtext language generator [22], with syntax transformed to an equivalent Java-like syntax with a number of useful syntactic sugar features such as variable definition, and taking advantage of Java's reflection mechanisms to make it easy to import a large variety of useful libraries and APIs for use in Protelis. All further code will thus be given in Protelis, rather than field calculus.

The level of abstraction is then raised by identifying a composable system resilient "building block" operators, which provide core functions of coordination as well as resilience and safety guarantees. Finally, these building blocks are composed into both general and domain-specific APIs, which may further exploit optimized equivalents of particular operators for improved performance in more restricted use cases.

4.1 Building Blocks for Resilient Coordination

We first begin to raise the level of abstraction from field calculus toward an effective programming environment for resilient distributed systems by identifying a system of highly general and guaranteed composable "building block" operators for the construction of resilient coordination applications. This new

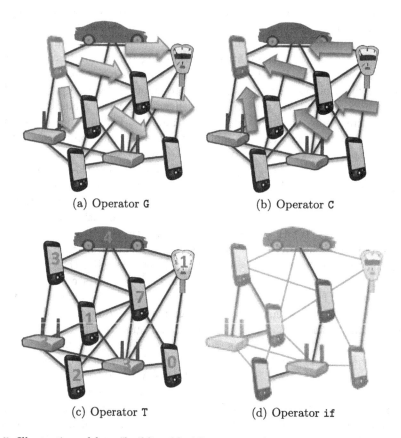

(a) Operator G   (b) Operator C

(c) Operator T   (d) Operator if

Fig. 5. Illustration of four "building block" operators for construction of resilient distributed services: information-spreading (G), information aggregation (C), aggregation over time (T), and partition into non-interacting subspaces (if).

layer (the middle layer in Fig. 3) is formed by careful selection of coordination mechanisms that are all *(i)* self-stabilizing, meaning that they can reactively adjust to changes in input values or the structure of the network, *(ii)* scalable to potentially very large networks, and *(iii)* preserve these resilience properties when the building blocks are composed together to form more complex coordination services. Critically, this means that it can be proven that any service built using only these "building blocks" will implicitly inherit such resilience [45].

One such set of operators has been identified already [8,45]: a set of three highly generalized coordination operators, G, C and T, along with field calculus' if and built-ins (Fig. 5). Each of these building blocks captures a family of frequently used strategies for achieving flexible and resilient decentralized behavior, hiding the complexity of using the low-level constructs of field calculus. The three building blocks are defined as:

- G(source,init,metric,accumulate) is a "spreading" operation generalizing distance measurement, broadcast, and projection, which takes four fields as inputs: source (a Boolean indicator field), init (initial values for the output field), metric (a function providing a map from each neighbor to a distance), and accumulate (a commutative and associative two-input function over values). It may be thought of as executing two tasks: first, computing a field of shortest-path distances from the source region according to the supplied function metric, and second, propagating values up the gradient of the distance field away from source, beginning with value initial and accumulating along the gradient with accumulate. For instance, if metric is physical distance, init is 0, and accumulate is addition, then G creates a field mapping each device to its shortest distance to a source.

- C(potential,accumulate,local,null) is an operation that is complementary to G: it accumulates information down the gradient of a supplied potential field. This operator takes four fields as inputs: potential (a numerical field), accumulate (a commutative and associative two-input function over values), local (values to be accumulated), and null (an idempotent value for accumulate). At each device, the idempotent null is combined with the local value and any values from neighbors with higher values of the potential field, using function accumulate to produce a cumulative value at each device. For instance, if potential is exactly a distance gradient computing with G in a given region R, accumulate is addition, and null is 0, then C collects the sum of values of local in region R.

- T(initial,floor,decay) deals with time, whereas G and C deal with space. Since time is one-dimensional, however, there is no distinction between spreading and collecting, and thus only a single operator. This operator takes three fields as inputs: initial (initial values for the resulting field), floor (corresponding final values), and decay (a one-input strictly decreasing function over values). Starting with initial at each node, that value gets decreased by function decay until eventually reaching the floor value, thus implementing a flexible count-down, where the rate of the count-down may change over time. For instance, if initial is a pair of a value v and a timeout t, floor is a pair of the blank value null and 0, and decay takes a pair, removing the elapsed time since previous computation from second component of the pair and turning the first component to null if the first reached 0, then T implements a limited-time memory of v.

```
def G(source, initial, metric, accumulate) {
  rep(dv <- [Infinity, initial]) {
    mux(source) {
      [0, initial]
    } else {
      minHood([nbr(dv.get(0)) + metric.apply(),
              accumulate.apply(nbr(dv.get(1)))])
    }
  }.get(1)
}
```

Fig. 6. Protelis implementation of operator G

Although there are only a few operators identified in [45], they are so general as to cover, individually or in combination, a large number of the common coordination patterns used in design of resilient systems. More importantly, when appropriately implemented in field calculus (e.g., Fig. 6), it has been shown that this system of operators, plus if and built-in operators, are elements of a "self-stabilizing language" where every program is a guaranteed to be self-stabilizing [45]. This means that distributed systems built from these operators enjoy a number of resilience properties: **stabilization:** if the input fields eventually reach a fixed state, the same happens for the output field; **resilience:** if some messages get lost during system evolution, or some node temporarily fails, this will not affect the final result; and **adaptability:** if input fields or network topology changes, the output field automatically adapts and changes correspondingly. These operators and their compositions are all also scalable for operation on potentially very large networks. Furthermore, this system of resilient operators can be readily expanded, simply by proving that any additional operators are also members of the self-stabilizing language, thereby proving that such an additional operator has the same resilience properties and can be safely composed with all previously identified operators.

4.2   Pragmatic Distributed Systems Engineering APIs

Building block operators are for the most part still too abstract and generalized to meet the pragmatic needs of typical applications programmers. To better meet these needs, various applications and combinations of "building block" operators can be captured into libraries, thereby forming a pragmatic and user-friendly API while still retaining all of the same resilience properties. Such libraries, both general and domain-specific, form the penultimate layer in Fig. 3, upon which application code (the highest layer) is actually written.

For example, a number of useful functions related to information diffusion and distributed action can be constructed from various configurations of operator G (along with built-ins). One such common computation is estimating distance from a set of source devices, which we have previously discussed as part of the field calculus example in Sect. 3. Implemented as an application of G, it may be expressed in Protelis as:

```
def distanceTo(source) {
  G(source, 0, () -> {nbrRange}, (v) -> {v + nbrRange})
}
```

Applying G in a different way implements another common coordination action, broadcasting a value across the network from a source:

```
def broadcast(source, value) {
  G(source, value, () -> { nbrRange }, (v) -> {v})
}
```

Other G-based operations include construction of a Voronoi partition and a "path forecast" that marks paths that cross an obstacle or region of interest.

Similarly, functions related to collective perception of information can be implemented using operator C, such as accumulating a summary of all the values of a variable in a region to a given sink device:

```
def summarize(sink, accumulate, local, null) {
  C(distanceTo(sink), accumulate, local, null)
}
```

or computing the variable's average or maximum value in that region. Likewise, state and memory operations may be implemented using operator T, such as holding a value until a specified timeout:

```
def limitedMemory(value, timeout) {
  T([timeout, value], [0, false],
    (t) -> {[t.get(0) - dt, t.get(1)]}).get(1)
}
```

These general API functions can then be combined together, just as in any other programming environment, to create higher level general libraries and more domain-specific libraries. For example, a common "higher-level" general operation is to share a summary throughout a region, which can be implemented by applying **broadcast** to the output of **summarize**. Likewise, in the domain of spatially-embedded services like the crowd-safety application discussed above, a useful building block is to organize an environment into dynamically defined "management regions," which can be implemented by combining state and partition functions.

A mixture of such libraries at various levels of specificity and abstraction thus forms the actual programming environment that a typical developer would use for engineering the distributed coordination aspects of a resilient distributed system, while implementing the purely local or cloud-based aspects of the service using more standard programming tools for those aspects of the system. Because the APIs are ultimately built on the foundations of resilient operators and the field calculus, however, it is guaranteed that any service developed in this manner also implicitly obtains the properties of resilience and safe composition from the lower layers of the abstraction hierarchy.

4.3   Improving Performance by Equivalent Substitutions

Finally, just as the performance of more conventional programs can be improved by changing the implementation of key libraries (e.g., changing a generic hash table implementation to one better balanced for an application's expected table size and access patterns), the performance of aggregate programs can be improved by substituting the generic building block operators by more specialized variants with better performance in particular contexts and patterns of use [45].

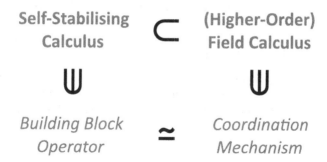

Fig. 7. Although field calculus can express any coordination mechanism, many useful mechanisms are difficult or impossible to express within a sublanguage that is known to be self-stabilizing. Any coordination mechanism that is asymptotically equivalent to a mechanism in the self-stabilizing subset, however, can be safely substituted without compromising safety or resilience guarantees. Figure adapted from [45].

Specifically, these substitutions make use of the mathematical relationship shown in Fig. 7: due to the functional composition model and modular proof used in establishing the self-stabilizing calculus, any coordination mechanism that can be guaranteed to self-stabilize to the same result as a building block operator can be substituted for that building block without affecting the self-stabilization of the overall program, including its final output. This allows creation of a "substitution library" of high-performance alternatives that can be used in certain circumstances and in those circumstances are more efficient or have more desirable dynamics. More formally:

Definition 1 (Substitutable Function). *Given functions* λ, λ' *with same type,* λ *is substitutable for* λ' *iff for any self-stabilizing list of expressions* $\overline{e}$, $(\lambda\ \overline{e})$ *always self-stabilizes to the same value as* $(\lambda'\ \overline{e})$.

The basic idea is that the property of self-stabilization specifies only the values after a function converges, so as long as two functions have the same converged values, they can be swapped without affecting any of the resilience properties based on self-stabilization. A building block operator with undesirable dynamical properties can thus be replaced by a more specialized coordination mechanism that improves overall performance without impairing resilience.

Three examples of substitution, given in [45], are:

- Distance estimation via G may converge extremely slowly when the network contains some devices that are close together [6]. Much faster alternatives exist, however, such as CRF-Gradient [6] and Flex-Gradient [4], and are known to self-stabilize to the same values as G distance estimation.
- Value collection with C is fragile: since it collects values over a spanning tree, even small perturbations can cause loss or duplication of values, with major transient impact on its results. When the accumulation is idempotent (e.g., logical AND) or separable (e.g., addition), this can be mitigated by accumulating across multiple paths.
- Low-pass filtering a signal is often useful for reducing noise. One common method, an exponential backoff filter, is substitutable with tracking a value via T, meaning that low-pass filters of this sort can be freely incorporated into programs without affecting their resilience.

When used in an application, such substitutions can markedly improve application performance. For example, consider an extremely simple distributed service for live estimation of crowd opinions of acts at a festival, implemented using

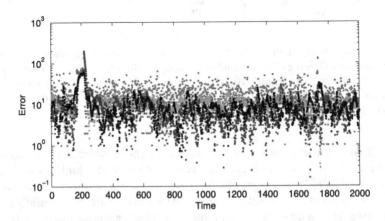

Fig. 8. Example crowd opinion feedback application is incrementally improved from its baseline performance (red) by first replacing T with an exponential filter (green), then C with multi-path summation (blue), and finally G with Flex-Gradient (black). Figure adapted from [45]. (Color figure online)

G to set up a potential field partitioning space into zones of influence for each act, C to sum a binary field of feedback, and T to track values:

```
(def add-range (v) (+ v (nbr-range)))

(def opinion-feedback (acts feedback)
  (T-filter
     (C (G acts 0 nbr-range add-range) sum feedback 0)))
```

In simulations of this scenario from [45], each incremental substitution of a generic function with a more optimized function improves the accuracy of the application: Fig. 8 shows how this application's performance can then be incrementally improved by first replacing T with an exponential filter, then C with multi-path summation, and finally G with Flex-Gradient.

Likewise, optimizations at lower layers of the framework have the potential improve the efficiency of field calculus implementations and the efficiency and simplicity of the implementation on particular devices and the interface with other applications and services. This layered approach to aggregate programming may thus serve as a framework for developing an efficient software ecosystem for engineering complex distributed systems, analogous to existing ecosystems for web or cloud development.

5 Application Examples

With the aid of appropriate domain-specific APIs, aggregate programming can greatly simplify the development and composition of distributed applications across a wide variety of domains. These can involve embedded devices and applications that are explicitly tied to space, but also can apply to more traditional location-agnostic computer networks. This section illustrates the breadth of possible applications by presenting examples across four domains: crowd safety at mass public events, UAV planning and control, construction of resilient enterprise systems, and network security.

5.1 Crowd Safety at Mass Events

One example, explored in [8], of an environment where aggregate programming is particularly applicable is at mass public events, such as marathons, outdoor concerts, festivals, and other civic activities. In these highly crowded environments, the combination of high densities of people and large spatial extent can often locally overwhelm the available infrastructure, causing cell phones to drop calls, data communications to become unreliable, etc. The physical environment is often overwhelmed as well, and the movement of high numbers of people in crowded and constrained environments can pose challenging emergent safety issues: in critically overcrowded environments, even the smallest incident can create a panic or stampede in which many people are injured or killed [43].

Fig. 9. A crowd safety service, restricted to run on personal devices (colored) in a simulation of approximately 2500 personal and embedded devices at the 2013 Vienna marathon, detects regions of potentially dangerous crowd density (red) and disseminates warnings to nearby devices (yellow). Figure adapted from [8]. (Color figure online)

Between smart-phones and other personal devices, however, the effective density of deployed infrastructure is much higher, since more people means more personal devices. Aggregate programming can be used to coordinate these devices, without the need for centrally deployed infrastructure, to provide services such as for crowd safety, to help identify and diffuse potentially dangerous situations. For example, crowding levels can be conservatively estimated by first estimating the local density of people as $\rho = \frac{|nbrs|}{p \cdot \pi r^2 \cdot w}$, where $|nbrs|$ counts neighbors within range r, p estimates the proportion of people with a participating device running the app and w estimates fraction of walkable space in the local urban environment, then comparing this estimate with "level of service" (LoS) ratings [24], taking LoS D (> 1.08 people/m^2) to indicate a crowd and LoS E (> 2.17 people/m^2) in a large group (e.g., $300+$ people) to indicate potentially dangerous density. Potential crowding danger can thus be detected and warnings disseminated robustly with just a few lines of Protelis code dynamically deployed and executed on individual devices [20,37]:

```
def dangerousDensity(p, r) {
  let mr = managementRegions(r*2, () -> { nbrRange });
  let danger = average(mr, densityEst(p, r)) > 2.17 &&
               summarize(mr, sum, 1 / p, 0) > 300;
  if(danger) { high } else { low }
}

def crowdTracking(p, r, t) {
  let crowdRgn = recentTrue(densityEst(p, r)>1.08, t);
  if(crowdRgn) { dangerousDensity(p, r) } else { none };
}

def crowdWarning(p, r, warn, t) {
  distanceTo(crowdTracking(p,r,t) == high) < warn
}
```

Figure 9 shows an ALCHEMIST [38] simulation of such a crowd safety service running in an environment of pervasively deployed devices: 1479 mobile personal devices, each following a smart-phone position trace collected at the 2013 Vienna marathon, as discussed in [3,36], and 1000 stationary devices, all communicating via once per second asynchronous local broadcasts with 100 meters range, with all devices participating in infrastructure services but the crowd safety service restricted to run only on the mobile personal devices. Building this program via aggregate programming ensures that it is resilience and adaptive despite its very short length, allowing it to effectively estimate crowding and distribute warnings while executing on a large number of highly mobile devices.

5.2   Humanitarian Assistance and Disaster Relief Operations

Humanitarian assistance and disaster relief operations are another example of an environment where distributed coordination is particularly valuable, due to existing infrastructure being damaged or overwhelmed. With appropriate mechanisms for distributed coordination, however, "tactical cloud" resources can substitute for fixed infrastructure in support of assistance and relief operations. For example, [40,44] present an architecture of "edge nodes" equivalent to a 1/2-rack of servers in sturdy self-contained travel cases, which can be effectively mounted and operated even in small vehicles such as HMMVWs or towed trailers. Continuing advances in computing capability mean that such edge nodes actually offer a startling amount of capability: 10 such units can be equivalent to an entire cargo-container portable data center. The challenge is how to effectively coordinate and operate mission critical services across such devices, particularly given that the communications network between nodes has limited capacity and changes frequently as nodes are moved around and also given that individual edge nodes may be taken offline at any time due to evolving mission requirements, failures, accidents, or hostile action. Aggregate programming can simplify the development of resilient services for the tactical cloud environment, whereas existing methods tend to push application development toward a "star" topology where edge nodes interact mostly indirectly by means of their communications with a larger infrastructure cloud.

Consider, for example, a representative service example of assisting in the search for missing persons following a major disaster such as tsunami. This is a good example of a distributable mission application, since it involves data at several different scales: missing person queries (e.g., providing a photo of a missing loved one) and responses (e.g., a brief fragment of a video log showing a missing person) are fairly lightweight and can be spread between servers fairly easily, while video logs (e.g., from helmet- and vehicle-mounted cameras) are quite large and are best to search locally.

An implementation of this coordination service requires less than 30 lines of Protelis [37] code: this implementation distributes missing person queries, has them satisfied by video logged by other teams, then forwards that information back toward the team where the query originated. Figure 10 shows a screenshot

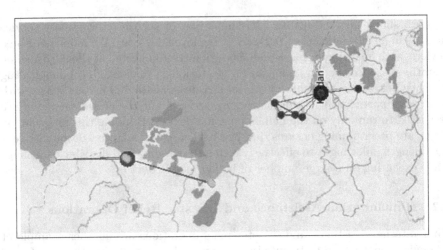

Fig. 10. Simulation of tactical cloud coordination in a humanitarian response scenario: tactical cloud nodes in survey team vehicles collectively help families find missing persons following a natural disaster: a query lodged with one team is opportunistically disseminated from its cloud node (red), to be compared against the video logs stored locally in each team's node. The desired information is located at a distant node (blue), then opportunistically forwarded to other nodes (green) until it can reach either the original source or some other node where the response can be received, thereby satisfying the query. (Color figure online)

from simulation of this scenario in the ALCHEMIST simulator [38]. In this scenario, a group of eleven survey teams are deployed amphibiously, then move around through the affected area, carrying out their survey mission over the course of several days. Each team hosts a half-rack server as part of their equipment, coordinating across tactical networks to collectively form a distributed cloud host for mission applications, such as searching video logs for missing persons and collating survey data. The distributed service implementation opportunistically disseminates queries, such that they end up moving implicitly by a combination of forwarding and taking advantage of vehicle motions to ferry data across gaps when there is no available connectivity. At each tactical cloud node, the query is executed against its video logs, and any matches are forwarded by the same opportunistic dissemination and marked off as resolved once the results of the service have been delivered to the person who requested assistance.

5.3   Resilient Enterprise Systems

Aggregate programming can also be applied to networks that are not closely tied to space, such as enterprise service networks, as in the work on distributed recovery of enterprise services presented in [15]. Management of small- to medium-scale enterprise systems is a pressing current problem (Fig. 11), since these systems are often quite complex, yet typically managed much more primitively than either individual machines (which are simpler and more uniform) or

Individual Server

Small- to Medium- Enterprise Network

Large Cloud Datacenter

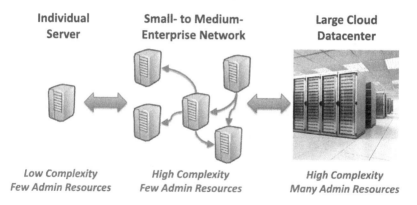

Low Complexity
Few Admin Resources

High Complexity
Few Admin Resources

High Complexity
Many Admin Resources

Fig. 11. Small- to medium-sized enterprises often have complex networks with many services and servers, but are not large enough to have significant administrative resources to devote to customization or to benefit from economies of scale. Figure adapted from [15].

large-scale datacenters (which can invest in high-scale or custom solutions). As a result, small and medium enterprises tend to have poor resilience and to suffer much more disruptive and extensive outages than large enterprises [1].

In [15], aggregate programming is used to address the common problem of safely and rapidly recovering from failures in a network of interdependent services, for which typical industry practice is to shut the entire system down and then restart services one at a time in a "known safe" order. The solution presented in [15], Dependency-Directed Recovery (DDR), uses Protelis [37] to implement a lightweight network of daemon processes that monitor service state, detecting dependencies (e.g., via methods such as in [13,29,39,42]) and controlling services to proactively bring down only those services with failed dependencies, then restart them in near-optimal time (Fig. 12). This system is realized with management daemons implemented Java, each hosting a Protelis VM executing the following simple coordination code:

```
// Collect state of monitored service from service manager daemon
let status = self.getEnvironmentVariable("serviceStatus");
let serviceID = self.getEnvironmentVariable("serviceID");
let depends = self.getEnvironmentVariable("dependencies");
let serviceDown = status=="hung" || status=="stop";

// Compute whether service can safely be run (i.e. dependencies are satisfied)
let liveSet = if(serviceDown) { [] } else { [serviceID] };
let nbrsLive = unionHood(nbr(liveSet));
let liveDependencies = nbrsLive.intersection(depends);
let safeToRun = liveDependencies.equals(depends);

// Act based on service state and whether it is safe to run
if(!safeToRun) {
```

```
if(!serviceDown) {
    self.stopService() // Take service down to avoid misbehavior
} else { false } // Wait for dependencies to recover before restarting
} else {
if(serviceDown) {
    self.restartService() // Safe to restart
} else { false } // Everything fine; no action needed
}
```

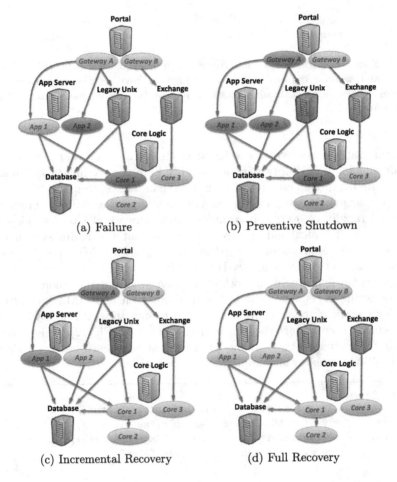

(a) Failure

(b) Preventive Shutdown

(c) Incremental Recovery

(d) Full Recovery

Fig. 12. Example of dependency-directed recovery in a service network, showing status *run* as green, *stop* as blue, and *hung* as red. Following failure of some set of services (a), other services that depend on them shut themselves down (b). As failed services restart, services that depend on them restart incrementally (c), until the entire service network has recovered (f). Figure adapted from [15].

With this program, any failure leads to a sequence of shutdowns, following dependency chains from failed services to the services that depend on them. Complementarily, when a service's dependencies start running again, that service restarts, becoming part of a wave of restarts propagating in parallel up the partial order established by dependencies.

Analysis of this system shows that it should produce distributed recovery in near-optimal time, slowed only be communication delays and the update period of the daemons. Experimental validation in emulated service networks of up to 20 services verifies this analysis, as well as showing a dramatic reduction in downtime compared to fixed-order restart, and allowing many services to continue running uninterrupted even while recovery is proceeding.

5.4   Network Security

For a final example, consider the value of effective and resilient coordination in network security. Improvements in virtualization technology have made it possible to trace and record the state evolution of an entire service or server, which can allow checkpointing of key points in process history, so that if attacks or faults are later detected the process can be "rewound" to a known-safe state and re-run with a dynamic patch or with the bad interaction edited out of the flow [16, 35]. Executing such mechanisms, however, requires that interactions be able to be tightly monitored and ordered, which is often quite difficult and costly for networked services.

Taking an aggregate programming perspective, however, we may recognize that when interactions between services can be monitored, as in many networked services, a partial order of events based on the sending and receiving of messages can be substituted for the total order otherwise required for checkpointing or rewind and replay. To enable this, each service in the network takes local checkpoints every time that it sends a message or processes a message that it has received. A send/receive pair between two interacting services may then be interpreted as a directed graph edge, from send to receive, and the union of these directed edges with directed edges expressing the local order of local checkpoints on each server forms an distributed acyclic directed graph that can be safely interpreted as a partial order over events. A distributed checkpoint can then be computed emergently using distributed graph inference to compute the closure of graph succession on a set of local events (e.g., a set of faults or attacks), rewind executed by coordinated deletion of this subgraph, and replay executed by re-executing the incoming edges to the subgraph. Critically, this does not require any sort of synchronization between services, as well as allowing recovery to take place asynchronously, with any service not affected by possible contamination able to run uninterrupted and other services being able to run again as soon as they themselves are free of possible contamination.

Using aggregate programming to implement this partial order approach, coordination for rewind and replay can be implemented in less than 100 lines of Protelis [37]. Figure 13 shows an example screenshot from a rewind and replay system running on a network of emulated services, in the process of editing out

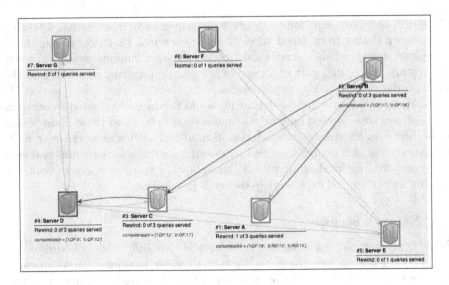

Fig. 13. Screenshot of distributed rewind and replay isolating and eliminating contamination (yellow machines) from an attack on a service network: following detection of an injected attack on a service (red box), potentially contaminated services (yellow box) suspend, trace potential contamination, and begin rewinding potentially contaminated interactions. Meanwhile, adjacent unaffected services (blue box) temporarily suspend operations to prevent spread of contamination while non-adjacent services (green box) continue to operate normally. (Color figure online)

an injected attack. Following detection of an injected attack on a service (e.g., via [16,35]), potentially contaminated services suspend, trace potential contamination, and begin rewinding potentially contaminated interactions. Meanwhile, adjacent unaffected services temporarily suspend operations as a "firebreak" against further spread of contamination, while non-adjacent services continue to operate normally.

6   Summary and Future Directions

This review has presented a summary of the aggregate programming approach to distributed systems engineering, including a review of its theoretical foundations in field calculus, how resilience can be guaranteed through composable "building blocks," and how these can be combined and refined to make effective APIs for engineering distributed applications across a wide range of domains. Overall, the aggregate programming approach offers the potential for complex distributed services to be specified succinctly and treated as coherent objects that can be safely encapsulated, modulated, and composed together, toward the ultimate goal of making distributed systems engineering as routine as ordinary single-device programming.

From this present state, four key directions for future work are:

- Further development of the theoretical foundations of aggregate programming, particularly with regards to mobile devices and the relationship between continuous environments and discrete networks of devices.
- Expansion of resilience results, including expansion of the set of building blocks and extension to a broader range of resilience properties, particularly regarding dynamical properties and feedback systems.
- Pragmatic improvements to the infrastructure and integration of aggregate programming, including expansion of libraries and APIs to more capabilities and more domains, integration with other pragmatic concerns such as security, optimizing usage of energy and other resources, and development of "operating system" layers for aggregate and hybrid aggregate/cloud architectures, as well as improvements to Protelis or other aggregate programming implementations.
- Developing applications of aggregate programming for a variety of problem domains, and transition of these applications into useful real-world deployments.

Our world is increasingly a world of computational aggregates, and methods such as these are the only way that we are likely to be able to keep engineering tractable, safe, and resilient in the increasingly complex interweaving of the informational and physical worlds, and our increasing dependence upon such distributed systems in the infrastructure of our civilization.

Acknowledgment. This work has been partially supported by the EU FP7 project "SAPERE - Self-aware Pervasive Service Ecosystems" under contract No. 256873 (Viroli), by the Italian PRIN 2010/2011 project "CINA: Compositionality, Interaction, Negotiation, Autonomicity" (Viroli), and by the United States Air Force and the Defense Advanced Research Projects Agency under Contract No. FA8750-10-C-0242 (Beal). The U.S. Government is authorized to reproduce and distribute reprints for Governmental purposes notwithstanding any copyright notation thereon. The views, opinions, and/or findings contained in this article are those of the author(s)/presenter(s) and should not be interpreted as representing the official views or policies of the Department of Defense or the U.S. Government. Approved for public release; distribution is unlimited.

References

1. Aberdeen Group: Why mid-sized enterprises should consider using disaster recovery-as-a-service, April 2012. http://www.aberdeen.com/Aberdeen-Library/7873/AI-disaster-recovery-downtime.aspx, Retrieved 13 July 2015
2. Anderson, D.P.: Boinc: a system for public-resource computing and storage. In: Proceedings of the Fifth IEEE/ACM International Workshop on Grid Computing, pp. 4–10. IEEE (2004)

3. Anzengruber, B., Pianini, D., Nieminen, J., Ferscha, A.: Predicting social density in mass events to prevent crowd disasters. In: Jatowt, A., Lim, E.-P., Ding, Y., Miura, A., Tezuka, T., Dias, G., Tanaka, K., Flanagin, A., Dai, B.T. (eds.) SocInfo 2013. LNCS, vol. 8238, pp. 206–215. Springer, Heidelberg (2013). http://dx.doi.org/10.1007/978-3-319-03260-3_18

4. Beal, J.: Flexible self-healing gradients. In: ACM Symposium on Applied Computing, pp. 1197–1201. ACM, New York, March 2009

5. Beal, J., Bachrach, J.: Infrastructure for engineered emergence in sensor/actuator networks. IEEE Intell. Syst. **21**, 10–19 (2006)

6. Beal, J., Bachrach, J., Vickery, D., Tobenkin, M.: Fast self-healing gradients. In: Proceedings of ACM SAC 2008, pp. 1969–1975. ACM (2008)

7. Beal, J., Dulman, S., Usbeck, K., Viroli, M., Correll, N.: Organizing the aggregate: languages for spatial computing. In: Mernik, M. (ed.) Formal and Practical Aspects of Domain-Specific Languages: Recent Developments, Chap. 16, pp. 436–501. IGI Global (2013). A longer version available at: http://arxiv.org/abs/1202.5509

8. Beal, J., Pianini, D., Viroli, M.: Aggregate programming for the internet of things. IEEE Comput. **48**(9), 22–30 (2015). http://jakebeal.com/Publications/Computer-AggregateProgramming-2015.pdf

9. Beal, J., Viroli, M.: Building blocks for aggregate programming of self-organising applications. In: Eighth IEEE International Conference on Self-Adaptive and Self-Organizing Systems Workshops, SASOW 2014, London, United Kingdom, 8–12 September, 2014, pp. 8–13 (2014). http://dx.doi.org/10.1109/SASOW.2014.6

10. Beal, J., Viroli, M.: Formal foundations of sensor network applications. SIGSPATIAL Spec. **7**(2), 36–42 (2015)

11. Beal, J., Viroli, M.: Space-time programming. Philos. Trans. R. Soc. Part A **73**, 20140220 (2015)

12. Beal, J., Viroli, M., Damiani, F.: Towards a unified model of spatial computing. In: 7th Spatial Computing Workshop (SCW 2014), AAMAS 2014, Paris, France, May 2014

13. Chen, X., Zhang, M., Mao, Z.M., Bahl, P.: Automating network application dependency discovery: experiences, limitations, and new solutions. In: OSDI, vol. 8, pp. 117–130 (2008)

14. Church, A.: A set of postulates for the foundation of logic. Ann. Math. **33**(2), 346–366 (1932)

15. Clark, S.S., Beal, J., Pal, P.: Distributed recovery for enterprise services. In: 2015 IEEE 9th International Conference on Self-Adaptive and Self-Organizing Systems (SASO), pp. 111–120, September 2015

16. Clark, S.S., Paulos, A., Benyo, B., Pal, P., Schantz, R.: Empirical evaluation of the a3 environment: evaluating defenses against zero-day attacks. In: 2015 10th International Conference on Availability, Reliability and Security (ARES), pp. 80–89. IEEE (2015)

17. Coore, D.: Botanical Computing: A Developmental Approach to Generating Interconnect Topologies on an Amorphous Computer. Ph.D. thesis, MIT (1999)

18. Damas, L., Milner, R.: Principal type-schemes for functional programs. In: Symposium on Principles of Programming Languages, POPL 1982, pp. 207–212. ACM (1982). http://doi.acm.org/10.1145/582153.582176

19. Damiani, F., Viroli, M., Beal, J.: A type-sound calculus of computational fields. Sci. Comput. Program. **117**, 17–44 (2016)

20. Damiani, F., Viroli, M., Pianini, D., Beal, J.: Code mobility meets self-organisation: a higher-order calculus of computational fields. In: Graf, S., Viswanathan, M. (eds.) FORTE 2015. LNCS, vol. 9039, pp. 113–128. Springer, Heidelberg (2015). http://dx.doi.org/10.1007/978-3-319-19195-9_8

21. Dean, J., Ghemawat, S.: Mapreduce: simplified data processing on large clusters. Commun. ACM **51**(1), 107–113 (2008)

22. Eysholdt, M., Behrens, H.: Xtext: implement your language faster than the quick and dirty way. In: OOPSLA, pp. 307–309. ACM (2010)

23. Finin, T., Fritzson, R., McKay, D., McEntire, R.: Kqml as an agent communication language. In: Proceedings of the Third International Conference on Information and Knowledge Management, CIKM 1994, pp. 456–463. ACM, New York (1994). http://doi.acm.org/10.1145/191246.191322

24. Fruin, J.: Pedestrian and Planning Design. Metropolitan Association of Urban Designers and Environmental Planners (1971)

25. Gentzsch, W.: Sun grid engine: towards creating a compute power grid. In: Proceedings of the First IEEE/ACM International Symposium on Cluster Computing and the Grid, pp. 35–36. IEEE (2001)

26. Giavitto, J.L., Godin, C., Michel, O., Prusinkiewicz, P.: Computational models for integrative and developmental biology. Technical report 72–2002, Univerite d'Evry, LaMI (2002)

27. Igarashi, A., Pierce, B.C., Wadler, P.: Featherweight java: a minimal core calculus for Java and GJ. ACM Trans. Program. Lang. Syst. **23**(3), 396–450 (2001)

28. Inchiosa, M., Parker, M.: Overcoming design and development challenges in agent-based modeling using ascape. Proc. Nat. Acad. Sci. U.S.A. **99**(Suppl. 3), 7304 (2002)

29. Lou, J.G., Fu, Q., Wang, Y., Li, J.: Mining dependency in distributed systems through unstructured logs analysis. ACM SIGOPS Operating Syst. Rev. **44**(1), 91–96 (2010)

30. Madden, S.R., Szewczyk, R., Franklin, M.J., Culler, D.: Supporting aggregate queries over ad-hoc wireless sensor networks. In: Workshop on Mobile Computing and Systems Applications (2002)

31. Mamei, M., Zambonelli, F.: Programming pervasive and mobile computing applications: the tota approach. ACM Trans. Softw. Eng. Methodologies **18**(4), 1–56 (2009)

32. Message Passing Interface Forum: MPI: A Message-Passing Interface Standard Version 2.2, September 2009

33. Nagpal, R.: Programmable Self-Assembly: Constructing Global Shape using Biologically-inspired Local Interactions and Origami Mathematics. Ph.D. thesis, MIT (2001)

34. Newton, R., Welsh, M.: Region streams: functional macroprogramming for sensor networks. In: First International Workshop on Data Management for Sensor Networks (DMSN), pp. 78–87, August 2004

35. Paulos, A., Pal, P., Schantz, R., Benyo, B., Johnson, D., Hibler, M., Eide, E.: Isolation of malicious external inputs in a security focused adaptive execution environment. In: 2013 Eighth International Conference on Availability, Reliability and Security (ARES), pp. 82–91. IEEE (2013)

36. Pianini, D., Viroli, M., Zambonelli, F., Ferscha, A.: HPC from a self-organisation perspective: the case of crowd steering at the urban scale. In: 2014 International Conference on High Performance Computing Simulation (HPCS), pp. 460–467, July 2014

37. Pianini, D., Beal, J., Viroli, M.: Practical aggregate programming with protelis. In: ACM Symposium on Applied Computing (SAC 2015) (2015)
38. Pianini, D., Montagna, S., Viroli, M.: Chemical-oriented simulation of computational systems with Alchemist. J. Simul. **7**, 202–215 (2013). http://www.palgrave-journals.com/jos/journal/vaop/full/jos201227a.html
39. Popa, L., Chun, B.G., Stoica, I., Chandrashekar, J., Taft, N.: Macroscope: endpoint approach to networked application dependency discovery. In: Proceedings of the 5th International Conference on Emerging Networking Experiments and Technologies, pp. 229–240. ACM (2009)
40. Simanta, S., Lewis, G.A., Morris, E.J., Ha, K., Satyanarayanan, M.: Cloud computing at the tactical edge. Technical report CMU/SEI-2012-TN-015, Carnegie Mellon University (2012)
41. Sklar, E.: Netlogo, a multi-agent simulation environment. Artif. Life **13**(3), 303–311 (2007)
42. Lgorzata Steinder, M., Sethi, A.S.: A survey of fault localization techniques in computer networks. Sci. Comput. Program. **53**(2), 165–194 (2004)
43. Still, G.K.: Introduction to Crowd Science. CRC Press, Boca Raton (2014)
44. Suggs, C.: Technical framework for cloud computing at the tactical edge. Technical report, US Navy Program Executive Office Command, Control, Communications, Computers and Intelligence (PEO C4I) (2013)
45. Viroli, M., Beal, J., Damiani, F., Pianini, D.: Efficient engineering of complex self-organizing systems by self-stabilising fields. In: IEEE International Conference on Self-Adaptive and Self-Organizing Systems (SASO), pp. 81–90. IEEE, September 2015
46. Viroli, M., Damiani, F., Beal, J.: A calculus of computational fields. In: Canal, C., Villari, M. (eds.) ESOCC 2013. CCIS, vol. 393, pp. 114–128. Springer, Heidelberg (2013)
47. Whitehouse, K., Sharp, C., Brewer, E., Culler, D.: Hood: a neighborhood abstraction for sensor networks. In: Proceedings of the 2nd International Conference on Mobile Systems, Applications, and Services. ACM Press (2004)

Author Index

Printed in the United States
By Bookmasters